Lubricant Analysis and Condition Monitoring

Lubricant Analysis and Condition Monitoring

R. David Whitby

CRC Press
Taylor & Francis Group
Boca Raton London New York

CRC Press is an imprint of the
Taylor & Francis Group, an **informa** business

First edition published 2022
by CRC Press
6000 Broken Sound Parkway NW, Suite 300, Boca Raton, FL 33487-2742

and by CRC Press
2 Park Square, Milton Park, Abingdon, Oxon, OX14 4RN

© 2022 Pathmaster Marketing Ltd.

CRC Press is an imprint of Taylor & Francis Group, LLC

ISBN: 978-1-032-15669-9 (hbk)
ISBN: 978-1-032-15670-5 (pbk)
ISBN: 978-1-003-24525-4 (ebk)

DOI: 10.1201/9781003245254

Typeset in Times
by codeMantra

I dedicate my second book to my Grandfather, Matthew Leach, and my first Chemistry Teacher, Dick Borrman, both of whom encouraged me to be curious and to learn new things constantly.

Contents

Preface

Almost all mechanical devices used in every industry require lubrication. Some machines, items of equipment or mechanical systems use very small amounts of lubricants, while others use huge quantities of oils and/or greases. The condition of lubricants affects the operation of machines and equipment, and the condition of the machines and systems affects the lubricants used in them.

The purpose of this book is to explain and discuss the benefits of identifying, planning, implementing and using lubricant and machine condition monitoring programmes to extend the lifetimes of both lubricants and machines, to achieve maximum productivity and profitability at the same time as reducing the impacts on waste and the environment.

The book describes the types of oils and greases used in modern machinery, equipment and systems and why it can be very important to monitor their condition. The book explains all the tests that can and should be used to monitor the condition of lubricants, machines and systems. It also describes tests that should not be used in a lubricant condition monitoring programme, and why.

The book's chapters explore and describe what and why to do specific actions and processes and how to achieve successful outcomes. The chapters aim to encompass all the information that users of lubricants and machines need in one handy volume.

Author

R David Whitby, BSc (Hons), CLP, is Chief Executive of Pathmaster Marketing Ltd., a business development consultancy for the international downstream oil, gas and energy industries, which he founded in 1992. Pathmaster Marketing has advised clients in the UK, France, Germany, Belgium, Denmark, Poland, Hungary, Russia, the US, Canada, Israel, Saudi Arabia, Iran, South Africa, Brazil, Singapore, Malaysia, Thailand and Australia on business planning, business strategy, market development and technology commercialisation. Specialist sectors include lubricants, fuels, new energies and speciality chemicals.

An Australian by birth, David began his career with British Petroleum, as a process chemist in a refinery in Western Australia. He worked for BP for 22 years in a number of management positions, including Marketing and Business Development Manager at Kalsep (an advanced separations company), Business Manager at BP Ventures, Project Leader for Industrial Lubricants at BP Research and Marketing Services Officer at Duckhams Oils.

David was Programme Director for Lubricants Courses at the Oxford Princeton Programme, and he ran the Advanced Lubrication Training Programme for the UK Lubricants Association. He has written numerous papers and articles on lubricants, has chaired and lectured to international conferences and directed over 120 training lubricants courses in more than 30 countries. He writes the bimonthly "Worldwide" column for *Tribology and Lubrication Tribology*, published by the US Society of Tribologists and Lubrication Engineers.

In addition to running Pathmaster Marketing, David was Non-Executive Chairman of Microbial Solutions Ltd., a start-up from the University of Oxford, from 2007 to 2015, and a Non-Executive Director of the Sonic Development Company Ltd., from 1998 to 2003. His first book, *Lubricant Blending and Quality Assurance*, was published by CRC Press in 2019.

David has lived in Woking, Surrey, United Kingdom, for more than 38 years and is married with two daughters and four grandchildren.

1 Introduction

1.1 PURPOSE

Almost all mechanical devices used in every industry require lubrication. Some machines, items of equipment or mechanical systems use very small amounts of lubricants, while others use huge quantities of oils and/or greases.

Examples of machines that use very little lubricants are personal computers (PCs) and laptops. In the past, many of these used a hard disc drive on which programs and data were stored. These hard discs had very small needle bearing that used a drop of a very high-performance lubricating oil. Now, PCs and laptops use solid-state storage media that do not need to be lubricated. However, both still need to use one or more cooling fans in order to keep the central processor and motherboard cool, and these still need tiny self-lubricated rolling bearings.

Examples of machines that use huge quantities of lubricants are ocean-going container ships and crude oil tankers, as well as very large trucks and machines used in mining operations. A large hydraulic system might use ten tonnes of hydraulic oil, while a 500-megawatt steam turbine used to generate electricity might use 300 tonnes of turbine oil. Other items of equipment, such as the gearbox in a helicopter or the aviation gas turbine in a passenger aircraft, are critical to the operation of the helicopter or aircraft.

A huge range of different oils are used in machines and equipment of all types. The types of oils are listed in Table 1.1. Some of these types are not used as lubricants, but are functional fluids without which the machines or equipment could not operate. They are included in Table 1.1 because they are associated with the lubricants industry, either through their derivation from crude oil or natural gas or through their derivation from the chemicals industry.

A slightly smaller range of greases, shown in Table 1.2, are used in numerous types of machines and equipment. All types of greases are used as lubricants, although they can have a number of other important functions, such as protecting against corrosion and ingress of water.

The range of industries in which these oils, greases and functional fluids are used is equally huge, as illustrated in Figure 1.1. The number of industries listed in Figure 1.1 is, in fact, a short list, as all of the headings have sub-headings and most of these have a number of sub-divisions. For example, although Research & Development (R&D) activities do not use big volumes of oils and greases, many of the items of equipment use small amounts of lubricants. Items such as ovens, refrigerators, mixers, robots, computers and testing machines all require lubrication in order to function.

It is axiomatic that the condition of a lubricating oil or grease is likely to affect the functioning or lifetime of a machine, item of equipment or mechanical system. An oxidised or dirty lubricant is likely to cause increased wear or corrosion. Conversely, the age, operating conditions, design, location or construction of a machine, item of

DOI: 10.1201/9781003245254-1

TABLE 1.1

Types and Applications of Lubricating Oils

Automotive Oils	Industrial Oils	Other Oils
Gasoline engine oils for cars, vans and taxis	Industrial gear oils	Aviation gas turbine oils
Diesel engine oils for cars, vans and taxis	Hydraulic oils	Aircraft engine oils
Natural gas engine oils for cars vans and taxis	Fire-resistant hydraulic fluids	Two-stroke marine engine oils
Hydrogen engine oils	Steam turbine oils for electricity generation	Four-stroke marine engine oils
Gasoline engine oils for trucks and buses	Gas turbine oils for electricity generation	Water-mix metalworking fluids
Diesel engine oils for trucks and buses	Water turbine oils for electricity generation	Metalworking oils
Natural gas engine oils for trucks and buses	Wind turbine oils	Metal-forming oils
Hybrid and electric vehicle oils	Air compressor oils	Metal-forming pastes
Two-stroke engine oils for motorcycles	Gas compressor oils	Metal-drawing oils and pastes
Four-stroke engine oils for motorcycles	Refrigerator and air conditioner oils	Metal-stamping oils and pastes
Multipurpose engine oils for farm machinery	Textile oils	Transformer and switchgear oils
Automotive gear oils	Wire rope lubricants	Electrical cable oils
Automatic transmission fluids	Heat transfer oils	Electrical cable pastes and compounds
Shock absorber oils	Heat treatment oils and fluids	

Source: Pathmaster Marketing Ltd.

TABLE 1.2

Types of Greases

Soap-Based Greases	Non-soap Greases
Lithium	Bentone (clay)
Lithium complex	Graphite (carbon black)
Calcium	Molybdenum disulphide
Calcium sulphonate	Polyurea (polymer)
Calcium complex	PTFE (polytetrafluoroethylene)
Aluminium	Silica
Aluminium complex	
Sodium	
Mixed soaps	

Source: Pathmaster Marketing Ltd.

Manufacturing	Processing	Construction
Aerospace	Chemicals	Houses
Automotive	Pharmaceuticals	Buildings
Clothing	Petroleum	Civil
Furniture	Paper	
	Rubber	**Utilities**
Fabrication	Surface coatings	Gas
Iron and steel		Electricity
Non-ferrous metals	**Agriculture**	Telephone
Composite materials	Food crops	
Textiles	Non-food crops	**Services**
Leather and fur	Plantations	Transportation
Floor coverings	Forestry	Restaurants
Electronics		Hotels
Machine tools	**Mining**	Food
Wood and wood products	Coal	Safety
Household goods	Minerals	Leisure
Consumer Goods Retailing	**R&D**	

FIGURE 1.1   Industry sectors.

Source: Pathmaster Marketing Ltd.

equipment or system is likely to affect the performance and lifetime of the oils and/or greases used as lubricants or functional fluids.

This is why monitoring the condition of both lubricants and machines is critically important to maximising the productivity, effectiveness, safety and profitability of both. An additional, and increasingly important, consideration is that lubricant and machine condition monitoring and predictive maintenance are likely to lessen waste and its impacts on the environment.

Fortunately, there are numerous sources of information and guidance about lubricant or machine condition monitoring and predictive maintenance. Most of these sources are in published magazines. For example, the magazine published by US Society of Tribologists and Lubrication Engineers (STLE), *Tribology and Lubrication Technology* (TLT), has regular articles about either lubricant condition monitoring or machine condition monitoring. Similarly, Noria Corporation publishes a monthly magazine *Machinery Lubrication*, which regularly contains articles about condition monitoring. Noria also publishes *Lube Tips*.

Noria also publishes a book, *Oil Analysis Basics*, a 192-page paperback that covers lubrication fundamentals, oil analysis and condition-based maintenance, contamination control, oil sampling methods, oil testing and analysis and targets, limits and data management. The second edition was published in 2010. However, this book does not appear to cover the monitoring of machines, equipment or systems in much depth.

Another book, *Machinery Oil Analysis: Methods Automation and Benefits*, a 506-page hardback published by the STLE in 2008, covers machinery lubrication, failure and maintenance concepts, machinery, fluid and filtration failure modes, oil sampling and testing and statistical analysis. However, this book does not appear to provide in-depth descriptions of oil and grease testing methods and explanations about which tests should and should not be used in a lubricant condition monitoring programme.

A third book, *Machinery Condition Monitoring, Principles and Practices*, published by CRC Press in 2017, is a "single source for practical machinery condition monitoring". The book focusses on the mechanical side of condition monitoring, with chapters on vibration and rotor dynamics, noise monitoring, thermography, signal processing and instrumentation and wear debris analysis, for example. However, it does not include lubricant analysis and lubricant condition monitoring.

Other, much shorter books are focussed on specific topics, such as *Wear Debris Analysis, Sourcebook for Used Oil Elements, Oil Sampling Procedures* and *In-Service Lubricant and Machine Analysis Diagnostics and Prognostics.*

Companies that offer lubricant testing and advice on lubricant condition monitoring have pages on their websites that describe what services they provide. Some of this information is also provided in the form of downloadable pamphlets. For example, Bureau Veritas publishes a 32-page pamphlet *The Basics of Oil Analysis*, currently in its fifth edition. Obviously, these types of web pages and pamphlets are intended as a basic introduction to the subject and not a comprehensive guide. Users of lubricants that would benefit from comprehensive guidance will be able to obtain it from a lubricant testing company once they have signed up to a three- or five-year contract.

While there are many sources of information about lubricant and machine condition monitoring, they are currently scattered in many different places. Users of lubricants in machines, equipment and systems may need to spend some considerable time searching for the precise information relevant to them. The purpose of this book is, therefore, to gather all the information that users of lubricants may need, in one place, and to ensure that the information is both up-to-date and forward-looking.

It is important to note that a large number of lubricant and equipment tests are described in the following chapters. Many of these tests can and should be used in lubricant and machine monitoring programmes. However, other tests should not be used in lubricant and machine monitoring programmes, generally because they take too long to produce a result and/or because they are too costly to be of practical value. Many of these expensive or time-consuming tests are very valuable for developing new or improved lubricants, but they are not suitable when an important test result is required within one or two days. This book will explain which tests should be used for condition monitoring and which should not, and why.

It is also important to note here, however, that the test methods described in this book and used for lubricant and/or machine condition monitoring programmes are only those that are either standard test methods published by national and international organisations that develop and specify tests for lubricants or those that have been widely accepted by lubricant users. There are many test methods that have been developed and used by individual companies for their own purposes. While these tests may be of significant value to those companies, they have not yet been demonstrated to be of value to a larger number of companies. There are too many of them to be included in this book, although some of them may gain sufficient acceptance to be included in later editions.

It is envisaged that this book will be of practical use to companies and organisations that use significant amounts of lubricants, have large or expensive machines, or use equipment that is critical to the safe or profitable operation of the company or organisation.

Many users of lubricating oils and greases, such as individual motorists or truck drivers, are unlikely to find much value in reading this book. Some of these users may be interested to read this book, to discover more of the technology that goes into the safe and reliable operation of machines, equipment and systems. Owners and operators of large fleets of trucks, buses, trains and planes should find the contents of this book to be particularly useful and valuable.

1.2   APPROACH

All lubricants deteriorate in use. Consequently, they have a finite useful life. This useful life almost always depends on the conditions in which a lubricant is being used, together with the type and application of the machinery, equipment or system which is being lubricated. Arduous conditions, heavy loads, low or high speeds, constant stop-start or shock-loads will all contribute to shortening the useful life of a lubricant. It is axiomatic, therefore, that lubricants affect machines and machines affect lubricants.

Monitoring how and why lubricants deteriorate in use will enable a determination of when a lubricant has reached the end of its useful life and therefore needs to be changed. The primary criterion for the need to change a lubricant is when its further deterioration would lead to a situation in which it would be unable to protect the machinery or equipment from damage. There are several separate or joint causes for this.

In engines, for example, abrasive and corrosive materials can cause bearing damage, or bore polish, by removing the cross-hatched honing marks which maintain the lubricant film, or in extreme cases, "scuffing" of piston and bore.

These effects are often interdependent and will cause further changes either directly or through catalytic effects. When these lubricant deterioration effects occur in such complex systems as lubricant formulations, then a structured approach is needed to understand and solve the problem.

A lubricant condition monitoring programme will help significantly in determining when a lubricant is likely to reach or has reached the end of its useful life. At the same time, monitoring the condition of the machine, equipment or system in which the lubricant is being used will add to the determination of when the lubricant needs to be changed.

Chapter 2 describes and explores the requirements and benefits of analysing lubricants and monitoring their condition. The reasons for lubricant deterioration are explained, and the methods for setting up and operating a lubricant condition monitoring programme are discussed.

The first requirement for the effective analysis of lubricants is to obtain a representative sample. This involves using proper sample bottles, selecting and using the appropriate methods and techniques for taking samples and determining the optimum sampling location and frequency.

The objective of sampling is to maximise data density while minimising data disturbance. Data density means gathering as much information in the sample as possible. Data density is maximised by sampling in the right location with the right equipment at the right time.

Most usually, the right location is a "live zone" within a machine, where oil is flowing in a turbulent manner. The sample will then contain all the useful information required for trend analysis, without losing any data resulting from either component fly-by or settling. The right equipment includes the use of special sampling valves, vacuum sampling pumps, disposable tubing and other accessories for taking samples as cleanly as possible. If the machine is critical to the company's operation or is not running as expected, the sampling frequency may be quite short, for example every two weeks. Otherwise, the sampling frequency could be as long as every six months. Taking consistent samples helps to plot meaningful trends of lubricant properties, lubricant performance and contamination levels.

Methods and equipment for the correct sampling of oils and greases are presented and discussed in Chapter 3. This includes ways to obtain the most representative samples and how to avoid problems with sampling different types of machines, equipment and systems.

Lubricants are formulated in a lubricant development laboratory, and they are then either blended in a blending plant, for oils and solid lubricants, or manufactured in a grease plant, for greases.

Lubricating oils are manufactured by taking appropriately selected base oils and blending them with appropriate additives, before delivering to an end user either in bulk or packaged in international bulk containers (IBCs), drums, plastic bottles or tin-plate cans. Contamination and base oil/additive degradation can occur at any stage of the blending, packaging, storage or transportation of finished lubricants.

While a lubricant manufacturer is responsible for insuring that the selection of base oils and additive package meets the required performance characteristics of the specific lubricant, the manufacturer is also responsible for insuring that the formulated lubricant is delivered to the end user without levels of degradation or contamination that might compromise the integrity of the fluid before it is used in a machine or equipment. Even before delivery to the end user, significant levels of contaminants, including heat, moisture and particles, can cause premature lubricant degradation in storage, at the manufacturing facility, at the blending plant or at the lubricant distribution warehouse.

Testing new lubricants upon receipt by a customer is essential to ensure the lubricant received is the lubricant ordered. This is critical for several reasons, for example to ensure that the oil received is the correct one, to establish a baseline for subsequent testing and monitoring of the oil condition and to establish the lubricant's level of cleanliness. Chapter 4 describes and discusses the need to test new lubricants prior to their use.

Once a baseline set of data has been established for a specific lubricant in a specific machine or item of equipment, the next stage in a lubricant analysis and condition monitoring programme is to take regular samples of lubricant from the machine and test them using the same tests as for the baseline.

Numerous test methods that can be used for testing both new lubricants and lubricants in use as part of a condition monitoring programme are discussed in depth in subsequent chapters. Chapter 5 focusses on the ways that can be used to test lubricants in use and how to report and assess the results.

Later chapters look at different types of equipment and machinery and how their operating condition can be monitored, using both lubricant testing and other analytical methods.

A huge number of tests can be, and are, used to evaluate the properties and performance of lubricating oils and grease in their numerous applications. Some of these tests are chemical, others are physical, some are mechanical and a few involve real-life operations.

However, not all of these tests are suitable for use in a lubricant condition monitoring programme. Chapter 6 describes and discusses those chemical tests that could or should be used to monitor the condition of oils and greases. Later chapters look at physical and mechanical tests for lubricant condition monitoring.

The international and national organisations that develop, publish, update and monitor these tests are summarised in Chapter 4. A number of original equipment manufacturers (OEMs) also develop, publish and update test methods, some of which can be useful in a lubricant condition monitoring programme. Those chemical tests that should (or sometimes could) be used to monitor each type of lubricant application will be discussed in later chapters.

The tests described in Chapter 6 cover only those that can or should be used to monitor the condition of oils and greases in operating machinery. Other chemical tests not described in Chapter 6 are used during the formulation and development of new or improved lubricants and/or during the blending of lubricating oils or the manufacture of greases.

Chapter 7 describes and discusses those physical tests that could or should be used to monitor the condition of oils and greases. The tests described in Chapter 7 cover only those that can or should be used to monitor the condition of oils and greases in operating machinery. Other physical tests not described in Chapter 7 are used during the formulation and development of new or improved lubricants and/or during the blending of lubricating oils or the manufacture of greases.

One of the main functions of a lubricant is to reduce mechanical wear. Closely related to wear reduction is the ability of lubricants of the extreme pressure (EP) type to prevent scuffing, scoring and seizure as applied loads are increased. As a result, a considerable number of machines and procedures have been developed to try to evaluate anti-wear and EP properties. In a number of cases, the same machines are used for both purposes, although different operating conditions may be used.

Wear can be divided into four classifications based on the cause: abrasive wear, corrosive (chemical) wear, adhesive wear and fatigue wear. To these four mechanisms of mechanical deterioration must be added seizure and pitting.

These mechanical deterioration types and the various machines that are used to evaluate the properties and performances of lubricating oils and greases used to eliminate, mitigate or control them are described and discussed in Chapter 8.

In the author's opinion, the only true test of a lubricant's performance is whether it functions satisfactorily for several years in the machinery for which it was formulated.

Lubricant specifications, laboratory tests, rig tests, engine tests and field trials are only useful as a guide or prediction of likely performance in service. In assessing the possible suitability of a specific lubricant for a specific application, it is critically important to remember to check that the predictive tests are the best ones for that application.

All of these laboratory, bench, rig and engine tests and field trials simply attempt to shorten the time taken to assess the properties or performance of lubricants, because the "real-life" performance of lubricants can take many years to establish. If actual performance in service was the only method by which lubricants could be evaluated, the development of new or improved products could take a very long time indeed. It is, however, very important to remember that laboratory, bench, rig, engine and even field tests are useful mainly for establishing lubricants that are unlikely to perform satisfactorily in practical operations, not those that definitely will perform satisfactorily for many years.

Laboratory tests, usually for physical or chemical properties, are usually quick and comparatively inexpensive. Bench and rig tests, which are intended to assess performance properties, such as anti-wear, corrosion inhibition or deposit formation, are performed in specially designed equipment or machines that are smaller and/or less complex than real engines or gearboxes. Consequently, they tend to be shorter and cheaper than engine tests. Engine tests are performed in special buildings, to try to provide as much repeatability and reproducibility as possible, and are more realistic, but are longer and more expensive than laboratory or rig tests. Field tests try to reproduce "real-life" operating conditions, but are lengthy and expensive.

The tests described in Chapter 9 are almost never used to monitor the condition of a lubricant in use. They are included in this book in order to describe why they are not used in a lubricant condition monitoring programme.

An engine oil obviously affects the operational efficiency of the engine, and monitoring the properties of the oil is obviously important. Conversely, maintaining those engine components that affect the lubrication process is also important.

With both gasoline and diesel engines, it is necessary to have clean combustion and crankcase ventilation air, which means that air cleaners and the positive crankcase ventilation (PCV) system must be serviced regularly. A clogged air cleaner, while it may be effective in cleaning the air, can restrict the volume of air reaching the engine enough to reduce power output significantly. To maintain power, drivers have a natural tendency to open the throttle, which only adds to the difficulty, as the extra fuel is not burned. Diesel engines are quite sensitive to such a condition and react by generating soot and building up engine deposits. Soot in engine oil can become very damaging. In addition to keeping the filters serviced, it is important that the piping connecting the filter to the engine be unobstructed and leak free. This is particularly important in installations having the air filter located at a considerable distance from the engine.

The ignition systems of spark ignition engines and the fuel injection systems of both gasoline and diesel engines should be in good working order and properly adjusted to assure the cleanest, most complete combustion possible. Malfunction or incorrect adjustment of these systems can result in increased amounts of unburned fuel in the cylinders and dilution or more rapid build-up of contaminants in the oil, as well as increased emissions.

Chapter 10 describes and discusses the methods that can be used to monitor the condition of both oils and engines of all types, including those used in cars, vans, taxis, motorcycles, trucks, buses, industrial and off-highway vehicles, power generation and ships. Hybrid and electric vehicles are also included.

To adjust the speed of an engine or motor to the required speed(s) of a machine, vehicle or item of equipment requires a transmission system of some type. The effective lubrication of these transmissions is just as important as the lubrication of the engines driving them and, usually, requires special lubricants.

Gears are used to transmit motion and power from one rotating shaft to another or from a rotating shaft to a reciprocating element. Gearboxes contain both gears and bearings, because the rotating shafts must rotate smoothly, whatever load is put on the gears. As gear teeth mesh, they roll and slide together. This combination of sliding and rolling occurs with all meshing gear teeth, regardless of the type of gears. The two factors that vary are the amount of sliding in proportion to the amount of rolling and the direction of slide relative to the lines of contact between the tooth surfaces.

Monitoring the oils and greases used to lubricate the different types of gears and bearings in transmissions can be critically important and may require different methodologies and tests to those used for engines. For example, if the gearbox in a helicopter fails, the aircraft cannot fly and it will fall to earth like a stone.

Chapter 11 describes and discusses the different types of gears and their lubrication, monitoring gears used in automotive applications and in industrial applications, and the different types of bearings and their lubrication. The lubrication and monitoring of wind turbine gears and bearings are also explored in depth.

Hydraulic systems, like many other industrial systems, are increasingly being operated at higher speeds, at higher pressures and with higher power outputs and shorter cycle times, to give higher productivity. Control valves have tighter clearances to improve performance and, to reduce cost, weight and space, reservoirs are shrinking in size. OEMs are designing systems with reduced noise levels and lower carbon footprints. This is the case across many industries, ranging from injection moulding and steel mills to off-road and construction equipment.

Since the hydraulic fluid is an integral part of the equipment's operational components, making all of these evolutionary changes in hydraulic system design places further demands on hydraulic oils.

End users share some responsibility in the increased stress on the hydraulic oils. Their attention to leak reduction inadvertently results in fewer top-up additions of new oil. Because of the economics of production, many hydraulic machines are operated by users at a higher production rate than originally designed.

Hydraulic oils and hydraulic systems and their issues, hydraulic oil performance, hydraulic fluid filtration and contamination control and methods to monitoring hydraulic oils and systems, including fire-resistant hydraulic fluids, are explained and discussed in Chapter 12.

Compressors and vacuum pumps are vitally important mechanical devices that are used to pressurise and circulate gases through processes, facilitate chemical reactions, provide inert gas for safety or control systems, recover and recompress process gases and maintain correct pressure levels by adding or removing gases or vapours from process systems. Compressors are also used to provide pneumatic (compressed air) power for construction and manufacturing operations. Gas compressors are used in almost all industries, including automotive, steel, chemical, mining, food, natural gas and petroleum production and processing and storage and energy conservation. Refrigerator compressors are used in refrigeration and air conditioning.

Steam and industrial gas turbines are used extensively in electricity generation, as prime movers for generators. They are also used for mechanical drive applications in many industries, to power centrifugal pumps, compressors, blowers and other machines. They continue to be used for shipboard propulsion. Gas turbines are used to power civil and military jets.

Water turbines are used for generating electricity in hydroelectric power stations. Lubricating, monitoring and maintaining compressors and turbines are obviously vitally important for a huge number of users of these machines.

The multitude of types of compressors and turbines and their applications are described and discussed in Chapter 13. It also looks at their lubrication, the types of lubricants used and the properties and performances required of those lubricants. The monitoring of lubricants and the maintenance of the machines are explored and discussed in depth.

Production engineering fluids are usually grouped into three main classes: metalworking fluids and pastes, heat treatment fluids and temporary corrosion protectives.

Metalworking fluids and pastes are engineering materials that optimise metalworking processes. Metalworking encompasses metal removal and metal deformation. Metalworking fluids used for metal removal are known as cutting and grinding fluids, while fluids and pastes used for drawing, rolling, bending or stamping processes are known as metal-forming fluids or pastes. The outcome of the two types of processes differs. The processes by which the machines make the products, the mechanics of the operations and the requirements for the fluids used in each process, are different.

Heat treatment, also known as quenching, is one or more operations involving the controlled heating and cooling of a metal in the solid state for the purpose of obtaining specific properties. Many types of heat-treating processes can be used to fulfill a wide variety of hardness and mechanical properties that may be required for metal components.

Temporary corrosion protectives provide corrosion protection for relatively short durations, during the storage or transportation of manufactured metal components or assemblies. "Temporary" refers to their ease of removal, not to the duration of the protection. They provide a water- and oxygen-resistant barrier by reason of their blanketing effect and natural or added corrosion inhibitors, which form an adsorbed layer on the metal surface.

The roles that each of these fluids and pastes plays in production engineering and the methods used to monitor and control them are described and examined in Chapter 14.

The ASTM defines a grease as "A solid to semifluid product of dispersion of a thickening agent in a liquid lubricant. Other ingredients imparting special properties may be included".

Greases are most usually used where a lubricant is required to maintain its position in a mechanism, particularly where opportunities for frequent re-lubrication may be limited or commercially unjustifiable. This may be due to the type of mechanism, the mode of action, the type of sealing or a need to minimise ingress of contaminants into the mechanism.

The main disadvantages of greases are their inability to remove wear debris rapidly and the lack of flushing action to remove wear particles or contamination from the mechanism. Conventional soap-based greases have a relatively limited operating temperature range, although more specialised greases can operate in more extreme conditions, but at a higher cost.

Approximately 90% of all rolling bearings are lubricated by grease. Monitoring greases is therefore quite important for many types of industrial machines. Chapter 15 explores and explains the specific requirements for monitoring the condition of greases and the machines in which they are used, in comparison with lubricating oils.

Lubricant condition monitoring and machine condition monitoring can be expensive, complex and time-consuming. They are not suitable for all mechanical equipment. Fortunately, for large and expensive machines and items of equipment, they can be extremely cost-effective.

Unfortunately, there are numerous significant issues that must be addressed in order to run an effective condition monitoring programme. Avoiding problems is very worthwhile and provides significant benefits when planning, implementing and running a condition monitoring programme.

Chapter 16 sets out why and how to define, plan and implement an achievable lubricant condition monitoring programme. It also describes and discusses the benefits of using the results of a lubricant condition monitoring programme to instigate effective predictive maintenance on all machines, items of equipment and lubricated systems.

2 Reasons for Analysing Lubricants and Monitoring Their Condition

2.1 INTRODUCTION

All lubricants deteriorate in use. Consequently, they have a finite useful life. This useful life almost always depends on the conditions in which a lubricant is being used, together with the type and application of the machinery, equipment or system which is being lubricated. Arduous conditions, heavy loads, low or high speeds, constant stop-start or shock-loads will all contribute to shortening the useful life of a lubricant. It is axiomatic, therefore, that lubricants affect machines and machines affect lubricants.

It is very important to remember that there is a fundamental difference between the service life of product and its shelf life. A product's shelf life is usually indicated by its manufacturer on the packaging to show the length of time the product can remain in its current packaging before being deemed unsuitable for use. The shelf life is usually indicated by a "use by" date or a "best before" date in the package. Many retail lubricants, for example automotive engine oils and automotive greases, do not have a manufacturer's shelf life, as it is assumed that they will be used well before the three to four year interval since they were manufactured. The author is aware of retail lubricants that are perfectly useable ten years after they were manufactured.

The service life of a product is determined by the application and conditions under which the product is being used. This is the subject of this book. Three examples of the differences between new and used lubricants are shown in Tables 2.1 to 2.3. The deterioration in each of the lubricants is obvious.

Monitoring how and why lubricants deteriorate in use will enable a determination of when a lubricant has reached the end of its useful life and therefore needs to be changed. The primary criterion for the need to change a lubricant is when its further deterioration would lead to a situation in which it would be unable to protect the machinery or equipment from damage. There are several separate or joint causes for this:

- It has become so oxidised that the increased acidity may lead to corrosion and/or the formation of sludge or varnish.
- Its viscosity has increased or decreased beyond a specification limit.

DOI: 10.1201/9781003245254-2

TABLE 2.1

New and Used Heavy-Duty Diesel Engine Oil

Property	New	Used
Appearance	Brown	Black
Viscosity at 40°C, cSt	71.3	61.8
Viscosity at 100°C, cSt	11.7	10.6
Viscosity index	160	162
Acid number, mg KOH/g	2.8	4.5
Base number, mg KOH/g	9.6	4.5
Water content, %wt	0.0	0.0
Soot content, %wt	0.0	1.2
Fuel dilution, %wt	0.0	4.0
Phosphorous, ppm	350	437
Zinc, ppm	400	602
Calcium, ppm	1,100	1,267
Boron, ppm	129	90
Magnesium, ppm	10	15
Sodium, ppm	<1	4
Iron, ppm	<1	9
Aluminium, ppm	<1	<1
Copper, ppm	<1	2
Lead, ppm	<1	8
Silicon, ppm	6	25

Source: STLE.

- It has become too laden with particulate dirt or metallic wear debris.
- Its additive pack has become depleted in one or more components.
- It has become contaminated with some other fluid, such as cooling water or machine cleaner.

In engines, for example, abrasive and corrosive materials can cause bearing damage, or bore polish, by removing the cross-hatched honing marks which maintain the lubricant film, or in extreme cases, "scuffing" of piston and bore.

These effects are often interdependent and will cause further changes either directly or through catalytic effects. When these lubricant deterioration effects occur in such complex systems as lubricant formulations, then a structured approach is needed to understand and solve the problem.

A lubricant condition monitoring programme will help significantly in determining when a lubricant is likely to reach or has reached the end of its useful life. At the same time, monitoring the condition of the machine, equipment or system in which the lubricant is being used will add to the determination of when the lubricant needs to be changed.

TABLE 2.2
New and Used Hydraulic Oil

Property	New	Used
Appearance	Bright	Dark brown
Colour, ASTM D1500	3.0	7.5 dil
Viscosity at 40°C, cSt	46.0	46.8
Viscosity at 100°C, cSt	6.72	6.78
Viscosity index	98	98
Acid number, mg KOH/g	1.0	2.1
Insolubles, mg/l	15	270
n-Pentane insolubles, %	0.0	2.5
Water content, %wt	0.0	0.0
Phosphorous, ppm	550	390
Zinc, ppm	560	402
Sodium, ppm	<1	21
Iron, ppm	<1	52
Aluminium, ppm	<1	10
Copper, ppm	1	17
Lead, ppm	<1	8
Chromium, ppm	<1	3
Tin, ppm	<1	2

Sources: STLE, Pathmaster Marketing Ltd.

2.2 METHODS TO DEFINE LUBRICANT SERVICE LIFE

In order to avoid the problems associated with lubricant deterioration in use, determining when to change a lubricant involves two extremes:

- A time- or distance-defined period of lubricant replacement. The three main types of these are, for example, 500 hours of operation, every year or 10,000 km or miles. These do not take into account the actual state of the lubricant. Instead, custom and practice show that the service interval set is sufficient to ensure that excessive wear does not occur. This is the precautionary principle. It does not require sampling and analyses or on-board sensors and is therefore low cost in terms of monitoring and maintenance. However, it may be high cost in terms of the remaining life, and hence, value of the lubricant, depending on the amount of lubricant in the machine. It can, therefore, tend to waste resources and not be beneficial to the environment.
- A quantitative determination of the state of the lubricant, achieved by sampling at regular intervals and monitoring various parameters to give a collective assessment of the condition of the lubricant. This is called "oil condition monitoring". The approach tends not to waste resources,

TABLE 2.3
New and Used Grease

Property	New	Used
Appearance	Bright	Dull
Texture	Smooth	Coarse
Colour	Mid-brown	Black
Penetration		
Unworked	310	450
Worked	319	459
NLGI number	1	000
Drop point, °C	181	143
Water content, %wt	0.0	0.2
Ash content, %wt	2.7	4.2
Calcium, ppm	3,000	5,900
Sodium, ppm	100	400
Potassium, ppm	<10	<10
Lithium, ppm	13,000	6,100
Lead, ppm	16,500	7,300
Iron	–	Major increase
Aluminium	–	Minor increase
Silicon	–	Traces

Source: STLE.

but can be high cost in terms of the testing required and, for smaller machines and items of equipment, may be impractical. The time interval of sampling should be, at most, half of the anticipated service interval. The database built up over time has value for the long term and is concerned with long-term trends in lubricant parameters such as wear metal concentrations, viscosity, particulate levels and many other factors that will be discussed in depth in the following chapters. For a full condition monitoring programme, the lubricant is replaced when its condition reaches a lower bound of aggregated parameters and it is judged to be, or close to being, unsuitable for its purpose of lubricating and protecting the mechanical system.

An interim position is to sum the overall performance of the system, be it engine or machine, from its last service interval by integrating power levels used in time intervals/distances travelled/time elapsed. The underlying assumption is that the level of performance and its time of operation are related to the degradation of the lubricant. For example, 100 km of unrestricted daytime high-speed driving on an autobahn in summer is assumed to degrade a lubricant more than 100 km of moderate speed urban driving in autumn or spring. Thus, the aggregates of high-power-level operation over time are weighted more than the same period of low-power operation.

2.3 CHEMICAL CAUSES OF LUBRICANT DETERIORATION

There are many chemical causes that can result in the deterioration of lubricants. The most common are:

- Oxidation.
- Thermal degradation.
- Additive depletion.
- Electrostatic spark discharge.
- Contamination.

Oxidation is any chemical reaction that involves the moving of electrons. When electrons are removed from a substance, it is said to have been oxidised. Normally, this is a reaction between oxygen and the substance. In the case of lubricants, the substances are organic molecules, such as hydrocarbons. When a hydrocarbon reacts with oxygen, the hydrocarbon loses some electrons and the oxygen gains those electrons.

Oxidation is the opposite of reduction, which is a reaction involving a gain of electrons. Consequently, a reduction reaction always occurs together with an oxidation reaction. Oxidation and reduction together are called redox (reduction and oxidation). Oxygen does not have to be present in a reaction for it to be a redox reaction. Oxidation is the loss of electrons, and reduction is the gain of electrons.

Oxidation of hydrocarbons and similar organic molecules can lead to an increase in viscosity and acidity and the formation of varnish, sludge and sediment. Additive depletion and a breakdown in the base oil can also result. Once an oil starts to oxidise, its acid number is likely to increase. In addition, rust and corrosion can form on the equipment due to oxidation. Increased viscosity can cause problems such as increased energy consumption, inability to operate as a coolant effectively and low fluid flow on start-up that will lead to increased wear. An increase in acidity is another cause of increased wear from acidic corrosion. The formation of sludge and varnish leads to problems of filter plugging. Oxidation is the single biggest factor in determining the useful service life of a lubricant.

An oxidation reaction goes through a free radical chain mechanism, consisting of three stages: initiation, propagation and termination. In the initiation reaction, free radicals are formed. Free radicals are atoms or molecules with unpaired electrons, making them highly reactive with hydrocarbon compounds to form hydroperoxides. In the propagation reaction, the hydroperoxides react with the oil, where more free radicals are created. In the presence of catalytic items such as water, wear metals and high temperatures, peroxides may split and then sustain the reaction. This propagation step continues to be carried out as there is a continuous feed of peroxides to promote the reaction. The termination step is due to the presence of one or more oxidation inhibitors, also known as antioxidants. Some oils have a natural oxidation inhibiting action while, more commonly, antioxidants are additives that are blended into the oil. These inhibitors work by breaking the oxidation chain reaction, decomposing the peroxides that are formed or deactivating the metal surfaces.

The temperature of the lubricant should be a primary concern. Because of friction, both between moving surfaces and internally within the lubricant, heat is generated.

Although a lubricant's principal function is to lubricate, a secondary function (among others) is to dissipate heat. As a consequence, the operating temperature of a lubricant is always higher than ambient.

The Arrhenius equation for chemical reactions, such as oxidation and thermal degradation, states that for every 10°C rise in temperature, the reaction rate approximately doubles. This means that if a lubricant has an oxidative lifetime of X at temperature Y°C, at a temperature of Y + 10°C, the lubricant's lifetime will be ½X. Keeping the lubricant as cool as possible when in use will extend its life and reduce the reaction of oxidative and/or thermal breakdown.

In addition to lubricating moving parts in machines, equipment and systems, oils are frequently required to dissipate heat. This means that oils will sometimes be heated above their recommended operating temperatures. This occurs mainly with turbine and compressor oils, but can also occur with some engines and gearboxes. Overheating can cause the lower boiling components of the oil to vaporise. Excessive overheating is likely to cause thermal degradation of the oil and/or decomposition of some of the additives in the oil. At temperatures greatly exceeding the thermal stability point of the lubricant, larger molecules will break apart into smaller molecules. This thermal cracking, often referred to as thermal breakdown, can initiate side reactions, induce polymerisation, produce gaseous by-products, destroy additives and generate insoluble by-products. Some of these changes will reduce the viscosity of the oil, while others will increase the oil's viscosity. Either of these will alter the ability of the oil to function effectively as a lubricant.

One type of thermal degradation of lubricating oil is known as micro-dieseling, also known as compressive heating or pressure-induced thermal breakdown. This is a process in which an air bubble transitions from a low-pressure region in a system to a high-pressure zone. When oil is pumped around a system, particularly in a hydraulic system, as the oil enters the pump, it is subjected to excessive low pressure (vacuum) and any dissolved air is released as small bubbles. Entrained air bubbles are even more of a problem.

As the oil travels to the high-pressure side of the pump, the air bubbles are compressed rapidly. This adiabatic compression can generate temperatures on the surfaces of the air bubbles in excess of 1,000°C. Oil molecules at the bubbles' surfaces will undergo rapid thermal and oxidative degradation, resulting in the formation of soot particles. As a result, the oil will darken and the soot particles are likely to start to cause higher rates of wear. It is therefore important to minimise air in oils used in high-pressure applications.

Additives are blended into final lubricant products to either enhance base oil properties, suppress base oil properties or impart new properties to the final product. Most additives are designed to be sacrificial in nature. They are used up during the service life of the lubricant. This makes the monitoring of additives an important part of any oil condition monitoring programme. Using oil analysis to monitor additive levels is important not only to assess the health of the lubricant but also to provide clues as to whether the additives are degrading faster than expected or what is causing the depletion of the additives. Numerous methods to monitor the additives in lubricants in use are discussed in Chapter 6.

Oil circulating systems are prone to electrostatic charges by friction caused due to oil flowing along the surfaces of the system. The strength of the static charge depends on the conductivity of the lubricant and the oil flow rate. The lower the conductivity and the higher the flow rate, the greater risk of electrostatic charging. This charging can accumulate and release a spark between 10,000°C and 20,000°C typically at sharp surfaces and most commonly in mechanical filters. The risk of electrostatic discharges increases when the lubricant is formulated with a hydrocracked or synthetic base oil, contains no polarising additives, flows through narrow pipes or contains high proportions of air bubbles. Although the sparks may be very small and for only fractions of a second, their cumulative effect is another source of thermal degradation of the oil.

Contaminants such as dirt, water, air and others can greatly influence the rate of lubricant degradation. Contamination of a lubricant generally has three sources: built-in, ingested and self-generated. Built-in contamination is a contamination that has been left in the component from the original manufacturing process. Ingested contamination is a contamination drawn in from outside the component, for example water, air, dirt and coolant. Self-generated refers to a contamination that is generated within the component such as wear metals and soot. All of these contaminants have the potential to accelerate the degradation of the lubricant. Dirt containing fine metal particles can be a catalyst that initiates and accelerates the degradation process of the lubricant. Air and water can provide a source of oxygen that reacts with the oil and leads to oxidation of the lubricant. Water can also help to deplete additives by reacting with them.

Solid particles are the most common particles found in an oil. There a few different ways that these particles are generated, including abrasive, adhesive and erosive wear due to fatigue. Wear particles that are equal to or slightly larger than the clearance space are the most damaging. Wear particles also form a chain reaction in that these particles become work-hardened, meaning they are harder than the original surface. If not removed by proper filtration, this can cause additional wear.

Water and air can provide a large amount of oxidation potential for reaction with the oil. The most common result of water contamination is rust. Water can also cause certain additives to become unstable, thereby reducing their effectiveness. In colder environments, water can cause a lack of lubrication due to freezing of the water. Air is found in four phases in oil: free air, dissolved air, entrained air and foam. Air contamination has the propensity to cause problems such as pump cavitation and micro-dieseling.

2.4   PHYSICAL CAUSES OF LUBRICANT DETERIORATION

A lubricant becomes physically unsuitable for further continued service use through a range of causes:

- Hard particulates from the thermal breakdown of hydrocarbons C_{30} and higher.
- Softer particulates from the thermal breakdown of C_{15} hydrocarbons from diesel fuels.

- Metallic materials, such as metallurgical cutting residues and welding repair particles or production grinding processes.
- Through defective sealing systems, which allow ingress of abrasive siliceous materials.
- Fuel condensing into the lubricant and reducing its viscosity, or together with condensed water, forming an emulsion of low lubricity value.
- Cooling water ingress into the lubricant system through defective seals.
- Textile materials such as (production line) cleaning cloths, contributing "lint", which compacts into obstructions of oilways.
- Infiltration through exhausted and inefficient oil filters.
- Filling through unclean filler pipes/tubes.
- Lubricant reservoirs open to the (unclean) atmospheres.
- Through overwhelmed air filters, for example in desert areas.

In some of these cases, it is the processes of chemical deterioration of the lubricant that have resulted in changes to their physical properties, such as viscosity, pour point, cold temperature flow or pressure–viscosity coefficient.

The hard particulates from the thermal breakdown of hydrocarbons C_{30} and higher are more commonly known as soot. These particles, unlike the softer soot particles formed from the thermal breakdown of C_{15} hydrocarbons, can contribute to higher rates of wear, particularly in reciprocating compressors and hydraulic systems. The softer soot particles can combine with water to produce grey or black sludges, mainly in the crankcase and oil passages in an engine. Soot can also adhere to varnish deposits and accelerate the build-up of carbon, if the varnish-forming tendencies of the oil are insufficiently controlled.

The occurrence of metallurgical cutting residues and welding repair particles or production grinding processes result from incomplete cleaning and flushing of such debris either during manufacturing or assembly of machines, equipment or systems. For many machines and systems, thorough cleaning of all components before assembly followed by thorough flushing of all lubrication systems is vitally important for ensuring a long service life for both the machine or system and the lubricants. This is particularly necessary for large steam and water turbines, which must be assembled from components inside the power plant in which they will operate. Metallic debris left in a machine or system is likely to cause catastrophic wear or seizure, stuck servo-valves and/or rapid filter blockage.

Defective sealing systems that allow dirt or dust into a lubricant will also result in higher rates of wear, stuck servo-valves and rapid filter blockage, in addition to helping to accelerate oil oxidation. Ingress of water into a lubricant, past defective seals, will help to accelerate rusting of steel components, to degrade additives and the generation of oil/water sludges. Lubricant reservoirs that are left open to the atmosphere are likely to allow the ingress of dust, dirt and/or moisture, particularly in dusty or humid climates.

Cleaning materials, such as cloths and rags, left in assembled machines and systems will generate fibres that will block oilways and filters. Blocked filters will be unable to remove machine generated wear or corrosion particles and may cause oil starvation of pumps and motors. Not only manufacturing debris should be thoroughly removed from machines, equipment and systems, but cleaning materials must be removed as well.

The debris of system wear, abrasive wear products from combustion processes and defective sealing materials are physical causes of lubricant deterioration. Another obvious physical cause of degradation is to add an incompatible lubricant to an existing formulation in an existing system. Although the base fluids may be miscible, their additives may be incompatible and precipitate, leaving the circulating fluid as a simple base oil system with little mechanical or tribological protection. In most cases, the physical causes of lubricant deterioration are simply related to good maintenance or the lack of its proper application.

2.5   DETERIORATION OF MACHINES AND EQUIPMENT

All machines, equipment and mechanical systems wear, but at different rates during their serviceable life. If nothing is done about the rate of wear, eventually the machine, equipment or system will fail. A plot of wear or system failure against time has a well-established pattern, which is frequently described as the "bath-tub" curve.

The plot, shown in Figure 2.1, does not describe "wear" or "failure" for individual systems, but is a statistical description of the relative wear or failure rates of a product group with time. Nominally identical units can fail at very different times, depending on a multitude of factors. However, with modern production methods and attention to the wear of components, most machines, equipment and systems should experience a "normal" lifetime. These serviceable lifetimes can range from 1 year, under very arduous operating conditions, to over 50 years under carefully controlled benign conditions.

Mechanical or system failures during the initial "running-in" period are always caused by material defects, design errors or assembly problems. Failures due to

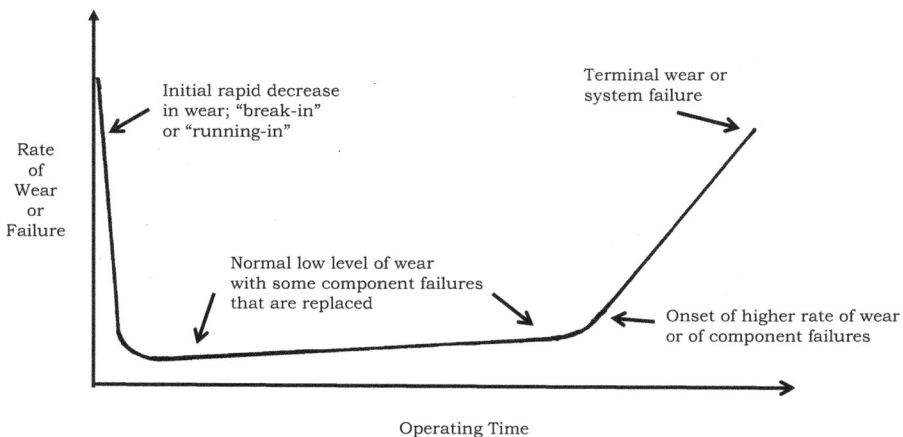

FIGURE 2.1   The "bath-tub" curve for wear or system failure.

Source: Pathmaster Marketing Ltd.

defective materials or design errors should be reported to the machine or equipment manufacturer or system designer. Failures due to incorrect assembly are obviously the responsibility of the people performing the assembly, whether they be internal or external to the system user.

Failures during the expected normal lifetime of the machine, equipment or system are usually considered to be random cases where stress exceeds strength. Eventual wearing out is a fact of life, due to either fatigue or material depletion by wear. It is therefore axiomatic that the useful operating life of a machine, item of equipment or mechanical system is limited by the component with the shortest life, unless that component can be repaired or replaced. This is where lubricant and machine condition monitoring can impact planned maintenance of components, machines and systems.

The "bath-tub" curve is used as an illustration of the three main periods of system wear or failure. In very many cases, wear and failure information is not recorded in a database, so that the initial, normal and terminal phases of system wear and failure can be measured and calibrated. The timescales for these phases usually vary between one system and another. Nominally identical machines or systems can experience different rates of wear or failure, although in many cases the differences are only slight. For example, one machine may be closer to an external doorway, leading to slightly wider ranges of operating temperature. Another machine may be closer to a source of vibration, such as a reciprocating compressor.

When condition monitoring is used to monitor the wear of a machine or system, a gradually increasing level of a wear metal in each sample of lubricant taken at regular intervals will indicate that the machine may have entered the final phase of its service life and its replacement or overhaul is becoming due. Arrangements to replace the machine, start ordering replacement parts and plan a maintenance schedule can be made before a failure or unexpected interruption of service occurs. This saves costs because the machine is worn, but not damaged and can be readily and economically overhauled or replaced in a planned operation that minimises service interruption. Knowledgeable replacement of worn systems or components is usually estimated to have a direct benefit:cost ratio of 10:1, rising to 20:1 when indirect costs of unexpected interruptions of service are included.

For many machines, particularly consumer goods, service life represents a commitment made by the item's manufacturer and is usually specified as a median. This is the time that any manufactured item can be expected to be "serviceable" or be supported by its manufacturer. As noted earlier, service life is not to be confused with shelf life, which deals with storage time, or with technical life, which is the maximum time during which the item can physically function. It also differs from "predicted life", or "mean time to failure". A manufacturer may estimate, by hypothetical modelling and calculation or extensive testing, a predicted life for a machine, item or system, for which it will honour warranty claims. This may also be useful for an end user when planning for mission fulfilment. The difference between service life and predicted life is most clear when considering mission time and reliability in comparison to mean time to failure and service life. For example, a missile system can have a mission time of less than 1 minute, a service life of 20 years, an active mean time to failure of 20 minutes, a dormant mean time between failures of 50 years and a reliability of 0.999999.

Consumers will have different expectations about service life and longevity, based upon factors such as use, cost and quality. Manufacturers will commit to a very conservative service life, usually two to five years for most commercial and consumer products. However, for large and expensive durable goods, the items are not consumable and service lives and maintenance activity will factor large in the service life. Again, a passenger aeroplane might have a mission time of 11 hours, a predicted active mean time between failures of 10,000 hours without maintenance (or 15,000 hours with maintenance), a reliability of 0.99999 and a service life of 40 years.

For an individual product, the component parts may each have independent service lives, resulting in several bath-tub curves. For example, a tyre will have a service life partitioning between the tread and the casing.

For equipment and machines that can be maintained, components that are known or accepted to wear out and will need to be replaced are associated with planning for the provision of spare parts. If there is no such planning, the useful life of the whole machine will be compromised. Again, a simple example is automotive tyres. Failure to plan for tyres wearing out would limit the service life of a car, van or truck to the extent of a single set of tyres.

Towards the end of any machine's useful life, maintenance costs start to become higher and higher and, at some point, the expense of maintaining the machine is higher than the cost of buying a new machine. For many items of industrial equipment, a new machine may offer significant additional benefits compared with the old machine. Benefits could include energy efficiency, enhanced reliability, greater automation and/or integration with other items of equipment.

Securing the maximum cost-effective life of each item of equipment should be the primary goal of a lubricant and machine condition monitoring programme.

2.6 RELATIONSHIPS BETWEEN LUBRICANT AND MACHINE DETERIORATION

It may not seem obvious at first, but the deterioration of lubricants affects the deterioration of machines and the deterioration of machines affects the deterioration of lubricants. A few examples will suffice to illustrate this.

Wear metal particles, particularly iron, copper, lead and aluminium, frequently act as catalysts to speed up the rate of oxidation of organic compounds, including the hydrocarbons and synthetic compounds present in lubricants. Filtering wear metal particles out of a lubricant will help to slow down the rate of oxidation and hence extend the lubricant's useful life.

Oxidation of lubricants, particularly oils, can result in the formation of sludge and/or varnish. Varnish on the surfaces of plain bearings is likely to result in a narrowing of the clearances in the bearing, reducing the flow of oil and thereby reducing the ability of the oil to transfer heat from the bearing. This may further reduce the clearances, resulting in an exponential effect that may ultimately cause lubricant starvation and bearing failure. Varnish on the internal surfaces of an electrohydraulic servo-valve in a hydraulic system may cause the valve to stick, rendering the hydraulic system inoperable.

Water contamination of a machine's lubricant may result in internal corrosion of steel surfaces. The resulting rust particles are likely to cause higher rates of wear, so keeping water out of a lubricant and ensuring that any free water separates from the lubricant rapidly are likely to help to extend the life of the machine. The same is true of dirt and dust contaminants.

Operating a machine at a higher power input or output than design is likely to lead to higher operating temperatures. These, in turn, are likely to result in higher rates of either or both oxidation and thermal degradation, thereby shortening a lubricant's useful life.

Vibration of a machine, or in one of its components, may result in increased metal-to-metal contact, again leading to higher rates of wear and higher operating temperatures (through increased friction), more metallic debris and the consequent problems outlined above.

2.7   ROOT CAUSE ANALYSIS OF PROBLEMS OR FAILURES

Identifying exactly why a machine, item of equipment or mechanical system experiences problems or fails is critically important to making sure that the same problem or failure does not occur again. Another machine, item of equipment or system may fail in the future, but the precise cause should be different from the last time.

Root cause analysis (RCA) is a process of discovering the root cause of a problem in order to identify appropriate solutions. RCA assumes that it is much more effective to solve and prevent underlying issues systematically, rather than just treating ad-hoc symptoms and making assumptions. RCA assumes that systems and events are interrelated. An action in one area triggers an action in another, and another and so on. By tracing back these actions, it is possible to discover where the problem started and how it grew into the current situation.

Consider, for example, a machine that stopped because it overloaded and the fuse blew. Preliminary investigation shows that the machine overloaded because a gearbox was not being lubricated correctly. Further investigation reveals that the oil circulating system for the gearbox had a pump which was not pumping adequately, hence the lack of lubrication. Investigation of the pump shows that it has a worn shaft. Investigation of why the shaft was worn discovers that there is a high level of silicon in the gear oil, leading to the means of preventing dirt from getting into the gear oil and into the pump. This damaged the pump's shaft.

The evident root cause of the problem is, therefore, that dirt was able to contaminate the lubrication system. Fixing this problem ought to prevent the whole sequence of events recurring. The real root cause could be a design issue if there is no filter to prevent dirt from entering the gear oil. Alternatively, the filter may not have been inspected to check that it had become overloaded with dirt and was no longer protecting the system, in which the real root cause is a maintenance issue.

Compare this with an investigation that does not find the root cause. Replacing the fuse, the gearbox or the lubrication pump will probably allow the machine to go back into operation for a while. But there is a risk that the problem will simply recur, until the root cause is dealt with.

RCA is now used in many situations, including manufacturing, industrial process control, telecommunications, information technology, rail, aviation and marine transportation, nuclear engineering, medicine, healthcare and accident analysis. It can be decomposed into four steps:

- The problem should be identified and described clearly. Effective problem statements and event descriptions (for example, as failures) are helpful and are usually required to ensure the execution of appropriate root cause analyses.
- A timeline should be established from the normal situation up to the time the problem occurred. RCA should establish a sequence of events for understanding the relationships between contributory (causal) factors, the root cause and the problem under investigation.
- Root cause(s) and other causal factors should be differentiated, for example by using event correlation. Correlating the sequence of events with the character, magnitude, location and timing of the problem, and possibly also with a library of previously analysed problems, should enable the investigator(s) to distinguish between the root cause, causal factors and non-causal factors. One way to trace down root causes consists in using hierarchical clustering and data-mining solutions. Another consists in comparing the situation under investigation with past situations stored in case libraries, using case-based reasoning methods.
- A causal graph between the root cause and the problem should be established. Investigator(s) should be able to extract from the sequences of events a sub-sequence of key events that explain the problem and convert it into a causal graph.

To be effective, RCA must be performed systematically. Typically, a team effort is required. The conclusions of the investigation and the root causes that are identified must be backed up by documented evidence. RCA generally serves as input to a remediation process whereby corrective actions are taken to prevent the problem from reoccurring.

In science and engineering, there are essentially two ways of repairing faults and solving problems:

- Reactive management: This involves reacting quickly after a problem occurs, usually by simply treating the symptoms. This type of management is implemented by reactive systems, self-adaptive systems, self-organised systems or complex adaptive systems. The aim is to react quickly and alleviate the effects of the problem as soon as possible.
- Proactive management: This consists in preventing problems from occurring. Many techniques can be used for this purpose, ranging from good practices in design to analysing problems that have already occurred in detail, and taking actions to make sure they never reoccur. Speed is not as important as the accuracy and precision of the diagnosis. The focus is on addressing the real cause of the problem as opposed to its effects.

More of the benefits of proactive management are described and discussed in Section 16.3, on predictive maintenance, in Chapter 16.

RCA is often used in proactive management to identify the root cause of a problem. It is customary to refer to the "root cause" in singular form, but one or several factors may in fact constitute the root cause(s) of the problem under study. A factor is considered the root cause of a problem if removing it prevents the problem from recurring. Conversely, a causal factor is one that affects an event's outcome, but is not the root cause. Although removing a causal factor can benefit an outcome, it does not prevent its recurrence with certainty.

RCA is performed with a collection of principles, techniques and methodologies that can all be leveraged to identify the root causes of an event or trend. The first goal of RCA is to discover the root cause of a problem or event. The second goal is to fully understand how to fix, compensate or learn from any underlying issues within the root cause. The third goal is to apply what has been learned from the analysis to systematically prevent future issues or to repeat successes. Analysis is only as good as what is done with it, so the third goal of RCA is important. RCA can also be used to modify core process and system issues in a way that prevents future problems.

2.8   FAILURE MODE AND EFFECT ANALYSIS

Failure Mode and Effect Analysis (FMEA) is a systematic method for identifying possible failures that pose the greatest overall risk for a process, product or service which could include failures in design, manufacturing or operation. It is also known as "Failure Modes and Effects Analysis", "Potential Failure Modes and Effects Analysis" and "Failure Modes, Effects and Criticality Analysis (FMECA)".

FMEA methodology is used to determine the chance of failure and the ensuing risks in developmental processes of services, products, production methods or process operations. The aim of FMEA is to define actions that reduce the possibility of failure. The multidisciplinary aspect of this tool ensures that a complete picture is identified regarding the quantifiability of risks. This means a hierarchy can be applied in the urgency of the risks. A useful tool that can help to do this is the Risk Analysis Tool.

The FMEA process depends on identifying:

- The failure mode(s): One or more ways in which a product, process or operation can fail, in terms of their possible deficiencies or defects.
- The effect(s) of failure: The consequences of one or more particular modes of failure.
- The cause(s) of failure: One or more of the possible causes of the observed mode(s) of failure.
- Analysis of the failure mode(s): Its or their frequency severity and possibility of detection.

FMEA can be used as an essential tool for improving both product and process design and operation. Sometimes the Design FMEA is referred to as DFMEA to differentiate it from the Process FMEA (PFMEA). Both use the same process.

Performing FMEA is a team operation. Everyone should feel fully involved in the process and in moving towards the goal. However, it is advisable to delegate certain responsibilities to specific people, so that monitoring and reporting can be effective. An effective FMEA should involve a line manager, an analyst and a reviewer. The line manager should assign someone to perform the FMEA (the analyst) and someone else to review it (the reviewer). The reviewer needs to have as much expertise and technical experience as the analyst. The analyst should describe the system under analysis, prepare system diagrams and use existing documentation to depict all major components and their performance criteria. The level of assembly may vary with the level of the analysis. He or she should then perform the FMEA as per the procedure described earlier. The written and documented results should then be presented to the reviewer, who should examine the FMEA for technical content and sign it if no significant problems are identified. If issues are found, they should be discussed with the analyst. Finally, the full FMEA documents should be presented to the line manager and potentially other managers as necessary.

Every industry is likely to have a slightly different approach to FMEA. With regard to lubricant analysis and condition monitoring, ASTM has published ASTM D7874 "Standard Guide for Applying Failure Mode and Effect Analysis (FMEA) to In-Service Lubricant Testing" which defines FMEA as an analytical approach to determine and address methodically all possible system or component failure modes and their associated causes and effects on system performance. Applying the FMEA process effectively requires an understanding of machine design requirements and equipment operating conditions, which lead to the identification of potential failure modes.

The ASTM D7874 guide is intended as a guideline for fluid analysis programmes and serves as an initial justification for selecting fluid tests and sampling frequencies. Plant operating experience along with the review and benchmarking of similar applications is required to ensure that lessons learned are implemented. Selection of proper fluid tests for assessing in-service component condition may have both safety and economic implications. Some failure modes may cause component disintegration, increasing the safety hazard. Thus, any fluid test that can predict such conditions should be included in the condition monitoring programme. Conversely, to maintain a sustainable and successful fluid monitoring programme, the scope of the fluid tests and their frequency should be carefully balanced between the associated risks and expected programme cost savings and benefits.

According to the guide, the failure modes monitored may be similar from one application to the next, but the risk and consequences of failure may differ. The analysis can be used to determine which in-service lubricant analysis tests would be of highest value and which would be ineffective for the failure modes of interest. This information can also be used to determine the best monitoring strategy for a suite of failure modes and how often assessment is needed to manage the risk of failure.

The guide describes a methodology to select tests to be used for in-service lubricant analysis. The selection of fluid tests for monitoring failure mode progression in industrial applications applies the principles of (FMEA). Although typical FMEA addresses all possible product failure modes, the focus of the guide is not intended to address failures that have a very high probability of unsafe operation as these should immediately be addressed by other means. The guide is limited to components selected for condition monitoring programmes by providing a methodology to choose fluid tests associated with specific failure modes for the purpose of identifying their earliest developing stage and monitoring fault progression. The scope of the guide is also focussed on those failure modes and their consequences that can effectively be detected and monitored by fluid analysis techniques.

The guide pertains to a process to be used to ensure an appropriate amount of condition monitoring is performed with the objective of improving equipment reliability, reducing maintenance costs and enhancing fluid analysis monitoring of industrial machinery. The guide can also be used to select the monitoring frequencies needed to make the failure determinations and provide an assessment of the strengths and weaknesses of a current condition monitoring programme. ASTM notes that the guide does not eliminate the programmatic requirements for appropriate assembly, operational and maintenance practices.

The first step in applying FMEA is selecting the components to test and identifying the possible failure modes that are associated with those components. For each component, the causes and effects of each failure mode are identified. Each failure mode is given a severity number (S) and an occurrence frequency number (O) to allow calculation of a criticality number ($S \times O$). The criticality number permits prioritisation among the different failure modes and allows determination of whether lubricant analysis can be applied to detect the failure mode.

If lubricant analysis can identify the failure mode, the next step is identifying the required test. A detection ability number (D) is used to rank how easily and reliably the failure mode can be detected using the chosen lubricant test.

To cross-check calculations, a comparison of the criticality number and the detection ability number should indicate that the failure modes with the highest criticality numbers also have the highest detection ability numbers. As a result, the likelihood of identifying failure modes with the selected fluid analysis test(s) is higher. Conversely, a mismatch between the two numbers could indicate a weakness in failure mode detection, requiring adjustments to enhance the programme.

By identifying and detecting equipment failure modes and predicting the rate of failure progression, effective condition monitoring programmes allow preventive action to be taken without unplanned downtime. This adds to predictive maintenance, which also uses other methods described and discussed in later chapters.

2.9 SUMMARY

A lubricant condition monitoring programme enables a user to have confidence in the quality of the lubricants in each machine and to help to analyse the condition of the machine in its current state. Lubricant condition monitoring can be used to determine when the lubricant needs to be changed, so that its inferior properties or performance does not impact the useful life of the machine.

Machine condition monitoring compliments lubricant condition monitoring, by providing additional information to evaluate whether and, if so, when the machine needs to be maintained, repaired or replaced.

The difficulty is that numerous facets of lubricant and machine analysis must be performed correctly, or the integrity and validity of the entire condition monitoring programme will be at risk.

3 Sampling Lubricants

3.1 INTRODUCTION

The first requirement for the effective analysis of lubricants is to obtain a representative sample. This involves using proper sample bottles, selecting and using the appropriate methods and techniques for taking samples and determining the optimum sampling location and frequency.

The objective of sampling is to maximise data density while minimising data disturbance. Data density means gathering as much information in the sample as possible. Data density is maximised by sampling in the right location with the right equipment at the right time.

Most usually, the right location is a "live zone" within a machine, where oil is flowing in a turbulent manner. The sample will then contain all the useful information required for trend analysis, without losing any data resulting from either component fly-by or settling. The right equipment includes the use of special sampling valves, vacuum sampling pumps, disposable tubing and other accessories for taking samples as cleanly as possible. If the machine is critical to the company's operation or is not running as expected, the sampling frequency may be quite short, for example every two weeks. Otherwise, the sampling frequency could be as long as every six months. Taking consistent samples helps to plot meaningful trends of lubricant properties, lubricant performance and contamination levels.

Data disturbance means allowing important information about the lubricant to escape. Minimising data disturbance depends on how well the sampling process is designed. A common mistake is failing to flush the sample equipment as part of the procedure for drawing the sample. Using disposable tubing and a vacuum sampling pump requires that the tubing must be flushed to get a genuinely representative sample. Flushing between five and ten times the dead volume from all sampling equipment is recommended. This ensures that any contaminants inside the tubing are cleared out and that the sample will be representative of the conditions inside the machine.

3.2 SAMPLE BOTTLES

Step 1 for obtaining a truly representative sample of a lubricant in a machine is to use the most appropriate sample bottles. Many companies believe, mistakenly, that any bottle or container in which to deliver the sample to a laboratory will be okay. This includes old glass bottles, empty milk bottles, used Coke bottles and any plastic bottle that has a screw cap. Numerous examples of completely unsuitable sample bottles are shown in Figure 3.1.

It is imperative that sample bottles are clean (preferably ultra-clean), dry and free of any material that may contaminate the sample. The characteristics and attributes

DOI: 10.1201/9781003245254-3

FIGURE 3.1   Typical examples of "bad" sample bottles.

Source: Noria Corporation, with permission.

Use	Material	Volume	Cleanliness	Typical Cost
General oil sampling for visual analysis	Polyethylene terphthalate PET (Transparent)	100 to 500 ml	Super-clean	£
General oil sampling	High density polyethylene HDPE (Opaque)	100 to 500 ml	Clean or super-clean	£
Hydraulic oil	PET or HDPE (Transparent or opaque)	100 to 500 ml	Clean or super-clean	££
Hydraulic oil for visual analysis	Glass (Transparent)	100 to 500 ml	Ultra-clean	£££

FIGURE 3.2   Oil sample bottle applications.

Source: Pathmaster Marketing Ltd.

of four types of suitable oil sample bottles are shown in Figure 3.2. Although some sample bottles are more suited for engine oils, gear oils and compressor oils, others need to be used for specific hydraulic and turbine oil samples, where ultra-cleanliness could be vitally important. In all cases, the cost of the sample bottle should not be an issue when the application requires it.

Oil sample bottles are available in only a few standard materials, most commonly plastic or glass. The material should be selected based on the type of fluid sampled and the cleanliness requirements. The most common plastic sample bottles are high-density polyethylene (HDPE), polypropylene (PP) or polyethylene terephthalate (PET). HDPE is opaque, which may be its main disadvantage. Not having the ability to clearly see the oil in the bottle prevents visual onsite analysis, which can be helpful in detecting water or heavy particle contamination. Conversely, PET is clear, but generally not suitable for samples at temperatures higher than about 90°C, for which PP is more suitable since it is able to withstand temperatures of up to 120°C. However, PET has greater compatibility than HDPE or PP with most industrial lubricants. Compared with glass bottles, both polyethylene-based bottles are relatively inexpensive, but they offer the benefits of excellent cleanliness levels and lubricant compatibility. Illustrations of all four types of sample bottles are shown in Figure 3.3. Glass bottles are available in clear and amber glass. Amber glass protects oil samples from sunlight, which might be important for some tests.

The size of the sample bottle should be based on the type of sample fluid, together with the number and type of tests to be conducted. For most standard oil analysis tests, oil samples are taken in a 100 or 120 millilitre (ml) bottle. For advanced, large numbers or exceptional tests, a 200 millilitre (ml) or 500 millilitre (ml) bottle may be required. An example of when a larger sample might be necessary would be for hydraulic fluid testing, especially aviation hydraulic fluid. Sample bottles can also come in smaller sizes for other applications.

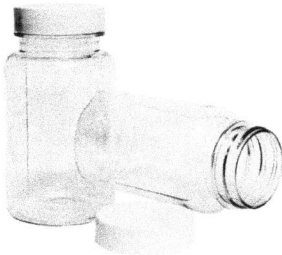

Polyethylene Terphthalate

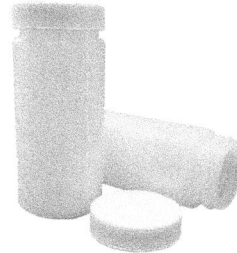

High Density Polyethylene

Polypropylene

Glass

FIGURE 3.3 Types of oil sample bottles.

Source: Pathmaster Marketing Ltd.

Lubrication engineers generally believe that particles too small to be seen by the unaided eye are the most destructive in three-body abrasive wear. These particles, which typically range in size from 5 to 15 microns (μm), have the capability of getting trapped between surface gaps, rather than either passing straight through the gap or staying outside the gap. For this reason, particle contamination analysis commonly presents data as three range numbers for particles greater than 4, 6 and 14 microns (μm).

Since these "invisible" particles are substantial in most samples of oil, it is essential that they are not present within the sample bottle prior to a sample being obtained. This means that drawing a sample into a washed-out beverage bottle will not be good enough. Even a sample bottle purchased with the lid and bottle in separate packages will not be sufficient.

The required sample bottle cleanliness will be based on the importance of the sample being taken and the sensitivity of the tests being conducted. Sample bottle cleanliness can be classed as ultra-clean, super-clean or clean. What this means is illustrated microscopically in Figure 3.4.

"Clean" oil sample bottles are defined as having less than 100 particles greater than 10 microns per millilitre (μm/ml) of fluid. This cleanliness level is the most common and least expensive. "Super-clean" oil sample bottles can be defined as having less than ten particles greater than 10 microns per millilitre (μm/ml) of fluid. "Ultra-clean" oil sample bottles are defined as having less than one particle greater than 10 microns per millilitre (μm/ml) of fluid. ISO 3722 also describes a certification procedure based on randomised testing for cleanliness.

The oil analysis laboratory will provide advice on the size of sample, and therefore sample bottle, required for each type of oil sample, together with the cleanliness requirements. One way to determine the necessary sample bottle cleanliness is to use the signal-to-noise ratio (SNR) method. The SNR is defined as the target cleanliness for the machine divided by the contamination identified for the bottle.

$$\text{SNR} = \frac{75^{(15/12)}}{7.5} = 10$$

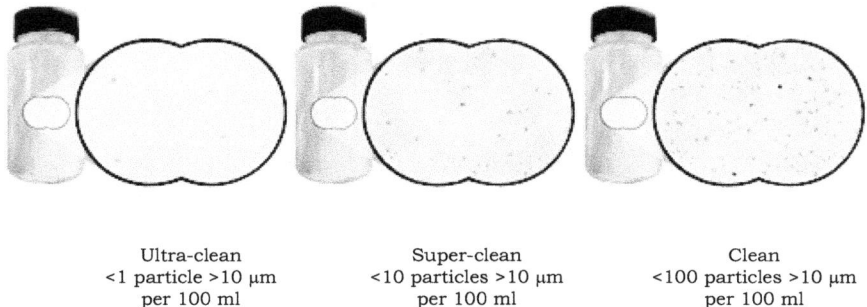

Ultra-clean	Super-clean	Clean
<1 particle >10 μm per 100 ml	<10 particles >10 μm per 100 ml	<100 particles >10 μm per 100 ml

FIGURE 3.4   Sample bottle cleanliness.

Source: Noria Corporation, with permission.

$$SNR = \frac{1000^{(19/16)}}{100} = 10$$

A high SNR is desirable. For example, an SNR of 5 would have a 20% variance of cleanliness accuracy, while an SNR of 10 would have a 10% variance. A higher SNR is achievable with fluids such as gear oils that do not require rigorous cleanliness levels. Fluids with greater cleanliness requirements, such as hydraulic oils, require cleaner sample bottles, preferably with an SNR value of 10 or higher.

When purchasing sample bottles, or using sample bottles provided by an independent oil analysis laboratory, it is very important to know that they are cleaned to the specifications required to meet the target cleanliness goals. If an oil is put into a dirty sample bottle, the test results will show that the oil in the system is dirty, when that may not be true.

More recently, new methods have focussed on improving sample bottles to improve not only their ease of use but also their ability to obtain a representative sample. One such example is the UCVD (Ultra-Clean Vacuum Device), which was invented and developed by Giuseppe Adriani of Mecoil Diagnosi Meccaniche Srl. It is widely known as the "SureSample" bottle. An example of this type of sample bottle is shown in Figure 3.5. It is designed to hold a pre-established vacuum. With the vacuum intact during distribution, the bottle is nearly void of all moisture and contaminants at an ultra-clean level. Once the bottle's nozzle is attached to a sample

Ultra Clean
Vacuum Device

FIGURE 3.5   Advanced sample bottle.

Source: Bureau Veritas, with permission.

tube and the other end inserted into a sample port, the nozzle is switched, allowing the vacuum to independently draw fluid into the bottle to the required level. This action is possible for nearly all viscosities, with higher viscosity fluids only requiring longer draw times. The method also may eliminate the need for a manual vacuum pump during sampling from any lubricating system, even non-pressurised systems. As a result, a sampling technician could potentially have several samples drawing into different bottles at the same time while being confident that the samples are not being contaminated in the process. There is, however, a danger in using this type of sample bottle with pressurised lubrication systems, particularly with hydraulic and turbine oils. Consequently, great care is required when sampling pressurised systems to ensure that the nozzle switch is turned very slowly and the oil allowed to be drawn in slowly.

3.3   METHODS FOR TAKING LUBRICANT SAMPLES

3.3.1   OIL SAMPLES

Several methods can be used for taking oil samples from equipment. Some are more effective than others. However, it is important that the method chosen can be used consistently by all those staff involved in taking oil samples. Each time a sample is taken, the end result should be the same regardless of the supervisor or technician taking the sample. Written procedures and specific training are vital to the success of an oil analysis programme.

With any sampling procedure, the sampling equipment should be flushed immediately before the sample is taken. A general rule for the amount to flush is ten times the estimated volume of the fluid pathway from the originating location to the sample bottle. This would include the sample valve with the pilot tube and the sample tube from the sample valve to the sample bottle.

3.3.1.1   Drop Tube Sampling

Drop tube sampling is an effective, low-cost way to obtain a sample using a vacuum pump. A diagram of the method is shown in Figure 3.6. However, when using this method there are a number of issues to consider.

In order to collect a sample, the machine, item of equipment or mechanical system must be opened, which will result in the oil being exposed to the environment. Opening a machine potentially allows significant amounts of airborne contamination to enter the oil and cause damage to the machine, equipment or system. Unless the sample is taken quite quickly, the airborne contamination may also contaminate the oil sample. In addition, the next oil sample may be slightly different from the current oil sample in terms of its water and/or silicon content, unless the machine's filtration system is operating correctly.

Using the drop tube method on a gearbox while it is running poses several problems. The plastic tubing may be pulled into the gearbox, which presents specific safety concerns for the person taking the sample. Other issues associated with drop tube sampling include large required flushing volume, difficulties in getting a consistent sample from the same location and problems with sampling high-viscosity

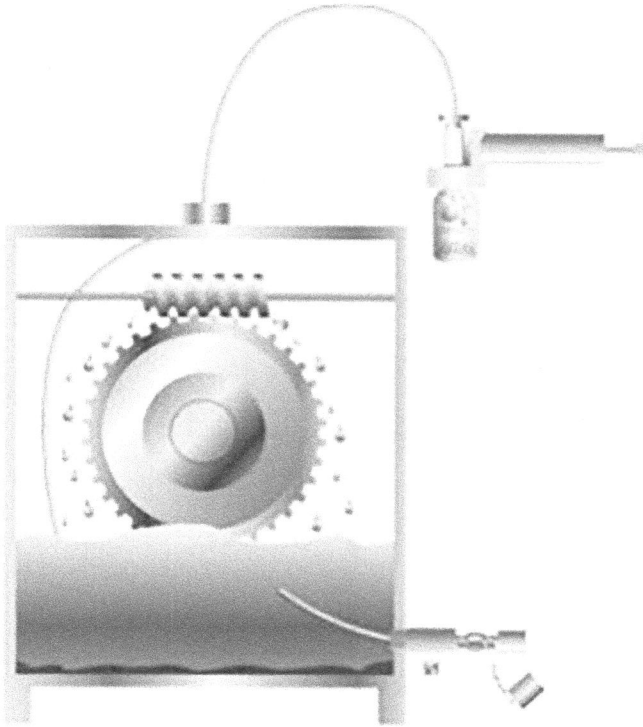

FIGURE 3.6 Drop tube sampling.

Source: Trico Corporation, with permission.

fluids. The key to an effective oil analysis programme is the ability to obtain an oil sample from a specific location while the machine is in operation and under normal load. Consequently, drop tube sampling should be avoided whenever possible.

3.3.1.2 Drain Port Sampling

The ideal location for drawing an oil sample from a reservoir or sump is to get it as close to the oil return-line or inlet strainer as possible. Another "rule of thumb" is to sample at 50% of the oil level. Sumps and reservoirs were designed to hold a large volume of oil, to dissipate heat and to allow air to rise and contaminants to settle. Therefore, the most concentrated contamination is on the bottom of the sump or reservoir, and the cleanest oil is towards the top. Although a drain plug might be considered a very convenient place to obtain a sample, if it is right at the bottom of the reservoir, it is highly unlikely to provide a representative sample even if large volumes of oil are flushed through the port before taking the sample. This is illustrated in Figure 3.7.

If the drain port is the only way to obtain a sample from a gearbox, hydraulic or turbine oil system reservoir, engine oil sump or compressor oil reservoir,

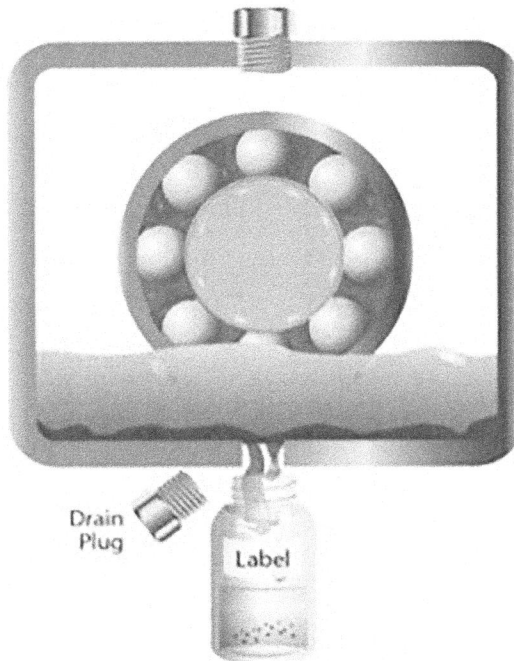

FIGURE 3.7   Drain port sampling.

Source: Trico Corporation, with permission.

commercially available pilot tubes can be installed on the bottom or the side of the reservoir or sump. These inward pilot tubes can be designed and installed to ensure that the sample is drawn in the most appropriate location of the sump or reservoir and that the sample is taken from exactly same location inside the system each time. This method is a more consistent and representative way of sampling oil than drop tube sampling.

3.3.1.3   Valve Sampling

Several designs of sampling valves are available commercially. Some are far superior to others. Sampling with valves as opposed to static sampling adds integrity and success to a condition monitoring programme.

Installing a sampling valve can help to maximise data density and minimise data disturbance. The sample obtained will provide the most representative information about the oil and the machine, including cleanliness, dryness, additive levels, wear metals, particles and more, as well as information that is uniform, consistent and unaltered in the sampling process. This can be very important when looking at trend analysis.

A most effective option, which is typically used on larger, pressurised systems, is a "Minimess®" style of sample valve, an example of which is shown in Figure 3.8.

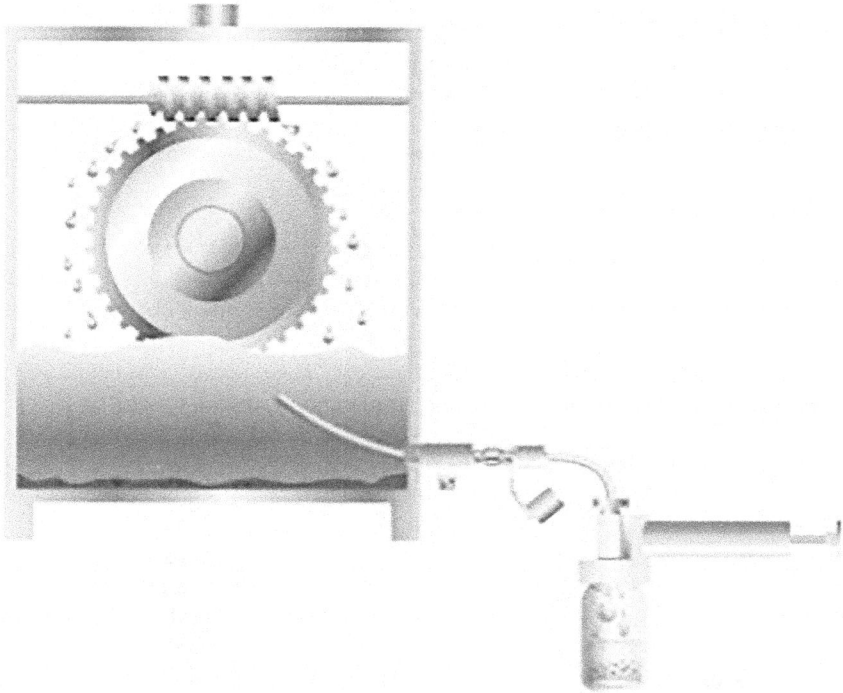

FIGURE 3.8   Minimess® sampling valve.

Source: Trico Corporation, with permission.

Various types of Minimess® style valves are available from a number of suppliers. These kinds of sample ports are check style, in which the valve is normally closed until the sample port adapter is threaded on. Sample ports are equipped with a dust cap that also has an o-ring for second-stage leak protection. The adapter has a hose barb on one side that accepts standard ¼ inch outside diameter plastic tubing. As the adapter is threaded onto the sample port, it unseats the check ball in the valve and allows the sample to flow into the sample bottle. These valves can be used on systems from 0 psi (assuming the line is flooded) to 5,000 psi.

A number of suppliers of Minimess® sample valves also supply oil sampling kits. One such kit, available from Hydrotechnik UK Ltd., contains:

- High pressure 630 bar Minimess® 1620 1.5 m sampling hose, vacuum Minimess® 1604 1.5 m sampling hose 1.5 m long and 100 millilitre (ml) plastic oil sampling bottle.
- 690 bar (10,000 psi) rated oil sampling tap with Minimess® 1620 hose connection and vacuum thief pump for tank/gearbox oil sampling via Minimess® 1604 sampling probe and vacuum hose.
- 1/4″, 3/8″ and 1/2″ bonded sealing washers.

- 1/4″, 3/8″, 1/2″ and 3/4″ BSP male/swivel female Minimess® 1620 sampling point tee adaptors.
- Minimess® 1620 sampling point to 1/4″ BSP male thread with captive seal, 1/4″ BSP female to 3/8″ BSP male thread adaptor and 1/4″ BSP female to 1/2″ BSP male thread adaptor.
- Minimess® 1604 sampling point with 6″ (150 mm) long sampling probe and 1/4″ BSP male screw-in thread with captive seal.

Installing a sampling valve such as a Minimess® valve on a reservoir or sump does not prevent the port from being used for other things. Adapters can be configured in a number of ways to allow for sight glasses, drain valves, quick connects, magnetic plugs and pressure gauges to be used. One arrangement that facilitates multiple functions is shown later in this chapter. Using T- and Y-adapters can create combinations that may be effective within nearly any desired configuration.

3.3.1.4 Safety Factors

Safety must be considered when taking samples from systems where the oil pressure is between 2,000 and 5,000 psi. Sample valves are available that offer superior quality and safety when taking an oil sample. For example, handheld pressure-reducing valves can be used in conjunction with the sample ports and adapters to reduce pressures of 5,000 psi to less than 50 psi. They also come in several adapter styles that allow for ease when installing them on a system. Another benefit to these types of sampling valves is that they hold a very small volume of static oil. This results in less oil flushing prior to taking a sample.

Care should also be taken when sampling oil at temperatures above 80°C. The sampling tubing and sample bottle must be able to withstand high temperatures, and suitable gloves and goggles should be used while taking the sample.

3.3.2 GREASE SAMPLES

Obtaining representative samples of grease is more difficult than for samples of oil.

Some lubricant analytical laboratories supply guidance and special kits for taking grease samples. One is OELCHECK GmbH, which sells a specially designed "Grease Sampling Kit", the components of which are shown in Figure 3.9.

Grease can be filled into a transparent tube with the aid of a large-volume syringe. The grease is only sucked part way into the tube, which has an approximate length of 20 cm, as shown in Figure 3.10. The syringe is not supposed to come in direct contact with the grease. The tube is first plugged onto the syringe cone, the surroundings of the extraction site are cleaned with a cleaning cloth and the grease is sucked into the plastic tube. Depending on the scope of the condition monitoring programme, an analysis is possible starting from approximately 1 g of grease (2 cm of grease in the tube). If the tube is filled completely (8 to 10 g of grease), more tests can be performed. After sampling, the tube is pulled off the syringe, it is bent in the middle and put into the sample bottle with the kink on top, as illustrated in Figure 3.11.

If an extraction of the grease with the syringe is not possible, the "Grease Sampling Kit" contains three spatulas of differing lengths. These can be used for grease in

Cleaning cloth

Sample tubes

Syringe

Spatulas

Comprehensive instructions are included in the kit

FIGURE 3.9 OELCHECK grease sampling kit.

Source: OELCHECK GmbH, Brannenburg, with permission.

FIGURE 3.10 Sampling grease with a syringe and tubing.

Source: OELCHECK GmbH, Brannenburg, with permission.

FIGURE 3.11 Grease sample.

Source: OELCHECK GmbH, Brannenburg, with permission.

FIGURE 3.12   Sampling grease with a spatula.

Source: OELCHECK GmbH, Brannenburg, with permission.

hard-to-reach bearings and inoperative components or if the sample has to be taken at the sealing lip or from pressure relief holes. The surroundings of the extraction site are first cleaned with the cleaning cloth provided. The grease is then scraped off at the representative spots with the spatula, as shown in Figure 3.12. Grease which may escape due to re-lubrication directly at the sealing point can be collected. The sample of grease on the spatula is then wiped off onto the sample bottle top or into the mouth of the sample bottle. The spatulas are provided in different versions and widths to allow a representative sampling.

If the grease is mixed in track at a slow rotation of the bearing before sampling, this will allow a diagnostically conclusive reference sample to be taken. If the bearing is to be dismantled or replaced, suck the used grease out of the area around the cage and between the rolling elements into the tube or remove the grease from this area with a spatula.

In the case of a slewing bearings (diameter <1,000 mm), extract the grease from the dedicated boreholes. The free space behind these holes is shut with special screws during operation. The points of withdrawal shall preferably be at the tracks in the main load area. Take several samples (perhaps three) from the bearer ring hole and from the retaining ring hole at a 120° angle.

ASTM standard D7718 was published in 2011 to provide an approved standard practice for sampling critical grease-lubricated equipment. The standard covers the method to obtain a trendable in-service lubricating grease sample from several different grease-lubricated equipment or machines. In some cases, it may be necessary to take more than one sample from a piece of equipment to obtain more trendable results. Examples of this could be a large bearing that does not fully rotate, such as a slew bearing, or one in which sufficient mixing does not otherwise occur. Samples taken in this way may need to be mixed to form a more homogeneous sample. This may also be true of other samples such as those taken from open face bearings. The standard covers methods for sampling greases from motor-operated valves, gearboxes, pillow-block bearings, electric motors, exposed bearings, open gears or failed grease-lubricated components.

A new method to obtain representative samples of grease for analysis involves using a device called a "Grease Thief", developed by MRG Corporation in the United States. For a grease-lubricated electric motor bearing, the Grease Thief screws into the bearing plug hole and grease is extracted into the device under variable pressure and force conditions, most usually during manual or automatic re-greasing. When a representative sample of grease has been collected, the Grease Thief is unscrewed, capped, put into a sample tube and sent to the laboratory for analysis. For other applications, such as pillow-block bearings, a Grease Thief sampling kit includes a spatula and syringe in order to collect a sample of grease exuding from the bearing, again during re-greasing. MRG Corporation supplies Grease Thief sampling kits and sampling instructions for use with wind turbines, robotics, electric motors, motor-operated valves and gearboxes and pillow-block bearings.

Previously, there was no comprehensive guidance for taking grease samples, which severely limited the use of grease analysis as a diagnostic tool. The use of the Grease Thief in the place of motor bearing drain plugs allows for proper grease purging while also capturing any purged grease for analysis.

In the past, vibration analysis would have been used to monitor the mechanical condition of a grease-lubricated electric motor. However, with Grease Thief sampling devices, it is possible to extract a reasonably representative sample of grease to detect mixing, grease degradation and wear particles.

For trend analyses, always take the samples of grease from the same location. A sample of new grease should be taken from the chosen location shortly before commissioning a bearing, for reference purposes.

3.4 SAMPLING LOCATIONS

In many factories, machines frequently do not have lubricant sampling points in the proper location(s) to obtain a truly representative sample. If a sample is not taken at the right location and in the correct way, it becomes increasingly difficult to ensure the representative nature of the oil analysis results. Successfully trending data becomes even less of a possibility.

The sample location should be selected on the basis of obtaining lubricant in a machine's most important lubrication zone, often called the "live zone". Typical examples of these locations are:

- For a rolling-element bearing, it would be the lubricant between the rollers and the race.
- For a gearbox, it would be the lubricant between meshing gear teeth.
- For a hydraulic system, it would be in the pipework from the pump.

Although it may not always be possible to take a lubricant sample from these locations, attempting to sample in close proximity or where the fluid returns back to a reservoir can be just as effective for maximising data density and minimising data disturbance.

3.4.1 SAMPLING DRY-SUMP CIRCULATING SYSTEMS

A dry-sump circulating system contains a central reservoir, a pump, various components to be lubricated, a filter and piping to connect them all together. A sample for this type of system should always be taken just downstream of components that may produce wear, since any wear particles produced will be among the most important properties of the oil.

Because several components are being lubricated, the "live zone" for sampling is in more than one location. This requires a sample to be taken after the lines converge on the return lines but before being passed through a return-line filter and back to the reservoir (in the case of a hydraulic system). This sampling location provides a good representative oil sample for the system as a whole before valuable wear data is removed by the filter. This is illustrated in Figure 3.13.

It is always best to take a sample where the fluid becomes turbulent. A turbulent sampling location is one where the fluid is forced to turn and tumble, such as on or just after a pipe elbow. This is important because without turbulence, heavier particles, such as wear debris, can be pushed past sample valve ports, especially when the velocity of the fluid is high. This particle "fly-by" phenomenon can result in false low particulate results. If there is sufficient pressure, a sampling tube with an adapter

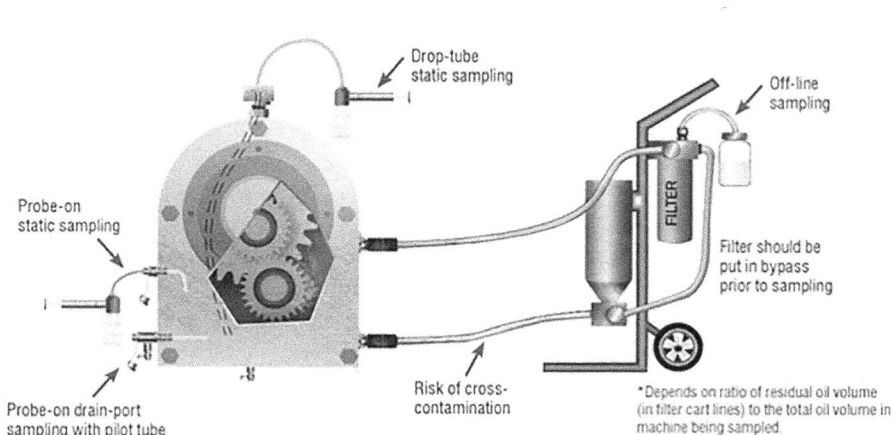

FIGURE 3.13 Oil sampling locations for a splash/bath lubricated machine.

Source: Noria Corporation, with permission.

affixed to a Minimess® valve can be used to draw the fluid into a certified clean bottle once the sampling lines have been flushed. If there is insufficient pressure, a vacuum sampler pump will be required.

When abnormal wear is evident in a primary sample taken after the lines converge, it will be necessary to obtain samples at secondary locations to determine where this wear is being produced. These sampling locations should be positioned immediately after the lubricant discharges from the component, for example in a bearing before the line branches into the return header.

3.4.2 SAMPLING WET-SUMP CIRCULATING SYSTEMS

Taking a representative oil sample from machine live zones can be challenging with a wet-sump system, such as a diesel engine or circulating gearbox in which a return or drain line is not accessible or does not exist. In this type of system, there will be either a pressurised supply-line or a kidney-loop filtration system.

With a system containing a pressurised supply line, the most appropriate sampling location may be after the pump but before the filter. Because the sample is taken at a pressurised point, the best sampling valve option is a Minimess® valve, which ensures no disturbance to the machine's operations.

If the wet sump is configured with a kidney-loop filtration system, a similar sampling location can be selected. For this configuration, a Minimess® valve for sampling can be installed after the pump and before the filter in the same manner that the supply-line system was equipped, since the system is also in a pressurised state. Many filter housings have convenient ports just upstream of the filter, which are often used for pressure gauges.

3.4.3 SAMPLING NON-CIRCULATING SYSTEMS

In systems without any circulation of lubricating oil, the primary means for obtaining a sample will be through the system's sump casing using a pilot tube from the sample valve to draw fluid from a turbulent, live-zone location. These types of systems will have a drain plug either near or at the bottom of the sump. This is where the sample port can be installed along with a T-adapter to allow for the addition of a ball valve for draining. Minimess® valves can be installed at an oil level port where a sight glass is typically situated. If this port is the most accessible, then a device that enables both sampling and sight glass functionality will be necessary.

During installation, the pilot tube should be carefully positioned so the tip is in an ideal live-zone location. Generally, this will be halfway up the oil level and a sufficient distance from any walls. It must be close to turbulent areas but no closer than 2 inches from any moving elements or walls within the system. For example, an ideal layout for a hydraulic oil reservoir is shown in Figure 3.14. The best hardware to use in conjunction with the Minimess® sample port would be a vacuum sampler. This helps to ensure that the sample never comes into contact with the environment.

3.4.4 MULTIPLE SAMPLING POINTS

The sampling locations described above can be called primary sampling locations. At these locations, the objective is to be able to draw a single sample that acts as a snapshot of the entire system.

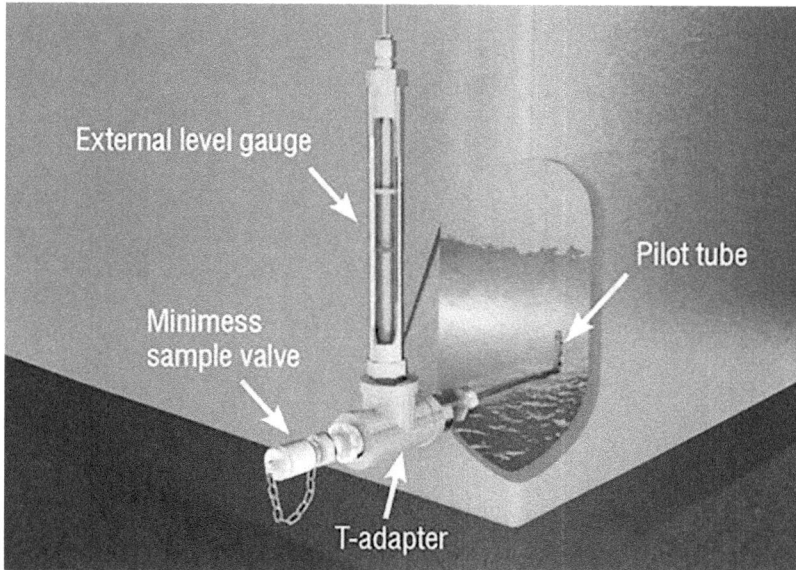

FIGURE 3.14 Oil reservoir sampling location with integrated sight glass.

Source: Noria Corporation, with permission.

Sometimes, however, samples taken from a primary sampling location miss a lot of valuable information. To capture this data, secondary sampling locations should be installed on many systems. The aim of a secondary location is to be able to pinpoint the cause of any fault seen on an oil analysis report. Unlike the primary sampling location, which provides an overall look at the entire machine, secondary sampling locations enable a focus on individual components inside the system. Some systems may require several secondary sampling locations.

Most circulating and hydraulic systems should have both a primary and secondary sampling locations to ensure that any identified failure mechanism can be tracked back to the component causing the problem. A secondary port can be used to help determine the source of wear debris or particles and, by installing sampling ports behind filters, how well the filter is removing particles can be monitored. While the primary port may get the most use, the secondary ports can be invaluable once a fault has been detected. An example of secondary sampling locations in a system is shown in Figure 3.15.

3.5 SAMPLING FREQUENCY

According to a recent survey by *Machinery Lubrication*, 64% of lubrication professionals rate machine criticality as the most important factor to consider when adjusting oil sampling frequency.

Determining sampling frequency is not easy. In essence, it is a trade-off. Sampling less frequently can risk missing a machine or lubricant failure, while sampling more

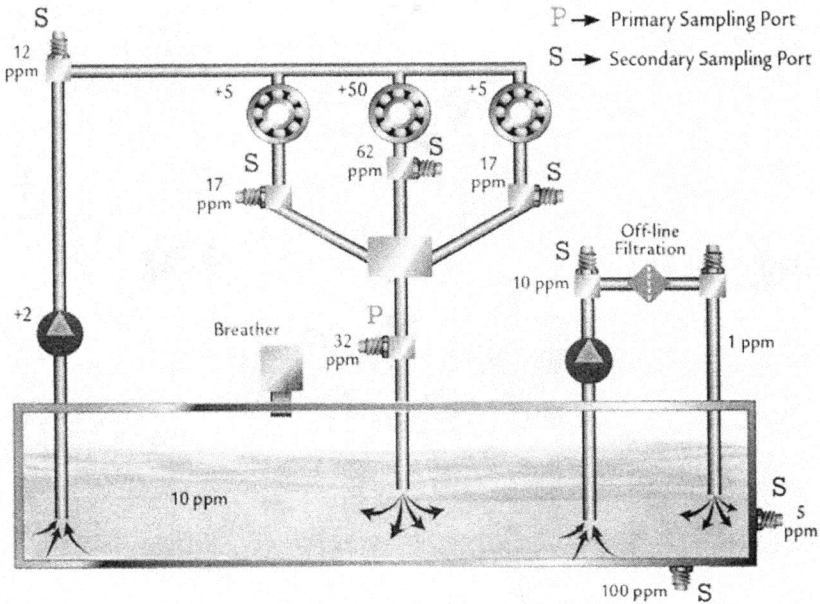

FIGURE 3.15   Examples of multiple sampling locations in a circulating oil system.

Source: Noria Corporation, with permission.

frequently can risk wasting time, money and lubricant. The determination can be done using one or more of the following variables:

- **Economic Penalty of Failure**: Samples should be taken more frequently when the cost of downtime, repair, rebuild, interruption to business and/or impact on product quality are high.
- **Fluid Environmental Severity**: Samples should be taken more frequently when demands placed on the lubricant by the environment and/or the machine are high.
- **Machine Age**: Based on the "bathtub curve", sampling should be taken more frequently when the machine is young (infant mortality) or old (surpassed maximum useful life). The bathtub curve was explained and discussed in the previous chapter.
- **Lubricant Age**: As with machine age, during the life of a lubricant in a machine, samples should be taken more frequently when the age of the oil is new or old.
- **Target Tightness**: Samples should be taken more frequently when oil analysis targets (such as ISO particle counts) are approaching or consistently near the flagging limits.

The relative importance of each of these variables is best determined by the people who are most familiar with the machine being sampled. Many of these questions asked about the machine are subjective. The most effective way to develop appropriate answers is to use historical information related to past machine failures, machine criticality, lubricant type, failure modes, lubricant change-outs and top-ups, adjustments to oil analysis targets over time, fluctuations in environmental conditions and other information.

The more that is known about each machine, the easier it is to calculate the optimum sampling frequency in relation to the key variables. Intelligent adjustments may need to be made to machine sampling frequency to optimise the value gained from lubricant analysis.

Guidance on sampling frequency for different lubricants in different machines, items of equipment and mechanical systems is described in the later chapters in this book.

3.6   SUMMARY

Correct sampling of lubricants is the first step in an effective lubricant analysis and condition monitoring programme. Use of the correct types of clean, super-clean or ultra-clean sample bottles is vital for obtaining truly representative samples of lubricants. Different sampling methods are required for oil and greases. Obtaining representative samples of greases is much more difficult than for oils.

Spending time and money on obtaining truly representative lubricant samples pays huge dividends in machine, equipment and mechanical system lifetimes and maintenance. Sampling and samples are the fundamental first steps for an effective condition monitoring programme.

4 Testing New Lubricants

4.1 INTRODUCTION

Lubricants are formulated in a lubricant development laboratory, and they are then either blended in a blending plant, for oils and solid lubricants, or manufactured in a grease plant, for greases.

Lubricating oils are manufactured by taking appropriately selected base oils and blending with appropriate additives, before delivering to an end user either in bulk or packaged in international bulk containers (IBCs), drums, plastic bottles or tin-plate cans. Contamination and base oil/additive degradation can occur at any stage of the blending, packaging, storage or transportation of finished lubricants.

While a lubricant manufacturer is responsible for insuring that the selection of base oils and additive package meets the required performance characteristics of the specific lubricant, the manufacturer is also responsible for insuring that the formulated lubricant is delivered to the end user without levels of degradation or contamination that might compromise the integrity of the fluid before it is used in a machine or equipment. Even before delivery to the end user, significant levels of contaminants, including heat, moisture and particles can cause premature lubricant degradation in storage, at the manufacturing facility, at the blending plant or at the lubricant distribution warehouse.

Testing new lubricants upon receipt by a customer is essential to ensure the lubricant received is the lubricant ordered. This is critical for several reasons, for example to ensure that the oil received is the correct one, to establish a baseline for subsequent testing and monitoring of the oil condition and to establish the lubricant's level of cleanliness.

4.2 ORGANISATIONS THAT DEVELOP AND SPECIFY TESTS FOR LUBRICANTS

The test methods used for both new lubricants and those in use will be discussed in depth in later chapters. A number of organisations develop test methods, establish their suitability to be used as test methods, establish their repeatability and reproducibility, monitor their use in practice and update them when necessary. Brief summaries of organisations that develop test methods and standards that can be used to monitor the condition of oils and greases follow, while further information can be found on their respective websites, the addresses for which are shown in Figure 4.1. The organisations described below do not include several that develop test methods, standards or specifications which are used to evaluate the performance of new lubricants in operating equipment and which are too lengthy or expensive to be of practical value for lubricant condition monitoring.

DOI: 10.1201/9781003245254-4

ASTM	American Society for Testing and Materials	www.astm.org
CEC	Coordinating European Council	www.cectests.org
IP	Institute of Petroleum (The Energy Institute)	www.energyinst.org
ISO	International Standards Organization	www.iso.org
SAE	Society of Automotive Engineers	www.sae.org
CEN	Comité Européen de Normalisation	www.cen.eu
DIN	Deutsches Institut für Normung	www.din.de
ANSI	American National Standards Institute	https://ansi.org.
AFNOR	Association Français de Normalisation	www.afnor.org
JASO	Japanese Automobile Standards Organization	www.jsae.or.jp

FIGURE 4.1   Test method organisations.

Source: Pathmaster Marketing Ltd.

4.2.1 AMERICAN SOCIETY FOR TESTING AND MATERIALS

The American Society for Testing and Materials (ASTM) is a globally recognised leader in the development and delivery of voluntary consensus standards. Over 12,000 ASTM standards are used around the world currently, to improve product quality, enhance health and safety, strengthen market access and trade and build consumer confidence. The organisation serves a broad range of industries, including metals, construction, petroleum, consumer products and many more.

More than 140 technical standards-writing committees have more than 30,000 of the world's top technical experts and business professionals representing 140 countries. They work in an open and transparent process and use ASTM's advanced IT infrastructure, to create the test methods, specifications, classifications, guides and practices that support industries and governments worldwide.

ASTM welcomes and encourages participation from anywhere in the world. Its open consensus process, using advanced Internet-based development tools for preparing standards, ensures worldwide access for all interested individuals. When new industries, such as nanotechnology, additive manufacturing and industrial biotechnology, aim to advance the growth of cutting-edge technologies through standardisation, many of them seek ASTM's assistance.

4.2.2 COORDINATING EUROPEAN COUNCIL

The Coordinating European Council (CEC) is an industry-based organisation that develops test methods for the performance testing of automotive engine oils, fuels and transmission fluids, using gasoline and diesel engines and associated bench test rigs. It also develops test methods for marine and large engine oils and two-stroke engine oils.

CEC maintains tests on an ongoing basis, concentrating on quality assurance (it forms part of the European Engine Lubricants Quality Management System

(EELQMS)) and maintains confidentiality amongst stakeholders. It also manages the provision of reference fluids (lubricants and fuels) for its tests.

CEC test methods are used extensively in Europe and throughout the world. It is based in Brussels and maintains a secretariat also in Brussels. CEC was established over 30 years ago and now has more than 1,500 participants from over 300 companies. Representatives are able to contribute in CEC affairs at all levels. The essential objective of the CEC is to develop performance test methods on behalf of the industries it represents in a timely, quality focussed and cost-effective manner and to ensure that such methods relate to market place demand.

4.2.3 Institute of Petroleum (The Energy Institute)

The Institute of Petroleum (IP) was a UK-based professional organisation founded in 1913 as the Institute of Petroleum Technologists. It changed its name to the Institute of Petroleum in 1938. The IP and the Institute of Energy (InstE) merged in 2003 to form the Energy Institute (EI). Both Institutes had a distinguished heritage developed over many years supporting their individual energy sectors.

The Energy Institute is a not-for-profit chartered professional membership organisation that brings together expertise to tackle urgent global challenges. Responding to the climate emergency while meeting the energy needs of the world's growing population calls for energy to be better understood, managed and valued. This ambition is at the heart of the EI's social purpose. It is a global, independent network of professionals spanning the world of energy, convening and facilitating debate, championing evidence and sharing fresh ideas, giving voice to issues of concern and where necessary challenging the industry it works with.

The EI delivers standards, guidance, training and qualifications that raise the bar in operations in almost all areas of the energy system, including upstream and downstream oil and gas, onshore and offshore wind power, fugitive methane reduction, battery storage, hydrogen, carbon capture, utilization and storage (CCUS) and integrated networks.

The IP developed test methods and standards for use in the petroleum industry, collected together under the general title "IP Standards for Petroleum and its Products" and published in several volumes. Part I was "Methods for Analysis and Testing", and Part IV was "Methods for Sampling". IP test methods and standards became so well-known and used worldwide over many years that the EI decided to continue to use these under their IP designations for lubricants, fuels, bitumen, waxes and other products derived from crude oil and natural gas.

4.2.4 International Standards Organization

The International Organization for Standardization, more commonly known as the International Standards Organization (ISO), is an independent, non-governmental organisation, the members of which are the standards organisations of the 165 member countries. ISO is the world's largest developer of voluntary international standards, and it facilitates world trade by providing common standards among nations. Through its members, it brings together experts to share knowledge and develop voluntary, consensus-based, market relevant international standards that support innovation and provide solutions to global challenges. ISO's secretariat is based in Geneva, Switzerland.

Sixty-five delegates from 25 countries met in London, in 1946, to discuss the future of International Standardization. In 1947, ISO officially came into existence with 67 technical committees (groups of experts focusing on a specific subject). More than 20,000 standards have been set to date, covering everything from manufactured products and technology to food safety, agriculture and healthcare.

Use of the standards aids in the creation of products and services that are safe, reliable and of good quality. The standards help businesses increase productivity while minimising errors and waste. By enabling products from different markets to be directly compared, they facilitate companies in entering new markets and assist in the development of global trade on a fair basis. The standards also serve to safeguard consumers and the end users of products and services, ensuring that certified products conform to the minimum standards set internationally.

Beyond the task of guiding thousands of documents through drafting, review, voting and publication, ISO also offers a range of services that support its strategic goals. Among these ISO works to help raise public awareness of standards and standardisation. It works with other organisations, such as the International Electrotechnical Commission (IEC) and the International Telecommunications Union (ITU), to create an annual World Standards Day. The day looks at how standards address the challenges that face society today. ISO also engages the wider public through its consumer committee on standards development (COPOLCO). It promotes the teaching of standardisation, by participating directly in a joint master's programme, helping its members to set up similar programmes and by maintaining a database of materials related to standards in education at all levels. In addition to raising awareness, ISO also helps its members through training and acting as a resource for standards-related research.

Full members (or member bodies) influence ISO standards development and strategy by participating and voting in ISO technical and policy meetings. Full members sell and adopt ISO International Standards nationally. Correspondent members observe the development of ISO standards and strategy by attending ISO technical and policy meetings as observers. Correspondent members that are national entities sell and adopt ISO International Standards nationally. Correspondent members in the territories that are not national entities sell ISO International Standards within their territory. Subscriber members keep up to date on ISO's work but cannot participate in it. They do not sell or adopt ISO International Standards nationally.

4.2.5 Society of Automotive Engineers

SAE International, previously known as the Society of Automotive Engineers (SAE), is a US-based, globally active, association and standards developing organisation for engineering professionals in various industries. SAE International's headquarters is in Warrendale, PA. The organisation's primary emphasis is on global transport industries such as aerospace, automotive and commercial vehicles. The name SAE International was established to reflect the broader emphasis on mobility.

SAE International has over 138,000 global members. Membership is granted to individuals, rather than companies. In addition to developing, publishing and monitoring standards, it also devotes resources to projects and programmes in science, technology, engineering and mathematics (STEM) education, in professional certification collegiate design competitions. SAE aims to be a leader in connecting and educating engineers while promoting, developing and advancing aerospace, commercial vehicle and automotive engineering. Its mission is to advance mobility knowledge and solutions for the benefit of humanity.

SAE International is a global association of more than 128,000 engineers and related technical experts in the aerospace, automotive and commercial vehicle industries. Its core competencies are life-long learning and voluntary consensus standards development. SAE International's charitable arm is the SAE Foundation, which supports many programmes, including A World In Motion® and the Collegiate Design Series.

According to SAE International, it is pre-eminent in serving its members and industry, by providing a neutral forum that convenes to address society's mobility needs, the most reliable and comprehensive collection of mobility engineering resources, science, technology, engineering and mathematics (STEM) education and professional development programmes. These are intended to inspire and build mobility's future workforce, consensus-based standards to advance quality, safety and innovation and a global community whose collective wisdom makes mobility more safe, clean and accessible. The organisation has an online database of more than 37,000 standards on quality, performance, safety cost optimisation of products and product life cycles.

4.2.6 Comité Européen de Normalisation

The Comité Européen de Normalisation (CEN), translated from the original French into English as The European Committee for Standardization, is a public standards organisation. Its mission is to foster the economy of the European Single Market and the wider European continent in global trading and the welfare of European citizens and the environment. It aims to do this by providing an efficient infrastructure to interested parties for the development, maintenance and distribution of coherent sets of standards and specifications.

CEN was founded in 1961. Its 34 national members work together to develop European standards (ENs) in various sectors to build a European internal market for goods and services and to position Europe in the global economy. CEN is officially recognised as a European standards body by the European Union (EU), the European Free Trade Association and the United Kingdom. The other official European standards bodies are the European Committee for Electrotechnical Standardization (CENELEC) and the European Telecommunications Standards Institute (ETSI).

More than 60,000 technical experts as well as business federations, consumer and other societal interest organisations are involved in the CEN network that reaches over 460 million people. In 1999, the European Parliament noted in a resolution that CEN, CENELEC and ETSI co-operate smoothly and that a merger of the three standardisation bodies would not have clear advantages. The standardisation organisations of the 30 national members represent the 27 member states of the EU, three countries of the European Free Trade Association (EFTA), the United Kingdom and other countries that are highly integrated into the European economy. An example of harmonised standards is that for materials and products used in construction and listed under the EU Construction Products Directive. The CE mark is a declaration by the manufacturer that a product complies with all relevant EU Directives. CEN (together with CENELEC) owns the Keymark, a voluntary quality mark for products and services. A product bearing the Keymark demonstrates conformity to European standards.

4.2.7 Deutsches Institut für Normung

The German Deutsches Institut für Normung (DIN) was founded in 1917 as the Normenausschuß der Deutschen Industrie (NADI, "Standardisation Committee of German Industry"). NADI was renamed Deutscher Normenausschuß (DNA, "German Standardisation Committee") in 1926 to reflect that the organisation now dealt with standardisation issues in many fields and not just for industrial products. In 1975, it was renamed again to Deutsches Institut für Normung and is recognised by the German government as the official national-standards body, representing German interests at the international and European levels. The acronym DIN is often incorrectly expanded as Deutsche Industrienorm ("German Industry Standard"), largely due to the historic origin of the DIN as NADI. NADI indeed published their standards as DI-Norm (Deutsche Industrienorm).

DIN is the independent platform for standardisation in Germany and worldwide. As a partner for industry, research and society as a whole, DIN plays a major role in helping innovations to reach the market in areas such as the digital economy or society, often within the framework of research projects.

More than 35,000 experts from industry, research, consumer protection and the public sector bring their expertise to work on standardisation projects managed by DIN. The results of these efforts are market-oriented standards and specifications that promote global trade and encourage rationalisation, quality assurance and the protection of society and the environment, as well as improving security and communication.

4.2.8 American National Standards Institute

The American National Standards Institute (ANSI) is a private, non-profit organisation that administers and coordinates the US voluntary standards and conformity assessment system for products, services, processes, systems and personnel. ANSI was founded in 1918 and works in close collaboration with stakeholders from industry and government to identify and develop standards- and conformance-based solutions to national and global priorities.

ANSI's headquarters are in Washington, DC, and its operations office is located in New York, NY. ANSI was originally formed when five engineering societies and three government agencies founded the American Engineering Standards Committee (AESC). In 1928, the AESC became the American Standards Association (ASA). In

1966, the ASA was reorganised and became United States of America Standards Institute (USASI). The present name was adopted in 1969.

ANSI does not develop standards, but provides a framework for developing fair standards and quality conformity assessment systems. The organisation works continually to safeguard their integrity. As a neutral venue for coordinating standards-based solutions, ANSI brings together private and public sector experts and stakeholders to initiate collaborative standardisation activities that respond to national priorities.

The organisation recognises that standards and technical regulations together impact up to 93% of global trade, so globally relevant standards and the conformance measures that assure their effective use help to increase efficiency, open markets, boost consumer confidence and reduce costs. ANSI is the US leader in fostering that potential for the benefit of businesses across every industry and consumers around the world. ANSI represents the interests of more than 270,000 companies and organisations and 30 million professionals worldwide.

ANSI also coordinates US standards with international standards so that American products can be used worldwide. It accredits standards that are developed by representatives of other standards organisations, government agencies, consumer groups, companies and others. These standards ensure that the characteristics and performance of products are consistent, that people use the same definitions and terms and that products are tested the same way. ANSI also accredits organisations that carry out product or personnel certification in accordance with requirements defined in international standards.

4.2.9 ASSOCIATION FRANÇAIS DE NORMALISATION

The Association Français de Normalisation (AFNOR) is the French national organisation for standardisation and the French member of ISO. The AFNOR Group develops its international standardisation activities, information provision, certification and training through a network of key partners in France who are members of the association. They are ACTIA (Association of Technical Cooperation for the Food Industry), ADEME (French Agency for Environment and Energy Management), ADEPT (Association for the Development of International Trade in Food Products and Techniques), COFRAC (French Accreditation Committee), CSTB (Scientific and Technical Center for Construction), CTI (Center Network Industrial Technology), INERIS (National Institute for Industrial Environment and Risks), LCIE (Laboratoire Central des Industries Électriques), LNE (Laboratoire National Metrology and Testing), UTAC (Union Technique de l'Automobile, Cycle and Motorcycle) and UTE (Union Technique de l'Électricité).

There are 1,600 members of the AFNOR Association, a workforce of 1,250, 40 offices around the world and 77,000 customers. The AFNOR Group designs and deploys solutions based on voluntary standards worldwide. The Group serves the general interest in its standardisation activities and provides services in such competitive sectors as training, professional and technical information and intelligence, assessment and certification. The AFNOR Group has 1,800 auditors and evaluators, 480 authors, 250 trainers and 19,300 professionals involved in standardisation.

AFNOR has four business units. AFNOR Normalisation is vested with a general-interest mission, as defined by the French Standardization Decree of 16 June

2009, to coordinate leadership of the French standardisation system. AFNOR Publishing is the official distributor of voluntary standards in France, offering a range of professional and technical information and intelligence solutions with carefully thought-out designs for easy online use. AFNOR Competencies proposes an end-to-end range of training courses for all organisations and people looking to better understand and implement the applicable regulatory, normative and technical context. AFNOR Certification provides certification and assessment services and engineering for products, systems, services and competencies and issues the AFAQ and NF marks and the European Ecolabel.

4.2.10   JAPANESE AUTOMOBILE STANDARDS ORGANIZATION

The Japanese Automobile Standards Organization (JASO) is an industry organisation and specification setting association for lubricants designated for use in a broad range of Japanese machines.

JASO is part of the Society of Automotive Engineers of Japan (JSAE), which involves members ranging from engineers to students united by the aspiration to encourage and participate in automotive research and development. JSAE currently has more than 50,000 individual members and over 500 corporate members. It endeavours constantly to facilitate productive exchange between members, including research presentations, workshops for domestic and foreign technicians, symposia, international conferences, exhibitions and tours. JSAE issues various publications, develops and publishes standards, confers JSAE engineering awards and provides training for engineers and research workers.

Standardisation activities are conducted after careful consideration of their benefit to the general consumer and the needs of society. In addition, standardisation also serves a wide range of roles, such as improving quality, safety and reliability, as well as increasing production efficiency and compatibility. This is aimed at creating positive economic effects.

The JSAE is composed of 70 different committees and aims to make major contributions with promoting standardisation in Japan and worldwide. This is achieved through numerous activities including deliberations on the creation of automobile-related international standards at ISO and IEC and establishing and revising organisation standards, JIS and JASO standards.

JASO's small engine specifications are broadly recognised and have been adopted globally by the industry as key performance standards for both 2-stroke and 4-stroke applications. The JASO standards are used by oil marketers and many original equipment manufacturers (OEMs) globally as an indication of the baseline performance upon which their performance requirements are built. The JASO emblem and associated specifications displayed on oil containers assist consumers with identifying the correct lubricants for motorcycles and other small engine equipment.

4.3   SOURCES OF DEFECTS WITH NEW LUBRICANTS

Reputable lubricant blending plants check the quality of the base oils and additives they use, both on receipt from their suppliers and during storage, prior to blending. They will test samples to look for off-specification raw materials. They will also

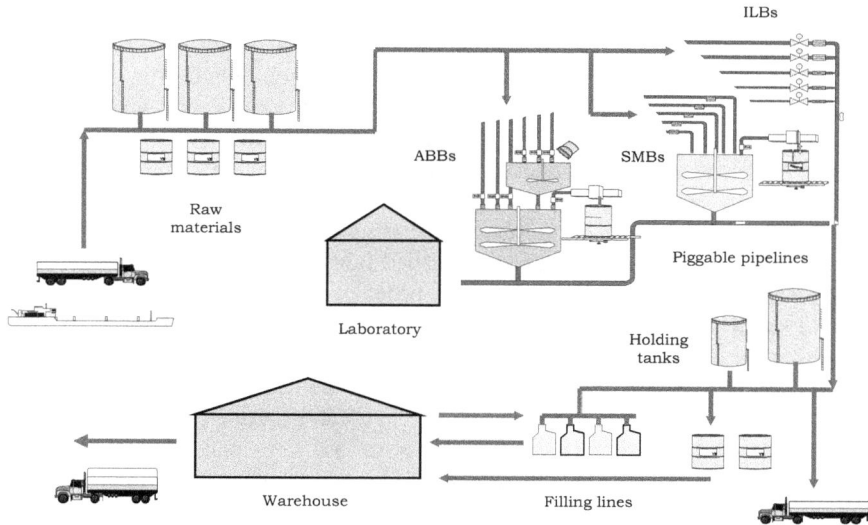

ILBs

ABBs

SMBs

Raw
materials

Piggable pipelines

Laboratory

Holding
tanks

Warehouse

Filling lines

FIGURE 4.2 Components of a lubricant blending plant.

**Source: Pathmaster Marketing Ltd., from an original drawing by Energy-Environmental
Engineering GmbH.**

require certificates of quality from their suppliers. After products are blended, they
will test them to ensure that they meet their specified quality requirements. Lubricant
blending plants are complex, with numerous products, formulations, components,
processes and equipment, as illustrated in Figure 4.2. Plants that manufacture greases
are equally complex and the production of greases involves more professional expert
know-how than exact science.

However, not all blending or grease manufacturing plants have the analytical
abilities to cover the range of potential non-conforming properties and contaminants
that can occur. For example, blending plants do not often have a particle counter
with which to check for particulate contamination. Emission spectrometers may not
be available to check for the contents of elements in additives or blended lubricants.
Some blending plants test each batch only for physical properties, such as viscosity,
viscosity index, flash point and pour point.

On occasions, errors occur during blending, particularly if the blending plant is
operated manually. Modern blending plants are completely computer-controlled, so
the addition of components, mixing, transfer, storage and packaging are fully automated. The mixing is done either in batch blenders or in-line blenders, generally
using mass flow meters or digital load cells. If these have not been calibrated correctly, the weights of the components may not be correct. If the blend is mixed at too
high a temperature, additive degradation can occur. Manual addition of an additive
may not be done correctly. In some cases, an additive may be left out completely
or added by mistake. In any of these cases, the resulting finished lubricant may be
off-specification. A good blending plant should pick up and correct these mistakes,
but sometimes they do not.

If a problem occurs with a lubricant in use, the lubricant manufacturer's formulation chemist(s) may decide to change the formulation slightly. Sometimes this happens without doing all the necessary laboratory, rig, bench or engine performance tests. This can contribute to the use of additives that are not compatible in service or which simply fail to properly dissolve in the base oil. The risk is greatest for blend-to-order requests or when formulations have been modified in a hurry.

Many blending plants use a number of blending vessels, in-line blenders and storage tanks for base oils, additives and finished lubricants. All these are connected using pipes, pumps and valves. Different products are supposed to be kept separate and small amounts of residues in vessels, tanks and pipes are supposed to be minimised or eliminated. Sometimes, however, cross-contamination of products does occur, almost always in error. The effects can be negligible, but in other cases, lubricant performance can be impaired. For lubricants shipped in bulk transportation, uncleaned tanks and tank compartments (or leaky compartments) can result in cross-contamination. It is always worthwhile checking that cross-contamination has not occurred.

The cleanliness of new lubricants varies considerably. New lubricants often exceed recommended cleanliness levels for in-service lubricants. Some blending plants use high-performance filters to clean new lubricants, others use only coarse filtration or no filtration at all. It is also common to find containers (drums, pails, bottles) on packaging lines and conveyors left open (exposed to atmospheric dust) for extended periods before they are filled and sealed. Bulk transport trucks may sit with their top bungs open before, during and after loading. All of these conditions can lead to both particle and moisture contamination.

Mislabelling of packaged lubricants is rare, but can sometimes happen due to human error. Because many different lubricants are handled, dispensed and packaged by manufacturing plants and distributors, a lubricant can accidentally be put into the wrong package or a package might be mislabelled.

4.4 PARTICULATE CONTAMINANTS

Several studies have indicated that the cost of excluding a gram of dirt is only about 10% of what it will cost once it gets into a lubricant. In some cases, when new oils from major manufacturers were tested, the ISO cleanliness codes (discussed later) have ranged from 14/11 (comparatively good) to 23/20 (not at all good). The average of these samples was 19/16, and several were 20/18 or 21/18.

However, responsibility for the cleanliness of new oils cannot be solely that of the lubricant supplier. Even if the supplier delivers lubricants below target levels of particulate or moisture contamination, poor practices at the user's site can compromise the efforts of the lubricant supplier to insure product integrity. Additionally, the cost to control contaminants throughout the manufacturing and delivery process must also be considered. While filtering oil to achieve a desired fluid cleanliness level is always possible, the cost to achieve this level of cleanliness, only to be compromised by poor practices once onsite may make this approach less than optimal. As an example, the typical cost associated with delivering a hydraulic fluid to a cleanliness rating of ISO 16/14/11 is typically around £0.15 per litre over and above the cost of the lubricant itself.

Upon delivery to the end user, the degree to which these contaminants should be excluded will be critically dependent on application. For example, a highly critical hydraulic application, with very tight tolerances and high pressures, will obviously require a significantly higher degree of contaminant exclusion than a high-viscosity open gear oil, to be used on a non-critical application.

In some cases, for example with military aviation hydraulic oils, the oil must be filtered, using very fine filters in a special delivery cart, as it is being pumped into the aircraft's hydraulic system. Nevertheless, whatever the application, it is still incumbent upon the lubricant manufacturer to provide some degree of contamination exclusion.

In order to balance cost versus benefit, it is generally good practice to set two levels of fluid cleanliness, one for new oil deliveries and one for in-service oils. Typically, the new oil cleanliness requirements should be set 1 to 2 ISO range codes higher than the in-service targets as a compromise between cost and acceptable new fluid cleanliness levels.

4.5   MOISTURE CONTAMINATION

Water is possibly a more harmful contaminant for stored lubricants than particulates. Although some solid contaminants, such as dust, can cause additive depletion through adsorption, in addition to promoting catalytic degradation at higher temperatures, the effects of water washing of additives and hydrolytic degradation make water a far bigger threat for stored lubricants. It is generally recommended to set maximum permissible water levels below the saturation point of the fluid at all ambient temperatures to which the fluid will be exposed during manufacture, storage, transportation and dispensing. For many industrial lubricants, this means that the water contents of new oils should be less than 100 ppm (0.01%) volume/volume.

It is well known that water can sometimes find its way into lubricants, particularly in hydraulic systems, steam turbines, enclosed gears and air compressors. The water might come from cooling coils, heat exchangers, steam condensers, washing of machines or condensation of atmospheric moisture. However, new lubricants can also experience moisture contamination.

A lubricant blending or grease manufacturing plant's managers, supervisors and operators understand that base oils may be delivered with some water and/or rust contamination, which may not be apparent when examined visually. Water in base oils can arise during the processes to manufacture them, whether from leaking heat exchangers or steam heating coils in storage tanks or during bulk transportation by tanker, barge, rail tank wagon or road tanker. Rainwater seeping into barges, rail tank wagons or road tankers should be avoided. Fortunately, water in base oils used to manufacture oils or greases will almost always be removed during manufacturing.

However, following manufacture, all packaged oils or greases need protection from rainfall, snow and condensation. Again, rainwater seeping into barges, rail tank wagons or road tankers being used to deliver blended oils in bulk needs to be avoided. The author is aware of several occasions in which seawater has accidentally entered tanks of finished lubricants being shipped in bulk in coastal tankers. In regions of high humidity in which where temperatures can drop to levels at or below the dew point, tanks, drums, bottle and cans of oil and drums, tins and cartridges of grease will need to be protected from changes in temperature. For example, drums of oil should be stored upside-down or on their sides if they are outdoors, as drums stored upright may "breathe in" moisture or rainwater, past the seals, as a result of the expansion and contraction of the air space above the oil due to changes in daily heat and cold. Drums of greases should be stored indoors for the same reason, since it is inadvisable to store them upside-down.

Most new lubricants have a moisture content of less than 50 parts per million (0.005%), although transformer and refrigerator oils are required to have even less. However, this should be checked before the oil or grease is put into service.

Water and rust can accelerate the deterioration of new oils and greases. Water in new oils and greases can react with additive to degrade them, which can result in the oil or grease being unfit for use. Water separates slowly, or not at all, from oil that has been oxidised or contaminated with dirt. In this respect, iron rust is a particularly unacceptable form of contamination.

4.6   ESTABLISHING A BASELINE FOR SUBSEQUENT TESTING AND MONITORING

To be able to conduct accurate lubricant condition monitoring, a baseline (starting point) sample should be taken. This will allow subsequent tests to be compared with the baseline test when the lubricant was new and will start the process of trend analysis, which is discussed in the next chapter. Without a starting point, it is not possible to know whether the lubricant's condition is stable, improving or deteriorating.

Once this baseline sample has been obtained, it should be kept as a reference. The lubricant supplier's blending plant should also have kept a reference sample. If there is a subsequent problem, these can be compared to check whether something may have happened during storage, transportation or delivery.

Tests (discussed in depth in later chapters) that can screen a variety of properties and performance characteristics all at once are an effective starting point. Good examples are elemental analysis, demulsibility and infra-red (IR) spectroscopy. Tests and properties that are essential to lubricant performance and machine reliability in the particular application should be prioritised. Good examples include viscosity, viscosity index, oxidation stability, cleanliness and dryness.

Tests that would quickly reveal a property relating to a specific quality concern, if any, should be considered first. It is advisable to keep testing streamlined and efficient. Expensive and/or time-consuming tests should only be done in exceptional circumstances, triggered by a non-conforming result from a screening test.

It is important that a consensus is developed with the lubricant supplier on which tests will be performed and what the condemning limits will be. When a lubricant user takes an active role in testing new lubricants and giving constructive feedback to the lubricant supplier, incremental improvements in lubricant quality are highly likely to result. Lubricant suppliers, particularly their blending plants and supply management chains, are encouraged to maintain close collaboration with customers on lubricant quality. Lubrication excellence is a collaborative process. Lubricant quality is a measurable property. If a performance property is important, it should be measured.

4.7 LUBRICANT STORAGE

Lubricants can be stored in tanks, drums, small containers or cans. The requirements in each case are that the material should be protected from contamination, should not deteriorate in storage, and that safety, health and environmental matters are addressed. For drums and small containers, it is important that the contents are easily identifiable from the markings and that all grades are easily accessible for shipping purposes.

Where these tanks, drums and small containers are stored is also very important for the protection of the lubricants. Although the design, facilities and operation of a warehouse share many of the features of warehouses in other industries, a few specific requirements are necessary.

4.7.1 Storage Vessels and Containers

4.7.1.1 Bulk Storage in Tanks

Fixed-roof tanks of various sizes may be used to store lubricants. Before use they should be thoroughly cleaned, be scale-free and may be coated internally with a proprietary protective coating that is of a type compatible with the lubricant to be stored. The tank should initially be dry, and filling arrangements must be such that ingress of water to the product is eliminated.

For viscous oils and in cooler climates, it will be necessary to provide some heating. For lighter oils, an outflow (suction) heater is sufficient, but if bulk heating is necessary, the heating coils must be sound and should be fed with hot oil or water preferably or fed with only low-pressure or exhaust steam to prevent local overheating.

The tanks should stand on an impervious concrete platform and be bunded (provided with a catchment area for the product, should the tank leak). A lookout must be kept for leaks at valves and flanges, and these should be eradicated as soon as possible. Lagging (insulation) of transfer lines is common, but the tanks themselves are not normally lagged for finished oils.

If horizontal-type tanks on piers are used, the same general advice applies, with the tanks being slightly sloping towards a draw-off point for samples. Where draw-off lines are common to different grades of oil, a pigging system should be installed.

4.7.1.2 Drum Storage

Larger volumes of finished lubricants are often stored in drums. This is the most difficult and potentially the most hazardous form of storage. Ideally, drums should be stored on their sides and with the bungs below the liquid level. This is to prevent water from collecting in the tops of the rims and being drawn in as the drums cool and keeping the bung seals moistened with product guards against leakage. However, storing large numbers of drums horizontally is hazardous unless a large amount of special-purpose racking is used, which can be costly. Drums stored one on top of the other should never be more than three drums high, and the ends must be securely chocked. Products must be stacked so that access to the lower layer is not required until the first layer has been removed. In no cases should drums be directly laid on the ground, but should be on battens or an impervious base. Provision should be made for stock rotation on as close to a "first-in/first-out" (FIFO) basis as can be achieved.

The safe storage period or "shelf-life" of products must be considered, and product should not be left for excessive periods at the bottom of a stack. Drums are frequently stored in fours on pallets. Full drums should normally be stored no more than two pallets high, although empty drums can be stored in higher piles if adequate care is taken in placing and removing the top layers. In particular, forklift trucks should have strong safety cabins and not be of the open type. Vertically stored drums must be protected from rain, and if they cannot be stored in a warehouse, they should be sheeted over with tarpaulins or plastic wrapped. In hot climates, the drums should be protected from direct sunlight by light-coloured screens or roofing.

Leakage must be prevented as much as possible. If taps are used on drums, then drip trays must be provided and care taken that the taps are functioning properly and are shut-off after use.

Small hand pumps are sometimes used to withdraw material from a drum placed on end, but again care must be taken that this does not leave a trail of oil when removed from the drum. To mop up accidental oil spills, proprietary crystalline materials are available which are much less hazardous than traditional sawdust.

Care must be taken when removing drums with forklift trucks that the drums are not accidentally punctured, thereby producing both an environmental problem and a health risk. Drums are often returnable or reusable, and therefore care should be taken at all times not to damage them unduly in handling operations.

4.7.1.3 Plastic Bottles and Tin-Plate Cans

These should always be kept under cover in a warehouse or other suitable building. Small containers will normally be packed in multiples in cardboard containers. Where the stock is for in-house use, provision must be made for unpacking and disposing of cartons and not allowing these to become oil soaked where they will present a hazard.

Because tin cans and plastic bottles are lightweight and much less robust than steel drums, they cannot be stacked on top of each other either very easily or very high. Even cardboard cartons of identically sized plastic bottles cannot be stacked more than three or four high. This means that pallets of plastic bottle storage cartons cannot be stacked on top of each other. This places limitations on the design and operation of lubricant storage warehouses.

4.7.2 Siting the Lubricants Store

The ideal site for a lubricants store in a warehouse or manufacturing plant has:

- A good reception area, with free access for vehicles and ample room for loading.
- Adequate space for filled packages (drums, cans and plastic bottles) of all grades and sizes of packed products.

- Adequate space for empty packages (drums and polycontainers) near the unloading point, so they can be delivered by delivery vehicles.
- A well-equipped loading dock, with direct access to the oil store.
- A location that minimises the work needed to get lubricants into and out of the store.

In large plants, there may be more than one storage warehouse needed for finished, packed lubricants. A main warehouse may then be used to stock several satellite warehouses.

4.7.2.1 Indoor Storage

Indoor storage of lubricants is desirable at all times. Since most companies cannot store everything indoors, however, they have to make the best use of the space available. They require easy access to the stock and freedom to use the packages in the order delivered. Three methods of storing packages are available and in common use; free stacking, palletisation and the use of racks.

In free stacking, the packages are placed on top of one another, and the safe height of the stack depends on its stability and the weight that the lower packages can support. The smallest packages – cans of one gallon (4 to 5 litres) or less and plastic bottles – are usually packed in stout fibreboard cases, or they can be bound together with tape for stability. Five-gallon (23 litre) drums and grease kegs are usually handled singly (unless palletised) and can be manhandled. The movement of drums calls for at least a hand trolley and skid and preferably a forklift device or mechanical hoist. The use of planks or slatted frames stabilises the stack and helps to prevent damage to the lower layers. If a forklift truck or similar is available, palletisation eases stacking, reduces handling risks and allows better access to the lower layers.

Small drums can be stacked on pallets in tiers seven-high, though such a practice is unusual except in a central store serving a number of secondary sites or where space limitations justify frequent "breaking" of the stacks. Drums can also be palletised, but more often they are stacked with interposed strips of timber.

Steel racks (of slotted or plain angle or of the clamped tubular type) have several merits; they allow space to be used to the best advantage, they ease stock handling and they encourage regular turnover. They should be installed with aisles wide enough to allow a forklift truck to be manoeuvred.

Only in very cold climates can indoor temperatures drop low enough to produce adverse effects in a lubricant. At the other end of the temperature scale, though, excessive heat due to the proximity of steam pipes, boilers, furnaces or flues should be avoided for grades containing volatile solvents. In many cases, because of insurance requirements or local fire regulations, it may be necessary to house such products, together with kerosene, white spirit and so on, in a store separate from lubricants. If one part of the store is hot, it should be reserved for oils of high viscosity.

The store should be dry at all times. Most containers are made of painted sheet steel or tin-plate, both of which will corrode if left in a damp condition long enough. Tin-plate in particular can rust through in only a few weeks, and the result is that both water and abrasive rust fall into the lubricant. In severe cases, rust can obscure the grade-name markings on containers.

Well-designed racks save space, ease handling and ensure first-in first-out procedure. In the absence of racks, palletising aids handling, but bottom layers tend to remain undisturbed.

If indoor storage is limited, it should be reserved for small packages and for lubricants affected by frost and heat. Take special precautions with outdoor stock; try to arrange for a short stock-life or frequent turnover.

It is very important not to use empty drums for road barriers or scaffold-pole support. In particular, NEVER use drums that have been used to store or dispense oils or greases as work-tables or trestles for welding or brazing work, because of the risk of explosions.

4.7.2.2 Outdoor Storage

Drums should not be stored upright outdoors unless they are upside-down, with the bungs at the bottom. As described previously, gaskets and o-ring seals around bungs are rarely air-tight. Any rainwater that collects on the tops of drums stored upright is usually sucked into the drum as it "breathes" (expansion and contraction of the drum's contents) due to daily changes in air temperatures.

Gaskets and o-ring seals around bungs in drums should ideally be wetted by the drum's contents. Drums stored horizontally should be placed so that their bungs are below the level of the oil, ideally at the "three o'clock" and "nine o'clock" positions.

Drums stored on their sides should be clear of the ground, perhaps on baulks of timber. They can be stacked three-high in this way, but must be carefully wedged to prevent movement. Steel sections are sometimes used instead of timber baulks.

All too often, where drums are stacked, the top ones are used and quickly replaced by new ones, so the lower ones remain undisturbed for years. For this reason alone, racking is to be preferred. Sloping racks, in which drums are loaded at one side and removed from the other (first in, first out) are ideal. Apart from ensuring regular replacement, a rack is convenient to load, is safe and makes for reasonable use of space. A corrugated-iron or plastic roof over the rack also provides a degree of protection against rain. Because of the rolling of drums, however, the ideal positions of the bungs cannot be maintained for long.

If small packages, drums and grease kegs have to be stored out of doors, they should be sheeted over, with free access for air, should be and examined regularly. The size of the stock should be adjusted to provide a quick turnover. On no account should the containers stand directly on the ground: rather they should be raised so that air can circulate beneath and around them.

If possible, outdoor storage sites should not be near dusty areas such as quarries and unmade roads, because thick dust on containers is likely to contaminate the contents when they are opened.

4.7.2.3 Storage of Special Types of Lubricant

Tanks for electrical and refrigerator oils are usually lined with amine-cured epoxy resin, and the air vents are protected with a silica-gel breather, to remove moisture. Self-sealing couplings are fitted to fill-pipes. During manufacture, some tanks for white oils are internally shot-blasted and immediately coated with rust-preventive oil. After erection on site, they are swabbed internally with the oil to be stored.

Desiccators are seldom needed in the storage of these oils, but as a precaution, the air vents on the tanks are protected with filters.

4.8   DISPENSING LUBRICANTS

When dispensing new lubricants, it is important to maintain accurate records. There should be written requisitions and authorised signatures, stipulating the quantity and grade of lubricant required and the application point (machine or equipment). The record should also list the date and time the lubricant was dispensed. It is easy now to record this information on a laptop or tablet, for later upload to a central register.

Containers of lubricants being dispensed must be marked clearly. This minimises the risk that the wrong lubricant will be put into a specific machine or item of equipment.

The tops of unopened drums and pails should be wiped clean before dispensing. This will help to ensure that any dirt or moisture does not get into the drum or pail. Dispensing grease is a problem; smaller packages are more convenient; grease dispensers are better.

The dispensing area should be kept separate from the main lubricant storage area. Nevertheless, the dispensing area should have adequate, but not excessive stocks, which should be checked regularly. All packages should come through the dispensing area, not direct from the storage area.

The dispensing area should be locked when unattended and should only be staffed by trained and authorised employees.

4.9   SUMMARY

Testing new lubricants, as they are received from a lubricant supplier, is the first step to achieving a consistent, reliable and meaningful lubricant and machine condition monitoring programme. The tests should determine whether the correct lubricant has been delivered in the correct pack and to establish the starting point (baseline) for the subsequent condition monitoring programme. Only tests that will be used in the condition monitoring programme should be selected, which excludes a large number of tests that are used to develop new lubricants or assess their performance against international, national or OEM's specifications.

The correct storage of lubricants is important to maintaining the properties and performance of the products, as well as to the cost-effectiveness of the complete supply chain. Lubricant storage is not simply about containers (tank, drums or bottles), but also about warehouses and their operation. Storing and dispensing new oils and greases correctly will ensure that they start their application in machines and equipment with the properties and performance that is intended.

5 Testing Lubricants in Use

5.1 INTRODUCTION

Once a baseline set of data has been established for a specific lubricant in a specific machine or item of equipment, the next stage in a lubricant analysis and condition monitoring programme is to take regular samples of lubricant from the machine and test them using the same tests as for the baseline.

Numerous test methods that can be used for testing both new lubricants and lubricants in use as part of a condition monitoring programme will be discussed in depth in subsequent chapters. Chapter 5 focusses on the ways that can be used to test lubricants in use and how to report and assess the results.

Later chapters will look at different types of equipment and machinery and how their operating condition can be monitored, using both lubricant testing and other analytical methods.

5.2 PROCEDURES FOR IN-SERVICE TESTING OF LUBRICANTS

5.2.1 FIELD TEST KITS

A number of companies market field test kits with which to measure several properties of lubricants in service. These kits were previously known as "spot tests", but have improved in reliability to be acceptable for continuing analyses where access to laboratory tests is limited, such as in small factories, on isolated sites, on outside and off-highway machines and on ships.

The aim of on-site oil analysis is not to replace a conventional test laboratory. Several key points need to be considered when selecting on-site tests:

- The frequency or total number of tests to be performed each month.
- The most critical and most time-sensitive data, for example with machines or equipment where there is little or no time between a potential failure starting and a functional failure occurring.
- The least skill-intensive kits and tests, so that non-technical staff do not have to become proficient at running complex scientific equipment.
- Kits, test methods, components and reagents whose costs are within both an initial and an annual budget.
- The comparative costs of off-site tests versus on-site tests.

Tests that can be done with on-site test kits or portable equipment include colour, kinematic viscosity, acid number (AN), base number (BN), particle counting, soot detection, water detection, foaming properties, demulsibility, voltametry, dielectric strength, ferrous corrosion resistance, fuel detection, FTIR (Fourier Transform Infra-Red) spectroscopy and analytical ferrography. Descriptions and

DOI: 10.1201/9781003245254-5

explanations of different test methods for measuring these properties are contained in Chapters 6 to 8.

Viscosity is readily measured by using a simple "falling ball" tube viscometer in the field on-site. Comparison with an identical apparatus, often in a "twin arrangement" containing a new sample of lubricant, gives a direct comparison of whether the used lubricant viscosity has comparatively increased or decreased by the respective times taken for the balls to descend in their tubes.

AN and BN kits are designed for use by non-technical staff in an on-site environment. These kits provide pre-measured, non-hazardous reagents in ready-to-use ampoules. A relatively accurate measure of acids or bases in the oil can usually be recorded within minutes. Although accurate, these test kits only approximate laboratory titration methods, such as ASTM D664, and the two should not be compared quantitatively.

The simplest method to determine particulate levels in a sample of a degraded lubricant is the blotter test, where a small volume of oil sample is pipetted onto a filter paper or some other absorbent material. This test can take various forms, either using a standard filter paper or a thin layer chromatographic (TLC) plate. The measurement concerned is the optical density (OD) of the central black spot. The higher the level of particulate, the denser (darker) the spot. The assumption is that the spread of the lubricant sample disperses carbon particulate within an expanding circle and that the OD of the carbonaceous deposit is a direct measurement of the mass of particulate present in that sample. The system can be quantified by use of a simple photometer, for field-based simple systems or a spectroreflectometer for laboratory measurements. Methods of automating these types of systems have included:

- Automated, accurate, constant volume pipetting of the oil samples.
- Video measurement of the oil sample blot on the filter paper, thus its OD.
- Data recording of these results.

Despite many attempts and applications, these advanced methods for determining particulate levels have not achieved widespread acceptance, possibly because of the increased complications built onto an initially simple test. Another, and major, problem is the heterogeneous nature of the samples presented for analysis, which give different responses, arising from:

- Different base oils, such as mineral oils and synthetic fluids used in modern lubricant formulations.
- Different formulations, such as the differences between hydraulic, automotive, aerospace and marine fluid formulations, with a high-dispersancy oil spreading its carbonaceous matter over a greater area than a low-dispersancy oil.

A further development of the blotter test is to use thin layer chromatography (TLC) plates, which are more uniform than paper. The intensity of the black spot from a 50 microlitre (μl) aliquot can be measured and, if its image is captured electronically, may be integrated across its area. But the black carbonaceous spot will also have a base oil ring extending beyond it, seen either as a change in white shade or a fluorescent area under UV illumination.

Several makes and models of particle counters are available currently. Most models are laser-based instruments and a couple are pore blockage (mesh obstruction) instruments. Although most models cost more than US$10,000, a particle counter should be high on the list of instruments to be purchased, when the budget is available. When performing ISO particle counts of any kind, it is crucial for the sample to be agitated aggressively using an industrial paint shaker. These two items need to be procured at the same time to ensure comparable results.

Water content can be measured in the field by mixing a lubricant sample with a carbide tablet in a sealed stainless-steel bomb. The measurement of water content is through the reaction of the carbide tablet (or calcium hydride in an alternative model) to generate gas pressure within the bomb. The pressure level generated is a measure of the water content of the sample. An alternative quick test for water content is the "crackle test", in which a small lubricant sample is heated suddenly. This can be done by rapidly inserting a hot soldering iron bit into the sample. If free water is present, a "crackling" noise will be heard, which is absent for dry samples. (The noise comes from the generation of steam in the sample.) Alternatively, a small drop of sample can be dropped from a syringe onto a "hot" laboratory hot plate, when again a crackle will be heard if the sample is wet. From experience, the limit of detection is taken to be 0.1% water.

The degree of oxidation can be measured by a simple colorimeter using a standard sample to measure colour, ASTM D1500. The trend compared with previous values is the important observation. If the change occurs early in the service of the lubricant, then the anti-oxidation reserve of the lubricant is being rapidly depleted or the lubricant is being contaminated. It is important to consider the change in colour in combination with values and changes determined for AN and viscosity for the same samples.

The instruments needed to perform a ferrogram (analytical ferrography) include a slide maker and a microscope. Although different slide makers are available, all produce a glass slide by holding ferrous particles in place with a magnet as the fluid is passed over. Typically, the microscope will have a 100X to 1,000X resolving power with a digital camera to capture ferrogram images. Together, these items are quite expensive. A significant amount of training and experience is required in order to achieve a successful on-site ferrography programme. These instruments are often used only by experienced on-site condition monitoring programmes with significant resources.

An example of an on-site test kit is explained in ASTM D7417 "Analysis of In-Service Lubricants Using a Particular Four-Part Integrated Tester". The test method covers the quantitative analysis of in-service lubricants using an automatic testing device that integrates four technologies: atomic emission spectroscopy, infra-red spectroscopy, viscosity and a laser particle counter. The test method is suitable for oils with kinematic viscosities from ISO 10 to ISO 320, so covers most hydraulic oils, gear oils, turbine oils and compressor oils. The test method may be used to establish trends in wear and contamination of in-service lubricants, but it may not give equivalent numerical results to current ASTM test methods. The integrated tester is used primarily to perform on-site analysis of in-service lubricants used in the automotive, highway trucking, mining, construction, off-road mining, marine, industrial, power generation, agriculture and manufacturing industries.

The integrated tester is prepared for analysis according to the operational manual and on-screen prompts. A sample of in-service lubricant is placed into the sample transport system and is analysed using available integrated devices. The application software guides the entire procedure, controls the transfer of the sample, stores data and generates on-screen and printed results with a printed generic recommendation of the lubricant's physical condition.

Further examples of on-site test kits are explained and discussed in Chapters 10 to 14.

5.2.2   IN-HOUSE LABORATORIES

Some commercial users of lubricants are sufficiently large to justify having their own analytical laboratory. These can range in size from quite small to comparatively large.

In most cases, an in-house laboratory will have all the basic test equipment, together with a few more specialised pieces of test and analytical equipment. In the infrequent cases where more advanced tests are required, the lubricant user will need to send samples to a larger, independent laboratory.

An in-house laboratory will need to have a lubricant sample storage area or room, sufficiently large to accommodate samples of new oils and greases for all types of lubricants used in the machines and equipment. These samples will need to be stored for a minimum of two years, and preferably three years. The sample storage area or room may also need to contain samples of in-service oils or greases from critically important machines or equipment, in case testing of these uncovers a problem and more extensive testing is required. The benefits of retaining samples of both new and in-service lubricants are discussed later in this chapter.

The in-house laboratory will also need to have a stock of appropriate new sample bottles, together with back-up sampling equipment, in case a vacuum pump, sampling valve or other item breaks.

5.2.3   INDEPENDENT LABORATORIES

There are many business services companies, worldwide, that offer testing and analysis services for a very wide range of industries, including lubricants. In addition to testing, the services often include inspection and certification so that client companies view the service provider as a problem-solving and business improvement partner. It is not unusual for one or more of the service provider's staff to be embedded within a client's manufacturing or management structure, working as an integral part of the problem-solving team, particularly for larger client companies.

For lubricants, the services provided can include:

- Provision of sample collection kits.
- Chemical, physical and microbiological analysis of oils and fluids.

- Chemical and physical tests on greases.
- Specialist testing, for example of electrical transformer oils for water content, dielectric strength, contaminants and acidity.
- Analysis of wear, additive and contaminant elements with expert interpretation of results.
- Interpretation of results against international standards.
- Web-based time trending of asset test results with 24/7 access and e-mail alerts.

Many independent service providers also offer oil and machine condition monitoring. Their accredited testing facilities for lubricating oils and greases and specialist chemists and engineers help to ensure the safe and cost-effective performance of the customer's equipment. This helps to prevent untimely failures, minimises long-term maintenance costs, enhances component life, reduces oil costs and maximises safety of equipment.

An increasing number of independent laboratories send test reports to client companies using e-mail. Additionally, client companies are able to access test report results on secure, individually password protected, web pages. This enables client companies to assess any actions required, effectively manage them, generate instructions and make document changes in one location. This results in cost savings through less administration time and paperwork. Asset test reports and trended data/information can be viewed in and exported directly from the software when a hard copy is needed. Interpretation of analytical results is the critical stage in the oil condition monitoring process.

Web-based asset management enables the independent laboratory to provide fleet and equipment profiling limits, wear rates and the incorporation of other data. Customer's own vibrational reports and interface capabilities, including work orders, oil changes, bearing changes and other information, can be added to the web pages. These events can be viewed directly on the trend graphs.

Sources of information about companies that provide lubricant testing and analysis services can be found at:

- **Europe**:
  European Lubricants Industry Directory (ELID):
  http://www.lube-media.com/directory. The directory is searchable and lists companies that offer lubricant testing.
- **North America**:
  Lubes 'n' Greases Directory:
  https://directory.lubesngreases.com. Click on Testing & Testing Equipment and then click on Testing. The directory was originally launched for North America, but has subsequently expanded to be worldwide. At the time of writing, it had information on a total of over 9,500 companies (in all categories) in most countries in North and South America, Europe, the Middle East, Africa, Asia and Oceania.
- **South America**: A comprehensive South American directory of lubricant testing companies did not appear to exist at the time of writing. South American companies are listed in the Lubes 'n' Greases directory.

- **Asia Pacific**: At the time of writing, there did not appear to be a comprehensive directory of lubricant testing companies for the Asia Pacific region. Standard business directories, such as Alibaba, do not appear to list companies that offer lubricant testing services.
- **Middle East and Africa**: A comprehensive Middle Eastern or African directory of lubricant testing companies did not appear to exist at the time of writing. Companies in these regions are listed in the Lubes 'n' Greases directory.

Several companies that provide lubricant testing and trend analysis have offices and laboratories in many countries worldwide. They include:

- **Intertek Testing Services (ITS)**: https://www.intertek.com. Intertek provides organisations around the world with quality control, research, testing, measurement and certification activities for industry, commerce, markets, institutions and governments. Industries that benefit from Intertek state-of-the-art laboratory services include consumer, petroleum, chemical, materials, energy, electronic, pharmaceutical, food, medical, minerals and more. Testing follows ASTM, ISO and many other industry standards.

  Intertek serves almost all countries in North, Central and South America, West and East Europe, the Middle East, Africa and Asia Pacific. It has more than 44,000 employees in 1,000 locations in over 100 countries. In addition to testing, inspecting and certifying products, Intertek is a Total Quality Assurance provider to industries worldwide, with innovative and bespoke assurance, testing, inspection and certification services to customers. Its services run 24 hours a day, 7 days a week.

- **ALS Global**: https://www.alsglobal.com. ALS Global works with clients to create cost-effective test packages applicable to their particular equipment and mechanical designs. Companies typically spend at least 10% of their revenues on maintenance and equipment repair. Standard testing programmes for equipment reliability are comprised of routine tests that can meet most service requirements.

  One of the many testing and analysis services offered by ALS Global is Oil, Fuel and Coolant Analysis. The company has offices or laboratories in the United States, Canada, Mexico, Argentina, Brazil, Chile, Peru, Colombia, Czech Republic, India, Singapore, Australia and Malaysia.

  ALS Global provides a basic preventative maintenance package designed to provide early detection of problems in the initial stages of development, excessive wear and the source of that component's wear, unwanted contaminants, such as dirt, water, coolant, nitration and incorrect oil, the lubricant's suitability for further service time, dilution of lubricants, misapplication of lubricants and the oil's physical properties.

- **Bureau Veritas**: https://oil-testing.com. Bureau Veritas operates oil analysis laboratories around the world, and its clients depend on them to deliver quality results and informative, actionable maintenance recommendations. Each laboratory provides customers with numerous advanced fluid testing and analysis services. The organisations' data analysts are US Society of Tribologists and Lubrication Engineers (STLE) Certified Lubrication Specialists and Oil Condition Monitoring Analysts with equipment-specific knowledge and more than 150 years of combined experience in machinery lubrication and maintenance.

The Bureau Veritas Group's oil analysis services were strengthened in 2014 with the acquisition of Analysts Inc., which was established in 1960 and was the first commercial laboratory to provide oil analysis test packages for specific equipment operations and the first to make maintenance recommendations to correct problems that oil testing and analysis identified. Bureau Veritas itself was established in 1828, aiming to become a global leader in Testing, Inspection and Certification (TIC), delivering high-quality services to help clients meet the growing challenges of quality, safety, environmental protection and social responsibility.

Bureau Veritas has lubricant testing offices or laboratories in the United States, China, Japan, Singapore, Dubai, Saudi Arabia, the Netherlands, Spain, Germany, Greece, Australia and South Africa.

- **SGS**: https://www.sgs.com. SGS is a leading inspection, verification, testing and certification company, headquartered in Switzerland. The company was established in 1878 and transformed grain trading in Europe by offering innovative agricultural inspection services. SGS was registered in Geneva as Société Générale de Surveillance in 1919. The current structure of the company, consisting of ten business segments operating across ten geographical regions, was formed in 2001. SGS is recognised as a global benchmark for quality and integrity. It has more than 89,000 employees, a network of more than 2,600 offices and laboratories around the world.

The company's core services are divided into four categories. Inspection and verification services involve activities such as checking the condition and weight of traded goods at trans-shipment, helping to control quantity and quality and meeting all relevant regulatory requirements across different regions and markets. Testing has a global network of facilities, which are staffed by knowledgeable and experienced personnel, and enables companies to reduce risks, shorten time to market and test the quality, safety and performance of products against relevant health, safety and regulatory standards. Certification enables companies to demonstrate that products, processes, systems or services are compliant with either national or international standards and regulations or customer-defined standards. Verification ensures that products and services comply with global standards and local regulations. SGS aims to combine global coverage with local knowledge, unrivalled experience and expertise in virtually every industry, to cover the entire supply chain from raw materials to final consumption.

SGS has offices and laboratories in almost every country in North and South America, Western and Central Europe, the Middle East, Africa and Asia Pacific. The company's services with lubricant testing can be seen on the SGS website by clicking on "Our Services", then "Construction" and then "Services Related to Machinery & Equipment". Contact details for the laboratories, not all of which are involved in testing lubricants, can be found on the company's website.

5.3 SAMPLE TRANSPORTATION

When samples of lubricants need to be sent to an independent testing laboratory, it is imperative that they are packaged and transported in such a way that the containers do not break or the contents do not become contaminated during transportation. In this regard, it is important to ensure that sample bottles are kept separate from each other when two or more are shipped together. This can be achieved either by wrapping each bottle in a protective bubble wrap or having separate compartments within the package for each bottle. Transportation of more than about six samples at one time may need several packages. Care should be taken that a package does not become too heavy to lift manually, unless a large package is put on a pallet for either a forklift truck or pallet trolley or can be moved using a sack-barrow.

If the independent laboratory is relatively close, say less than 50 km by road, the samples can be sent in protected boxes in a van or small truck. Some companies use specialist couriers. Using a postal system is not advisable, unless a special contract has been agreed. Glass sample bottles are usually suitable for on-site and in-house testing, while plastic sample bottles are likely to be better for transportation to a more distant external laboratory.

Where the independent laboratory is further away, or overseas, special rail or air freight arrangements may be required. This applies particularly to marine lubricants. Although air freight is comparatively expensive, it has the advantage of being comparatively quick.

The time taken (in days) from point of sampling, to receipt of the sample by the laboratory, to testing and to the availability of the results may not matter for some machines or equipment, but may be very important for others. For marine lubricants, for example, the samples should be taken just before the ship reaches port and should be dispatched immediately after docking. The results should be back to the ship before it sails again, usually no more than two days, in case a lubricant needs to be changed or an item of equipment needs to be maintained or repaired.

When setting up a new condition monitoring programme, items of equipment that are critical to the operation of the plant, facility or system should be identified. Lubricant samples from these items of equipment may need to be tested and the results available very quickly, in case something is amiss and needs urgent action. The consequences of delayed action could be loss of production, loss of sales and revenues, damage to the plant or facility, damage to the environment or even injury to people. Procedures for rapid transportation to the test laboratory, immediate testing and web-based delivery of test results should be established as part of the condition monitoring programme.

Consideration may also need to be given to the types of packaging for lubricant sample bottles. For example, with a system of regular and routine sending of samples to an independent laboratory, procedures for the return of empty sample bottles and packages might be negotiated as part of the contract. Return and re-use of sample bottles and packaging will be beneficial for the environment. In such cases, wooden or plastic boxes for transportation will be much better than cardboard boxes, which may last for only a few journeys.

5.4   LUBRICANT ANALYSIS FLAGGING LIMITS

Each report of a lubricant analysis may have between 10 and 40 parameters for which an acceptable range of results has been defined. Evaluating the results will depend on the type of oil, its formulation, how it should be maintained and the operating conditions of the machine or equipment in which the lubricant is being used.

Standard methodologies for defining or establishing the "normal" ranges, together with the identification and revision of limits and a reliable system for recognising failure modes, are the basic components of an acceptable lubricant analysis programme.

Most laboratories offer comments and recommendations in the form of "flags" that are patterned after green, yellow and red traffic lights. Yellow- and red-flagged test results indicate that a threshold or limit has been exceeded and that further action is required by the end user. Green-flagged information can be archived for trending analysis. With computer and web-based recording and reporting of test results, it is easy to show green-flagged results in either black or green, yellow-flagged results in either very dark yellow or orange and red-flagged results in bright red. Modern colour printers can achieve the same indications.

Some lubricant properties, such as particle counts, only have upper limits. Other properties, such as oxidative stability, only have lower limits. Properties, such as viscosity, that measure stability, have both upper and lower limits.

There is no universal approach for setting alarm limits. Also, some properties and statistics in the laboratory report are not critical to the specific piece of equipment. As a result, not every lubricant analysis parameter needs an alarm limit.

Because oil analysis is as much expertise as formulae and there are so many considerations involved, most laboratories do not publish limit information. The issue for end users is who sets the range or limit and who to believe; the lubricant formulator, the equipment manufacturer, the laboratory, the industry or a test method specification organisation.

In 2011, the American Society for Testing and Materials (ASTM) published ASTM D7720 "Standard Guide for Statistically Evaluating Measurand Alarm Limits when Using Oil Analysis to Monitor Equipment and Oil for Fitness and Contamination".

The guide was published because "Alarm limits are used extensively for condition monitoring using data from in-service lubricant sample test results". There are many bases for initially choosing values for these alarm limits. Many questions need to be addressed. These include:

- Are those limits right or wrong?
- Are there too many false-positive or false-negative results?
- Are they practical?

The guide teaches statistical techniques for evaluating whether alarm limits are meaningful and if they are reasonable for flagging problems requiring immediate or future action.

The guide is intended to increase the consistency, usefulness and dependability of condition-based action recommendations by providing machinery maintenance and monitoring personnel with a meaningful and practical way to evaluate alarm limits to aid the interpretation of monitoring machinery and oil condition as well as lubricant system contamination data.

The scope of the guide is:

- It provides specific requirements to statistically evaluate measurand alarm thresholds, which are called alarm limits, as they are applied to data collected from in-service oil analysis. These alarm limits are typically used for condition monitoring to produce severity indications relating to states of machinery wear, oil quality and system contamination. Alarm limits distinguish or separate various levels of alarm. Four levels are common and are used in the guide, though three levels or five levels can also be used.
- A basic statistical process control technique described in the guide is recommended to evaluate alarm limits when measurand data sets may be characterised as both parametric and in control. A frequency distribution for this kind of parametric data set fits a well-behaved two-tail normal distribution having a "bell" curve appearance. Statistical control limits are calculated using this technique. These control limits distinguish, at a chosen level of confidence, signal-to-noise ratio for an in-control data set from variation that has significant, assignable causes. The operator can use them to objectively create, evaluate and adjust alarm limits.
- A statistical cumulative distribution technique described in the guide is also recommended to create, evaluate and adjust alarm limits. This particular technique employs a percent cumulative distribution of sorted data set values. The technique is based on an actual data set distribution and therefore is not dependent on a presumed statistical profile. The technique may be used when the data set is either parametric or nonparametric, and it may be used if a frequency distribution appears skewed or has only a single tail. Also, this technique may be used when the data set includes special cause variation in addition to common cause variation, although the technique should be repeated when a special cause changes significantly or is eliminated. Outputs of this technique are specific measurand values corresponding to selected percentage levels in a cumulative distribution plot of the sorted data set. These percent-based measurand values are used to create, evaluate and adjust alarm limits.
- The guide may be applied to sample data from testing of in-service lubricating oil samples collected from machinery (for example, diesel engines, pumps, gas turbines, industrial turbines, hydraulic systems) whether from large fleets or individual industrial applications.
- The guide may also be applied to sample data from testing in-service oil samples collected from other equipment applications where monitoring

for wear, oil condition or system contamination is important. For example, it may be applied to data sets from oil-filled transformer and circuit breaker applications.

- Alarm limit evaluating techniques, which are not statistically based, are not covered by the guide. Also, the techniques of the standard may be inconsistent with the following alarm limit selection techniques: "rate-of-change", absolute alarming, multi-parameter alarming and empirically derived alarm limits.
- The techniques in the guide deliver outputs that may be compared with other alarm limit selection techniques. The techniques in the guide do not preclude or supersede limits that have been established and validated by an Original Equipment Manufacturer (OEM) or another responsible party.
- The standard does not purport to address all of the safety concerns, if any, associated with its use. It is the responsibility of the user of the standard to establish appropriate safety and health practices and determine the applicability of regulatory limitations prior to use.

The complete guide can be downloaded from www.astm.org. In addition, all the larger independent test laboratories will have a copy of the guide.

5.5   TREND ANALYSIS

Trend analysis is a method of analysing time series data (information in sequence over time) involving comparison of the same test result over a significantly long period, to identify the general pattern of a relationship between associated factors or variables. This can then be used to project the future direction of the pattern. In engineering and lubrication, trend analysis is the widely used practice of collecting data and attempting to identify what is happening in an item of equipment or system.

Although trend analysis is often used to predict future events, it can also be used to estimate uncertain events in the past.

When testing lubricants in service, trend analysis and monitoring can help guard against costly equipment damage and repairs. It provides valuable early warning if the lubricants and fluids in expensive equipment are experiencing contamination or degradation. Lubricant degradation or contamination are sure signs of potential or impending equipment wear, damage or failure.

Used oil testing measurements should be monitored for adverse and unexpected trending, include wear metals, AN, viscosity, water and other key analysis. To capture these early warning trends, used oil samples are taken from equipment and engines and tested at periodic time intervals. Oil condition trend analysis can identify problems in the early stages of development, allowing time for corrective action to be taken before valuable equipment experiences costly repair, down-time or failure.

Used oil condition trend analysis helps extend operations, reduce shutdowns and mitigate down-time for large engines, power-trains, pumps, gears, bearings and other expensive system components. Additional ferrography, failure analysis, chemical analysis, forensic analysis and materials analysis expertise can help troubleshoot problems and pinpoint root causes.

Examples of trends in test results for oils in use are shown in Figures 5.1 to 5.7. Example graphs for trends for four elements are shown in Figure 5.1. All four trends show the same pattern of an upward trend, which suggests that the equipment is showing a few signs of some deterioration, although the graphs for lead and silicon contents are relatively flat compared with those for iron and chromium. The graph shown in Figure 5.2 is for a series of analyses on a pump turbine guide bearing. The trend lines for coarse iron and coarse copper indicate some type of wear in the bearing.

Three types of trend graphs can be plotted, as shown in Figures 5.3 to 5.5:

- Linear trend.
- Logarithmic trend.
- Exponential trend.

In practice, almost all trend graphs are plotted linearly, as this is more than satisfactory to reveal trends, as shown in Figure 5.2. Trends that have to be plotted either logarithmically or exponentially reveal that a really major change is happening or has happened. In either case, immediate corrective action is required. Alternatively, it may be too late to avert serious problems.

In statistics, trend analysis often refers to techniques for extracting an underlying pattern of behaviour in a time series which would otherwise be partly or nearly completely hidden by noise. If the trend can be assumed to be linear, trend analysis can be undertaken within a formal regression analysis, as described in trend estimation.

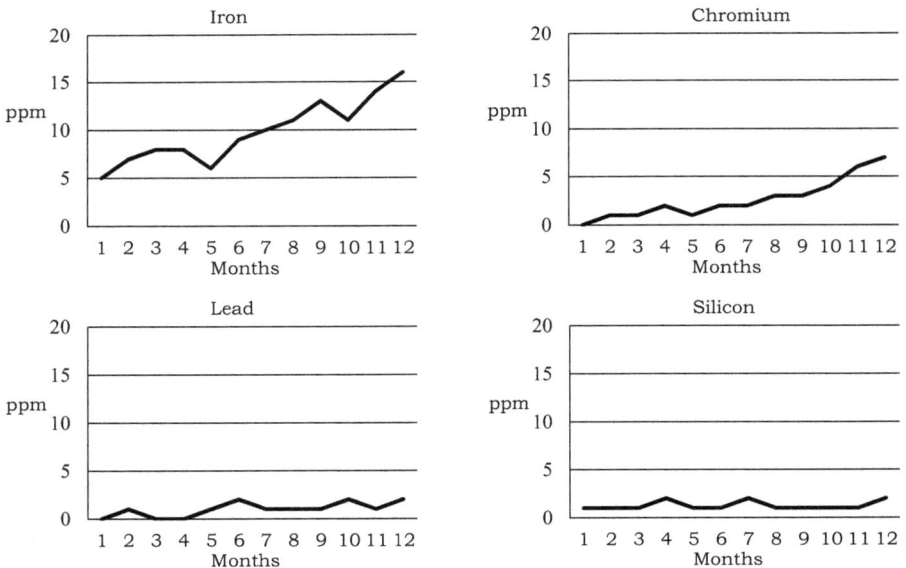

FIGURE 5.1   Examples of trends.

Source: Pathmaster Marketing Ltd.

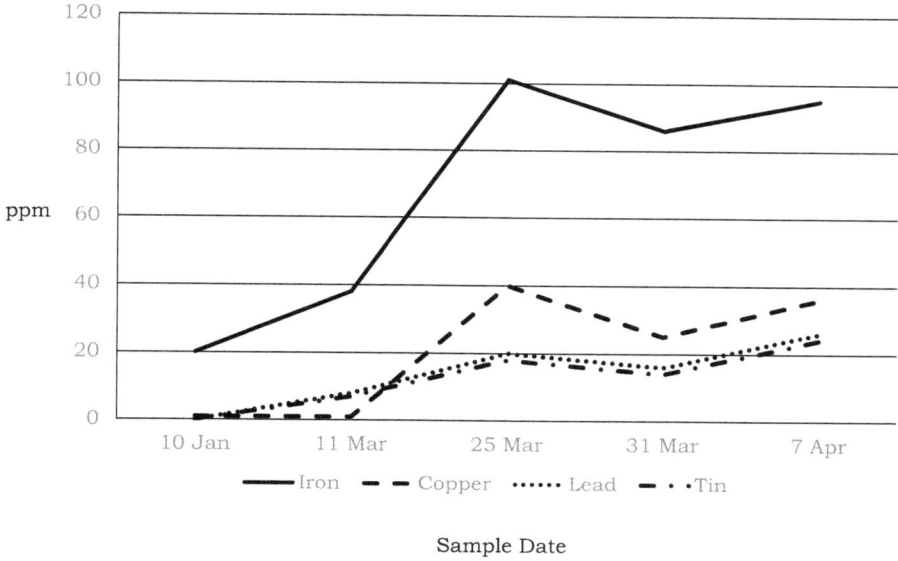

FIGURE 5.2   Pump turbine guide bearing wear trend.

Source: Pathmaster Marketing Ltd, from data obtained from Spectro Inc.

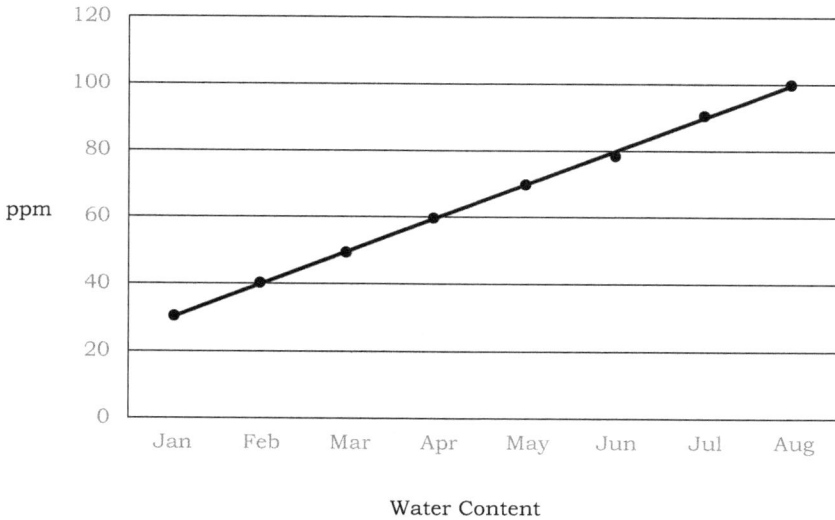

FIGURE 5.3   Example of a linear trend.

Source: Pathmaster Marketing Ltd.

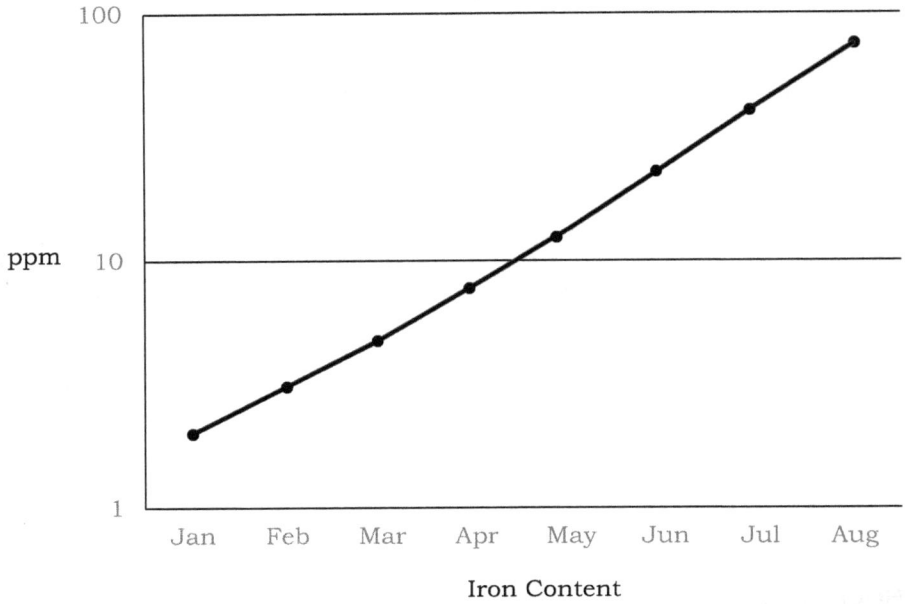

FIGURE 5.4   Example of a logarithmic trend.

Source: Pathmaster Marketing Ltd.

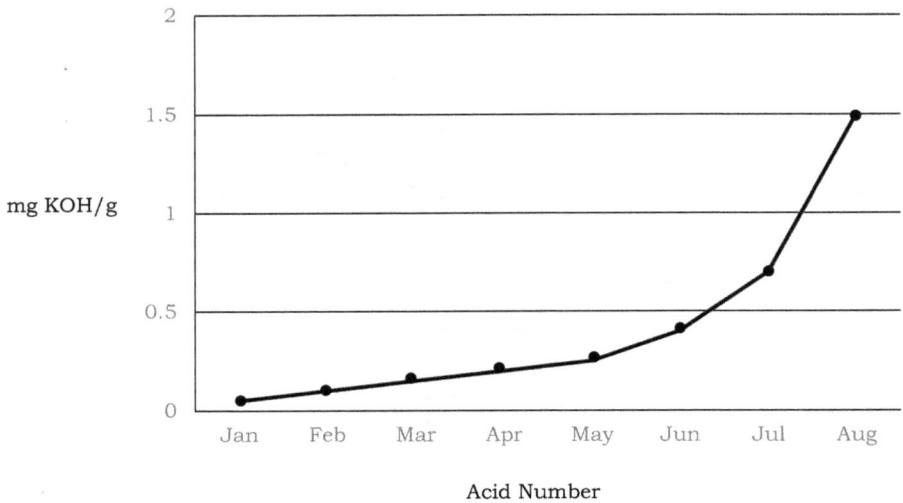

FIGURE 5.5   Example of an exponential trend.

Source: Pathmaster Marketing Ltd.

If the trends have other shapes than linear, trend testing can be done by nonparametric methods, for example the Mann–Kendall test, which is a version of Kendall rank correlation coefficient. For testing and visualisation of non-linear trends, smoothing can also be used.

Analysing trends may not be straightforward. Many tests for lubricants have issues with repeatability and reproducibility. The repeatability of a test method is defined by ASTM as the difference between the results obtained using the standard test method performed on the same sample, by the same operator, in the same laboratory, using the same apparatus, on the same day (although this latter condition depends on the specific standard test method). The reproducibility of a test method is the difference between the results of tests obtained using the standard test method performed on two identical samples, by different operators, in different laboratories using different apparatus. ASTM also defines an intermediate precision condition in which different operators use the same apparatus in the same laboratory to test the same sample using the standard test method on different days.

As a consequence of the issue of test repeatability, some laboratories will include error bars with the results obtained on each sample, to reflect the limited certainty of the test result. The inclusion of error bars will give a trend graph of the type shown in Figure 5.6. The example shown in the graph indicates that the trend appears to be upwards, but with some uncertainty as to how much and whether corrective action might be required. This uncertainty can become clearer if a regression analysis is performed on the data, as illustrated in Figure 5.7. The linear regression line shows

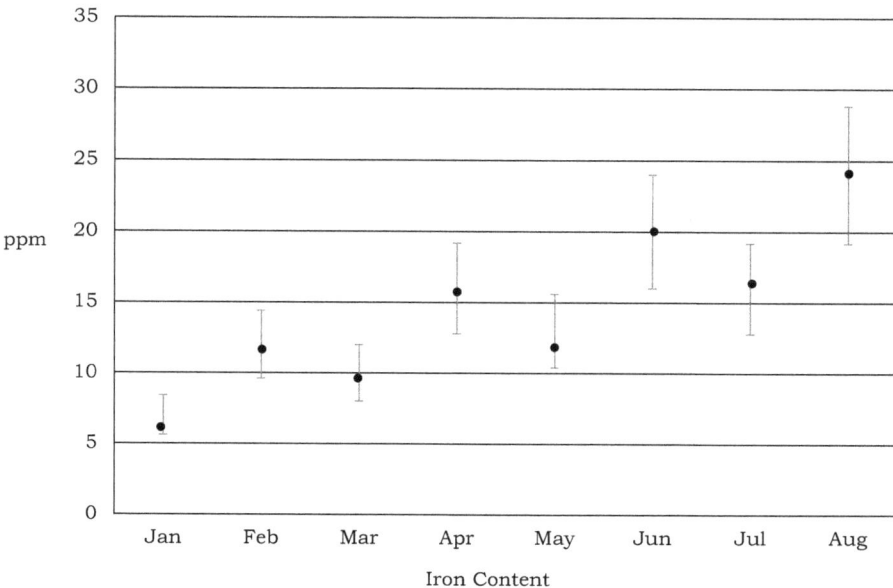

FIGURE 5.6   Error bars in a trend graph.

Source: Pathmaster Marketing Ltd.

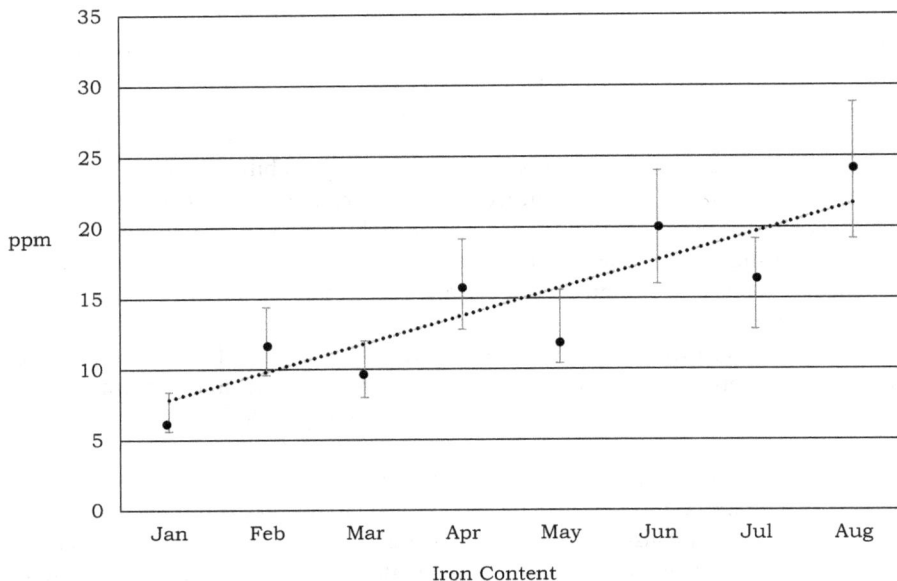

FIGURE 5.7 Regression analysis to show a linear trend.

Source: Pathmaster Marketing Ltd.

that the trend is clearly upwards and that the next sample (or the one after that) may breach a flagging limit.

A helpful guide as to which elements that are or might be present in in-service oils and greases and may need to be subject to trend analysis is provided by ASTM in standard method D6224. The list of elements and where and how they arise is summarised in Figure 5.8.

5.6   LUBRICATION AND LUBRICANT PROBLEMS AND THEIR RESOLUTION

Complaints can arise internally within a company that uses lubricants, for example that a machine is not operating satisfactorily with the lubricant that has been supplied. Conversely, they can be external complaints by a customer to the supplier of the lubricant. While attitudes and approaches may differ somewhat, the basic rules for dealing with complaints are essentially the same in both cases. For maximum coverage, the subject will be treated as though it is the latter form of complaint and some ground rules will be discussed as to how the lubricant supplier can handle it.

A policy for handling complaints should be developed and communicated to all people involved in the lubricant suppler. Rapid action, replacement of material and compensation for damage can often prevent the loss of a customer who has complained. However, too-ready acceptance of responsibility for a problem can become expensive, particularly if compensation for mechanical damage and loss of business

Element	Symbol	Wear Metal	Additive	Contam- inant	Sources
Aluminium	Al	x		x	Pistons, bearings, dirt
Antimony	Sb	x	x	x	Journal bearings, greases
Barium	Ba		x		Old type detergents
Boron	B		x	x	Additives, coolants
Cadmium	Cd	x			Journal bearings, plating
Calcium	Ca		x	x	Additives, water, greases
Chromium	Cr	x			Compression rings, cylinders
Copper	Cu	x	x		Bearings, cages, bushings, valve guides, coolers, pumps
Iron	Fe	x		x	Cylinders, shafts, gears, bearings, housings, cases, rust
Lead	Pb	x		x	Journal bearings, platings, pumps, paint, solder
Magnesium	Mg		x	x	Additives, sea water
Manganese	Mn	x			Shafts, valves
Molybdenum	Mo	x	x		Additives, compression rings
Nickel	Ni	x			Rolling bearings
Phosphorous	P	x	x		Additives, surface finish
Potassium	K			x	Coolants
Silicon	Si	x	x	x	Dirt, additives, alloyed with Fe
Silver	Ag	x			Wrist pins, solder
Sodium	Na		x	x	Additives, coolants, water, greases
Tin	Sn	x			Journal bearings, rolling bearing cages, solder
Titanium	Ti	x			Turbine blades, paint
Vanadium	V	x			Turbine blades, valves
Zinc	Zn		x	x	Additives, galvanised steel

FIGURE 5.8   Sources of elements in oils and greases.

Source: ASTM D6224.

is concerned. A compromise is to immediately replace any lubricant which is suspected of being either contaminated or off-specification, but to resist suggestions that mechanical damage is directly associated with lubricant quality until an exhaustive investigation has taken place. Field staff, whether sales managers, salespeople or customer support engineers, investigating complaints must know precisely what offers they can make to an aggrieved customer.

It is most important that complaints be properly documented, and it is well worthwhile to develop special forms for the purpose. The best way of recording this information is on a mobile device, such as a tablet, laptop or smartphone. Details of oil grade, delivery date, batch number and customer storage conditions need to be ascertained, plus information on the equipment type, duty and age or condition. A sample of the oil which is the subject of a complaint needs to be obtained, and a sample of used oil if the complaint concerns used oil condition or equipment malfunction. The taking of samples requires particular care and attention. A representative average sample must be obtained, or specific top, middle and bottom samples in the case of non-homogeneous material. A copy of the equipment manufacturer's recommendations, or their contact address and telephone number, should also be obtained.

It is usually clear by this time whether the complaint is relatively trivial or if it has potentially serious implications. In the case of contaminated unused oil, the only testing necessary may be a visual examination without specific identification of any contamination or sediments. If the sample appears normal, it should be compared with the retained sample which was taken at the blending plant before delivery. If they appear identical, correctness of grade can be further confirmed by tests such as

BN, sulphated ash or metals analysis. If desired, contamination can be checked for by flash point, water content and filtration of the sample.

If the unused oil appears to be within specification, and the complaint concerns its performance in a machine, the first action is to check whether the grade is one recommended for that machine and the service conditions. For example, with a heavy-duty diesel engine, the laboratory should examine used oil samples for contamination, fuel diluent, insolubles, AN or BN and elemental contents. (These tests are explained and discussed in Chapters 6 and 7.) From these results, it will be possible to assess the overall condition of the oil, the mechanical condition of the engine and/or the length of time the oil has been in service. A judgement will then have to be made based on experience as to whether the problem is one of oil quality, machine condition or the service conditions.

The usual causes of complaints are:

- **Wrong grade of oil supplied or used**: Easily detected in the laboratory, if not from package or drum labels.
- **Contamination**: Usually detectable by appearance or smell; can take place at the blending plant, in transit (bulk) or at the customer's premises.
- **Wrong oil recommendation**: To be assessed by a sales engineer after considering engine or equipment type and service.
- **Poor engine or equipment condition**: Can be assessed by examination of the used oil for fuel dilution, sludge, wear metals and other factors.
- **Extreme service conditions**: Detectable from insolubles contents, viscosity, wear metals and so on, provided that maintenance procedures are known. The user may also provide information on the type and severity of service.
- **Poor maintenance**: Can produce symptoms similar to either or both of the above cases and has to be assessed more from a knowledge of the customer's operations.
- **Mechanical failure**: Often the basic origin of the complaint and does not need specific confirmation. However, the parts should be examined as carefully as possible by a trained technical service engineer to try to determine a sequence of events.

Complaints of mechanical failure, blaming the detergency of the lubricant as the primary cause, are frequently unjustified. An engine maintained in good condition and operated within its normal duties should operate successfully without problems on the specified quality of lubricant. International standards of lubricant quality contain sufficient safety margins that normal quality variations will produce no detrimental effects. (An exception was the specification of low-temperature flow qualities after standing at very low temperatures. New tests on cold cranking and low-temperature flow had to be developed after a series of engine failures due to low-temperature operability problems.)

A common cause of problems is poor maintenance, which can lead to either filter blocking, or in other cases excessive fuel dilution and consequent loss of oil viscosity. Diesel injector maintenance is particularly important for a heavy-duty diesel engine, and in the past many mechanical problems blamed on the lubricant have turned out to be caused by dribbling injectors producing excessive fuel dilution.

Complaints of short overhaul lifetimes due to poor lubricant performance are harder to deal with. Assuming that a maker's recommended quality of lubricant is in use, either the equipment is old and/or poorly maintained or the service conditions are more severe than the user realises. (It is a common misapprehension that engines spending much time idling stress lubricant less than when working hard. The opposite may often be the case.) One solution here may be to provide a higher-quality lubricant for a trial period at no increased cost and to monitor the effect with the user's co-operation.

It is worth noting that, in the author's experience, approximately three-quarters of all customers' complaints about lubricants have not been caused by poor quality lubricants.

5.7 SUMMARY

A condition monitoring programme requires regular routine testing of lubricating oils and greases. This can be done using either field test kits, an in-house laboratory or an independent laboratory or a combination of all three. Numerous companies offer to test lubricant samples for smaller users of lubricants. A few such companies have offices and laboratories in every region and have acquired considerable expertise in assisting clients with condition monitoring, inspection and quality assurance.

Transporting lubricant samples to laboratories requires close attention to methods to avoid sample contamination or breakage. In some cases, the time to take the sample, deliver it to the laboratory and receive the test results can be essential to the effective monitoring of critically important machines, equipment or systems.

Analysing the results of the routine tests, using flagging limits and trend graphs, will enable machines, equipment and systems to be monitored effectively, to determine both lubricant and machinery service lifetimes. Condition monitoring will also detect the cause(s) of both lubrication and lubricant problems. It will also help to determine the root cause(s) of problems and, therefore, how to fix them. It is worth noting that, in the author's experience, approximately three-quarters of all customers' complaints about lubricants have not been caused by poor quality lubricants.

6 Chemical Tests for Lubricants

6.1 INTRODUCTION

A huge number of tests can be, and are, used to evaluate the properties and performance of lubricating oils and grease in their numerous applications. Some of these tests are chemical, others are physical, some are mechanical and a few involve real-life operations.

However, not all of these tests are suitable for use in a lubricant condition monitoring programme. This chapter will describe and discuss those chemical tests that could or should be used to monitor the condition of oils and greases. Later chapters will look at physical and mechanical tests for lubricant condition monitoring.

The international and national organisations that develop, publish, update and monitor these tests were summarised in Chapter 4. A number of original equipment manufacturers (OEMs) also develop, publish and update test methods, some of which can be useful in a lubricant condition monitoring programme. Those chemical tests that should (or sometimes could) be used to monitor each type of lubricant application will be discussed in later chapters.

The tests described in this chapter cover only those that can or should be used to monitor the condition of oils and greases in operating machinery. Other chemical tests not described in this chapter are used during the formulation and development of new or improved lubricants and/or during the blending of lubricating oils or the manufacture of greases.

6.2 CHEMICAL TESTS FOR OILS

6.2.1 ACID NUMBER

The acid number (AN) of an oil is synonymous with neutralisation number. The AN of an oil is the weight in milligrams (mg) of potassium hydroxide required to neutralise one gram of oil and is a measure of all the materials in an oil that will react with potassium hydroxide (KOH) under specified test conditions. The usual major components of such materials are organic acids, soaps of heavy metals, intermediate and advanced oxidation products, organic nitrates, nitro compounds and other compounds that may be present as additives.

Since a variety of degradation products contribute to the AN value, and since the organic acids present vary widely in corrosive properties, the test cannot be used to predict corrosiveness of an oil under service conditions.

Tests were developed to provide a quick determination of the amount of acid in an oil by neutralising it with a base. The amount of acid in the oil was expressed in terms

DOI: 10.1201/9781003245254-6

of the amount of a standard base required to neutralise a specified volume of oil. This quantity of base came to be called the neutralisation number of the oil.

The two most commonly used tests for acidity (AN) are ASTM D664 (equivalent to IP 177 and ISO 6619) and ASTM D974 (equivalent to IP 139, ISO 6618, DIN 51558T1 and AFNOR T60-112). With method ASTM D664/IP 177/ISO 6619, acidity is determined by potentiometric titration. The sample is dissolved in a mixture of toluene and iso-propanol containing a small amount of water. With the methods ASTM D974, IP 139 and ISO 6618, the acidity is determined by colour-indicator titration, again with the sample being dissolved in a mixture of toluene and iso-propanol containing a small amount of water. In both methods, the total AN (although this was previously abbreviated to TAN) is determined by the quantity of standard KOH solution required to titrate 1 gram of oil. The strong acid number is the quantity of KOH required to titrate the strong acids extracted from 1 gram of oil. The test results are given as mgKOH/g.

6.2.2 BASE NUMBER

Three methods are used to determine the base number (BN) of lubricating oils, whether new or in-service. The first two are ASTM D2896 "Standard Test Method for Base Number of Petroleum Products by Potentiometric Perchloric Acid Titration" (equivalent to IP 276 and ISO 3771 tests) and ASTM D4739 "Standard Test Method for Base Number Determination by Potentiometric Hydrochloric Acid Titration" (equivalent to IP 417 and ISO 6619 tests). Both methods have been developed as newer alternatives to methods ASTM D664 and ASTM D974, which were used previously to determine the BN of a lubricant.

New and used lubricants can contain basic constituents that are present as additives. The relative amounts of these materials can be determined by titration with acids. The base number is a measure of the amount of basic substance in the oil, always under the conditions of the test. It is sometimes used as a measure of lubricant degradation in service, although condemning limits must be established empirically.

ASTM D2896 covers the determination of basic constituents in petroleum products by titration with perchloric acid in glacial acetic acid. The method has two procedures, A and B, which use different titration solvent volumes and sample weights. A round robin on a series of new and used oils and additive concentrates has shown that the two procedures give statistically equivalent results.

Appendix X2 in the method provides the use of an alternative solvent system which eliminates the use of chlorobenzene. The use of the alternative solvent gives statistically equivalent results, but the test precision is worse. Paragraph X2.5.5 provides guidance when comparing results using the two different solvents.

Constituents in the lubricant that may be considered to have basic characteristics include organic and inorganic bases, amino compounds, salts of weak acids (soaps), basic salts of polyacidic bases and salts of heavy metals. The test method can be used to determine BNs >300 mgKOH/g, although the precision statement has been obtained only on BNs ≤300 mgKOH/g.

ASTM D4739 uses a weaker acid (alcoholic hydrochloric acid) to titrate the bases than ASTM D2896, and the titration solvents (a mixture of iso-propanol, toluene and chloroform with a small amount of water) are also different. As a result, ASTM

D2896 will titrate salts of weak acids (soaps), basic salts of polyacidic bases and weak alkaline salts of some metals. They do not protect the oil from acidic components due to the degradation of the oil. ASTM D2896 may produce a falsely exaggerated BN. ASTM D4739 will probably not titrate these weak bases but, if so, will titrate them to a lesser degree of completion. It measures only the basic components of the additive package that neutralises acids. Conversely, if the additive package contains weak basic components that do not play a role in neutralising the acidic components of the degrading oil, then ASTM D4739 results may be falsely understated.

Particular care is required in the interpretation of the BN of new and used lubricants. When the BN of a new oil is required as an expression of its manufactured quality, ASTM D2896 is preferred, since it is known to titrate weak bases that ASTM D4739 may or may not titrate reliably.

When the BN of in-service or at-term oil is required, ASTM D2896 is preferred because in many cases, especially for internal combustion engine oils, weakly basic degradation products are possible. ASTM D2896 will titrate these, thus giving a false value of essential basicity.

The third test is ASTM D5984 "Base Number in Lubricants by Colour Indicator Titration", which is newer than the previous methods. This is a semi-quantitative field test method used on new and used lubricating oils. In-service oils may contain basic constituents present as additives or as degradation products formed during service. A decrease in the measured BN often is used as a measure of lubricant degradation. ASTM D5984 uses reagents that are considered less hazardous than those used in other alternate BN methods. It uses pre-packaged reagents for field use in cases in which laboratory equipment is unavailable and quick results are at a premium. The test covers BNs from 0 to 20. Oils with higher BNs can be analysed by diluting the sample or using a smaller sample size. Results obtained by the test method are similar to those obtained by ASTM D2896.

The sample is dissolved in iso-octane and alcoholic hydrochloric acid. The solution is mixed with sodium chloride solution, and the aqueous and organic phases are allowed to separate. The aqueous phase is then decanted off and titrated with sodium hydroxide solution using methyl red indicator. When the solution colour changes from magenta to yellow, the BN is read off the side of the titrating burette.

When testing used engine lubricants, it should be recognised that certain weak bases are the result of the service rather than having been built into the oil. ASTM D4739 can be used to indicate relative changes that occur in oil during use under oxidising or other service conditions regardless of the colour or other properties of the resulting oil. The values obtained, however, are intended to be compared with the other values obtained by the method only; BNs obtained by ASTM D4739 are not intended to be equal to values by other test methods. Although the analysis is made under closely specified conditions, the test method is not intended to, and does not, result in reported basic properties that can be used under all service conditions to predict the performance of an oil. For example, no overall relationship is known between bearing corrosion or the control of corrosive wear in an engine and BN.

As a consequence of the differences between ASTM D664, ASTM D2896 and ASTM D4739, when considering oil condition monitoring, it is advisable to use ASTM D2896, IP 276 or ISO 3771 as the preferred test method for BN.

6.2.3 OXIDATION RESISTANCE

Oxidation of lubricating oils depends on the temperature, amount of oxygen contacting the oil and the catalytic effects of metals. If the oil's service conditions are known, these three variables can be adjusted to provide a test that closely represents actual service. However, oxidation in service is often an extremely slow process, so the test may be time-consuming. To shorten the test time, the test temperature is usually raised and catalysts added to accelerate the oxidation. Unfortunately, these measures tend to make the test a less reliable indication of expected field performance. As a result, very few oxidation tests have received wide acceptance, although a considerable number of tests are used by specific laboratories that have developed satisfactory correlations for them.

One oxidation test that is widely used is ASTM D943 "Oxidation Characteristics of Inhibited Steam-Turbine Oils". ISO 4263 is equivalent to ASTM D943. This test is commonly known as TOST (turbine oil stability test). While TOST is intended mainly for use on inhibited steam turbine oils, it has been used for hydraulic and circulation oils and for base oils for use in the manufacture of turbine, hydraulic and circulation oils. The test is operated at a moderate temperature (95°C, 203°F). Iron and copper catalyst wires are immersed in the oil sample, to which water is added. Oxygen is bubbled through the sample at a prescribed rate. The test is run either for a prescribed number of hours, after which the neutralisation number of the oil is determined, or until the neutralisation number reaches a value of 2.0. The result in the latter case is than reported as the hours to a neutralisation number of 2.0.

Objections to ASTM D943 are that extremely long test times, often on the order of several thousand hours, are required for stable oils, and that the only criterion for acceptability is the neutralisation number. Severe sludging and deposits on the catalyst wires can occur with some oils without excessive increase in the neutralisation number. A modification of the procedure, called procedure B, overcomes some of these latter objections by requiring a determination of the sludge content. Consequently, ASTM D943 tests are not suitable for an oil condition monitoring programme.

Two oxidation tests that are used primarily in Europe are the IP 280 procedure for turbine oils and the DIN 51506 (PNEUROP procedure of the Comité Européen des Constructeurs de Compresseurs et d'Outillage) test for compressor oils.

In the IP 280 procedure, often referred to as the CIGRE test (for Conference Internationale des Grandes Reseaux Electriques a Haute Tension), oxygen is passed through a sample of oil containing soluble iron and copper catalysts. The sample is held at 120°C (248°F), and the test time is 164 hours. During the test, volatile acids formed are absorbed in an absorption tube. At the end of the test, the ANs of the oil and the absorbent are determined and combined to give the total acidity; the sludge is determined as a weight percent. These may then be further combined to give the total oxidation products. Compared with ASTM D943, the IP 280 test requires a relatively short, fixed test time, and the amount of sludge formed during the test is an important criterion of the evaluation. Where the limits for satisfactory performance in the test are properly set, some believe that good correlation with performance in modern turbines is obtained. A concern of the procedure is the use of an oil-soluble catalyst which does not recognise the benefits of oils that specifically resist catalyst dissolution in service.

In the DIN 51506 test, a sample of oil containing iron oxide as a catalyst is aged by being held at 200°C (392°F) for 24 hours while air is bubbled through it. At the end of the test, the evaporation loss and the Conradson carbon residue (CCR) of the remaining sample are determined. The evaporation loss is significant only in that it must not exceed 20%, so the main criterion is the CCR value. The test is believed to correlate to some extent with the tendency of oils to form carbonaceous deposits on compressor valves. Although the DIN 51506 test takes less time than the ASTM D943 test, it is not really suitable for an oil condition monitoring programme.

The ASTM D2272 "Standard Test Method for Oxidation Stability of Steam Turbine Oils by Rotating Pressure Vessel", which is technically equivalent to the IP 229 test, is widely known as the "RBOT" or "RPVOT" test. The test uses an oxygen-pressured cylinder to evaluate the oxidation stability of new and in-service oils in the presence of water and a copper catalyst coil at 150°C. This procedure is often used as a screening test and as a quality control test because of the speed at which results can be obtained. A 50 gram (g) sample of the test oil, 5 millilitres (ml) of water and a copper wire catalyst are placed in a small cylinder and pressurised to 90 psi with oxygen at room temperature (25°C). The cylinder is then placed in a 150°C bath and rotated at 100 rpm. The pressure increases as the vessel is heated, reaches a maximum value and then drops as oxidation occurs. Once a 25 psi drop from the maximum pressure is observed, the amount of time from the vessel being placed in the bath through the drop is reported. An unmodified oil typically will run less than 30 minutes, and a high-quality formulated oil can run in excess of 1,000 minutes.

The test is not intended to be a substitute for ASTM D943 or be used to compare the service lives of new oils of different compositions. It can also be used to assess the remaining oxidation test life of an in-service oil.

Appendix X1 of ASTM D2272 describes a new optional turbine oil (unused) sample nitrogen purge pre-treatment procedure for determining the percent residual ratio of RPVOT value for the pre-treated sample divided by RPVOT value of the new (untreated) oil, sometimes referred to as a "% RPVOT Retention". This nitrogen purge pre-treatment approach was designed to detect volatile antioxidant inhibitors that are not desirable for use in high-temperature gas turbines.

The ASTM D4742 "Standard Test Method for Oxidation Stability of Gasoline Automotive Engine Oils by Thin-Film Oxygen Uptake (TFOUT)" test evaluates the oxidation stability of engine oils for gasoline automotive engines. The test, run at 160°C, uses a high-pressure reactor pressurised with oxygen together with a metal catalyst package, a fuel catalyst and water in a partial simulation of the conditions to which an oil may be subjected in a gasoline combustion engine. The test method can be used for engine oils with viscosity in the range from 4 to 21 mm^2/s (cSt) at 100°C, including re-refined oils. The test method is not a substitute for the engine testing of an engine oil in established engine tests, such as Sequence IIID, and is much too long for use in a lubricant condition monitoring programme.

Another oxidation stability test that is much too long to be of value in a lubricant condition monitoring programme is ASTM D5846 "Oxidation Test for Hydraulic and Turbine Oils Using the Universal Oxidation Test Apparatus".

The DKA test (CEC L-48-95) accesses the tendency of transmission lubricants to deteriorate by oxidation under specified conditions. It applies to fully formulated transmission lubricants (ATF and gear oils). Samples of oil are subjected to oxidation conditions by heating to 160°C (or other specified temperature) and passing air through it at a specified flow rate during a period of 192 hours. Two different apparatus can be used, referred to as A and B. Samples are evaluated for sludge residue on the test glassware. The change in viscosity at 40°C and 100°C, and the difference in AN between fresh oil and oxidised oil are measured. Again, this test is too long to be suitable for an oil condition monitoring programme.

The ASTM D4310 "Standard Test Method for Determination of Sludging and Corrosion Tendencies of Inhibited Mineral Oils" is a modification of ASTM D943. It is used to determine the tendencies of inhibited mineral oil-based steam turbine lubricants and mineral oil-based anti-wear hydraulic oils to corrode copper catalyst metal and to form sludge during oxidation. The test is conducted in the presence of oxygen, water and copper and iron metals at an elevated temperature. This test is also used for testing circulating oils having a specific gravity less than that of water and containing rust and oxidation inhibitors. The test uses a special piece of glassware known as an oxidation cell. A 300 ml sample of the test oil, 60 ml of water and a catalyst (a 225 mm braided low-carbon steel-copper coil) are placed in the oxidation cell which is heated in a bath to 95°C. Oxygen is delivered to the system at a rate of 3 litres per hour. The test is run for 1,000 hours (41.67 days). Upon completion of the test, the weight of the insoluble material (sludge) that is formed and the total amount of copper in the oil, water and sludge phases are reported.

During round robin testing, copper and iron in the oil, water and sludge phases were measured. However, the values for the total iron were found to be so low (that is, below 0.8 mg), that statistical analysis was inappropriate. The results of the co-operative test programme are available from ASTM.

Procedure A of the test method requires the determination and report of the weight of the sludge and the total amount of copper in the oil, water and sludge phases. Procedure B requires the sludge determination only. The AN determination is optional for both procedures.

The ASTM D2893 "Standard Test Method for Oxidation Characteristics of Extreme-Pressure Lubrication Oils" test methods have been widely used to measure the oxidation stability of extreme-pressure lubricating fluids, gear oils and mineral oils. Two procedures can be used. The changes in the lubricant resulting from these test methods are not always necessarily associated with oxidation; some changes may be due to thermal degradation.

Procedure A is run at 95°C, and Procedure B is run at 121°C, both in temperature-controlled heating baths. For both Procedures, more than one test oil can be tested at a time, provided the heating bath is able to accommodate the required number of test tubes. 300 ml of each test oil is placed in a separate test tube, each of which is then fitted with corks and air delivery tubes, so that the lower ends of the tubes are within 6 mm of the bottoms of the test tube. The test tube is immersed in the heating bath so that the heating medium is at least 50 mm above the level of the oil sample. Dried air is flowed through the test tube at a rate of 10 litres per hour.

The temperature of the oil samples and the rate of air flow are checked every hour and any necessary adjustments made. The test starts once the oil samples have reached the desired temperature. When using multi-cell baths, one way of checking the temperature of the oil samples can be to use a dummy cell in the bath, similar to the way it is used in ASTM D943. The test is stopped after 312 hours (13 days), and the samples are tested for kinematic viscosity at 100°C (using equivalent test methods ASTM D445, IP 71, ISO 3104, DIN 51562 or AFNOR T60-100) and precipitation number (using test method ASTM D91). These results are compared with those obtained for new oils, and the percentage increase in viscosity and change in precipitation number are reported. The test is much too long to be useful in an oil condition monitoring programme.

The ASTM D6186 pressure differential scanning calorimetry (PDSC) test method is used to determine the oxidation induction time of lubricating oils subjected to oxygen at 3.5 MPa (500 psig) and temperatures between 130°C and 210°C. The test method is faster than other oil oxidation tests and requires a very small amount of sample. A 3 mg sample of test oil is weighed into a new sample pan and placed in a test cell. The cell is heated at a rate of 100°C/min to a specified test temperature and then pressurised with oxygen to 3.5 MPa. The pressure is maintained using a flow rate of 100 ml/min. The test is run for 120 minutes or until after the oxidation exotherm has occurred. The oxidation induction time is defined as the time from when the oxidation valve is opened to the onset time for the oxidation exotherm. The onset time is extrapolated from the thermal curve. If more than one oxidation exotherm is observed, then the largest exotherm is reported. This relatively short test is ideal for monitoring the condition of in-service oils.

Another PDSC test is CEC L-85-99, in which a sample of candidate oil is heated in a PDSC unit to a defined temperature, then held isothermally at that temperature for up to 2 hours, to determine the oxidative induction time (OIT). In this test, 2 mg of sample is heated between 50°C and 210°C and then held at that temperature for up to 2 hours in a closed system at 100 psi (6.9 bar) overpressure. The OIT, expressed in minutes, is the onset time observed from achieving the isothermal temperature. The test was developed for use in the ACEA E5 heavy-duty diesel engine oil specification, in which the straight pass/fail criterion is set at 65 minutes for candidate oils.

A test that can be used to assess the oxidation of in-service lubricants is ASTM D7214 "Oxidation of Used Lubricants by FTIR Using Peak Area Increase Calculation". The peak area increase (PAI) is representative of the quantity of all the compounds containing a carbonyl function that have formed by the oxidation of the lubricant, including aldehydes, ketones, carboxylic acids, esters and anhydrides. The PAI gives representative information on the chemical degradation of the lubricant that has been caused by oxidation. This test was developed for transmission oils and is used in the CEC L-48-A-00 test as a parameter for the end of test evaluation. The test method, however, is not intended to measure an absolute oxidation property that can be used to predict performance of an oil in service. The test also may be used for other oils than in-service transmission oils.

Fourier transform infrared (FTIR) spectra of the fresh and the used oils are recorded in a transmission cell of known path length. Both spectra are converted to absorbance mode and then are subtracted. Using this resulting differential spectrum, a baseline is set under the peak corresponding to the carbonyl region around 1,650 and 1,820 cm^{-1}, and the area created by this baseline and the carbonyl peak is calculated. The area of the carbonyl region is divided by the cell path length in millimetres, and this result is reported as peak area increase. A few circumstances can interfere with the test result. Very viscous oils, use of an ester as a base oil or high soot contents may require dilution of the sample and a specific area calculation as described in the method. The results of the test may be affected by the presence of other components with an absorbance band in the zone of 1,600 to 1,800 cm^{-1}. Low PAI values may be difficult to determine in those cases.

6.2.4 ANTI-OXIDANT CONTENT

The amounts of oxidation inhibitors in some new and in-service industrial oils can be measured using two methods. ASTM D6810 "Hindered Phenolic Anti-Oxidant Content in Non-Zinc Turbine Oils by Linear Sweep Voltammetry" is applicable to new or used Type HL turbine oils (those that do not contain zinc dialkyldithiophosphate or zinc diarylyldithiophosphate (ZDDP) oxidation inhibitors) in concentrations from 0.0075%wt up to concentrations found in new oils. This is achieved by measuring the amount of current flow at a specified voltage in the produced voltammogram.

A measured quantity of sample is dispensed into a vial containing a measured quantity of alcohol-based electrolyte solution and containing a layer of sand. When the vial is shaken, the hindered antioxidants and other solution-soluble oil components present in the sample are extracted into the solution, and the remaining droplets suspended in the solution are agglomerated by the sand. The sand/droplet suspension is allowed to settle out, and the hindered phenol antioxidants dissolved in the solution are quantified by voltammetric analysis. The results are calculated and reported as a weight percent of antioxidant or as a millimole of antioxidant per litre of sample for prepared and fresh oils and as percent remaining antioxidant for used oils. The results of samples taken at regular intervals can thus be used for trend analysis in a lubricant condition monitoring programme.

A more comprehensive test is ASTM D6971 "Standard Test Method for Measurement of Hindered Phenolic and Aromatic Amine Antioxidant Content in Non-zinc Turbine Oils by Linear Sweep Voltammetry" which covers the determination of hindered phenol and aromatic amine antioxidants in new or used type non-zinc turbine oils. The test measures the same concentrations as ASTM D6810. The test is not designed or intended to detect all of the antioxidant intermediates formed during the thermal and oxidative stressing of the oils, which are recognised as having some contribution to the remaining useful life of the used or in-service oil. Nor does it measure the overall stability of an oil, which is determined by the total contribution of all species present. Before making a final judgement on the remaining useful life of the used oil, which might result in the replacement of the oil reservoir, it is advised to perform additional analytical techniques in accordance with the guidelines in standards ASTM D6224, ASTM D4378 and ASTM D2272 (which are discussed in depth in later chapters), having the capability of measuring the remaining oxidative life of the used oil.

The test is applicable to non-zinc type turbine oils as defined by ISO 6743 Part 4, Table 1. These are refined mineral oils containing rust and oxidation inhibitors but not anti-wear additives. The test is also suitable for manufacturing control and specification acceptance.

A measured quantity of sample is dispensed into a vial containing a measured quantity of acetone-based electrolyte solution and a layer of sand. When the vial is shaken, the hindered phenol and aromatic amine antioxidants and other solution-soluble oil components in the sample are extracted into the solution, and the remaining droplets suspended in the solution are agglomerated by the sand. The sand/droplet suspension is allowed to settle out and the hindered phenol and aromatic amine antioxidants dissolved in the solution are quantified by voltammetric analysis. The results are calculated and reported as mass % of antioxidant or as mmol of antioxidant per litre of sample for prepared and fresh oils and as a % remaining antioxidant for used oils.

When a voltammetric analysis is obtained for a turbine oil inhibited with a typical synergistic mixture of hindered phenol and aromatic amine antioxidants, there is an increase in the current of the produced voltammogram between 8 and 12 seconds (or 0.8 and 1.2 V applied voltage) for the aromatic amines, and an increase in the current of the produced voltammogram between 13 and 16 seconds (or 1.3 and 1.6 V applied voltage) for the hindered phenols in the neutral acetone test solution. Hindered phenol antioxidants detected by voltammetric analysis include, but are not limited to, 2,6-di-tert-butyl-4-methylphenol, 2,6-di-tert-butylphenol and 4,4'-methylenebis (2,6-di-tert-butylphenol). Aromatic amine antioxidants detected by voltammetric analysis include, but are not limited to, phenyl alpha naphthylamines and alkylated diphenylamines.

When ASTM D6971 is used for in-service oils, it is commonly referred to as the Remaining Useful Life Evaluation Routine (RULER) test. The test is quick and of low cost, so is ideal for oil condition monitoring, particularly for ashless turbine, hydraulic and industrial gear oils.

6.2.5 VARNISH FORMATION POTENTIAL

ASTM D7843 "Standard Test Method for Measurement of Lubricant Generated Insoluble Colour Bodies in In-Service Turbine Oils using Membrane Patch Colorimetry" can be a guide to end users on the formation of lubricant-generated, insoluble deposits. The results from the test are intended to be used as a condition monitoring trending tool as part of a comprehensive programme, as outlined in standards such as ASTM D4378 and ASTM D6224.

The test extracts insoluble contaminants from a sample of in-service turbine oil onto a patch, and the colour of the membrane patch is analysed by a spectrophotometer. The test is known as the membrane patch calorimetry (MPC) test. The results are reported on a scale from 0 to 100, within the CIE LAB scale. Ratings of 40 or higher are considered cause for alarm. An illustration of increasing MPC ratings is shown in Figure 6.1.

As oils age in operation, lubricant degradation products are formed. Many of these products are acidic and negatively affect the acid level of the oil. Other degradation products have high molecular weights and are insoluble in the oil. Lubricating

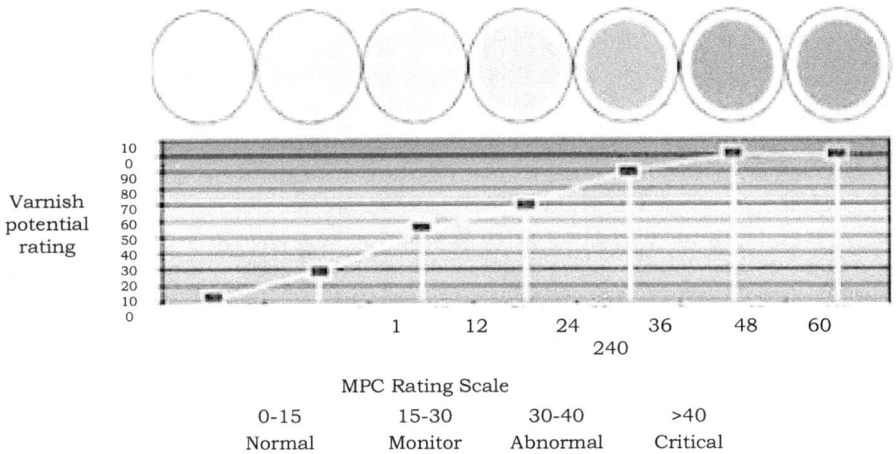

FIGURE 6.1   Examples of membrane patches.

Source: Pathmaster Marketing Ltd.

oils have some degree of "solvency" that allows for the suspension of these insoluble products for a period of time. As the level of insolubles builds up in the oil, eventually these products drop out, and harmful varnish and sludge form on components.

Until recently, oil analysis testing was unable to determine the potential for varnish and sludge build-up in lubricated systems. Traditional oil analysis testing measures parameters, such as the AN of an oil, in order to assess its degradation. In many cases, the AN test was the sole indicator of the necessity of an oil change.

However, a problem with the original ASTM D7843 test method was identified by Professor Akira Sasaki in 2014.[1] Because the spectrophotometers recommended in the standard were not designed to address the specific needs of the test, a significant portion of the varnish sub-micron particles trapped inside the membrane volume are not visible and are often missed during measurement. When the membrane is lit from the back, it leads to different results when compared with the way it looks when lit from the front. In 2015, Professor Sasaki[2] developed a new colorimetric patch analyser (CPA) which can measure the colour of contaminants on the patch surface using reflected light and the colour inside the patch by using transmitted light.

In addition, one of the most common concerns when using a solvent to dilute an oil in a laboratory test is that the solvent may dissolve part of what is to be measured and distort the test results. *Machinery Lubrication* magazine published a new index in order to address this concern in 2014.[3] In the alternative method, the volume of oil sample is passed through membranes in accordance with ASTM D7843, but no solvent is used. Although the proposed method can take considerable time because of the need to pass thick oil through the membrane, it will eliminate the possible effects of the solvent. The new index is also measured by a spectrophotometer and is called iMPC. The final index will be the number obtained from the division of iMPC into MPC. The effect of this on monitoring hydraulic and turbine oil condition is discussed in Chapters 12 and 13, respectively.

With MPC, oil analysts have a method to help to identify oils that have a high varnish potential, giving an operator time to schedule an oil change before harmful varnish build-up can cause problems with sensitive components and the failure of critical equipment.

When used as part of a regular oil analysis programme, MPC can provide confidence in determining proper oil change intervals. In the long term, this can mean the difference between cost savings as the result of safely extending oil drains and major equipment damage and failure as the result of harmful varnish build-up.

The test is not appropriate for turbine oils that contain dyes. It can also be used for hydraulic and compressor oils.

6.2.6 Corrosion Resistance

The rust-protective properties of lubricating oils are difficult to evaluate. Rusting of ferrous metals is a chemical reaction that is initiated almost immediately when a specimen is exposed to air and moisture. Once initiated, the reaction is difficult to stop. Thus, when specimens are prepared for rust tests, extreme care must be taken to minimise exposure to air and moisture so that rusting will not start before the rust-protective agent has been applied and the test begun. Even with proper precautions, rust tests do not generally show good repeatability or reproducibility. (A test method is considered to be repeatable when two analysts using the same apparatus and reagents in the same laboratory get the same results. A test method is considered to be reproducible when two analysts using the same apparatus and reagents in different laboratories get the same results. Test results obtained by different analysts often differ, sometimes by small amounts, but sometimes considerably.)

Most laboratory rust tests involve polishing or sandblasting a test specimen, coating it with the oil to be tested, then subjecting it to rusting conditions. Testing may be in a humidity cabinet, by atmospheric exposure or by some form of dynamic test. In the latter category is ASTM D665 "Rust Preventive Characteristics of Steam Turbine Oil in the Presence of Water". The IP 135, ISO 7120, DIN 51585 and AFNOR T60-151 test methods are identical to ASTM D665. In this test, a steel specimen is immersed in a mixture of distilled or synthetic seawater and the oil under test. The oil and water mixture is stirred continuously during the test, which usually lasts for 24 hours. The specimen is then examined for rusting. Examples of freshly polished, mildly rusted and severely rusted steel specimens are shown in Figure 6.2.

The ASTM D6557 "Standard Test Method for Evaluation of Rust Preventive Characteristics of Automotive Engine Oils" test was designed as a replacement for ASTM D5844. The test was designed to measure the ability of an engine oil to protect valve train components against rusting or corrosion under low-temperature, short-trip service. ASTM D5844 was correlated with vehicles in that type of service prior to 1978. Correlation between the two test methods has been demonstrated for most, but not all, of the test oils evaluated.

The test covers a Ball Rust Test (BRT) procedure for evaluating the anti-rust ability of fluid lubricants. The procedure is particularly suitable for the evaluation of automotive engine oils under low-temperature, acidic service conditions.

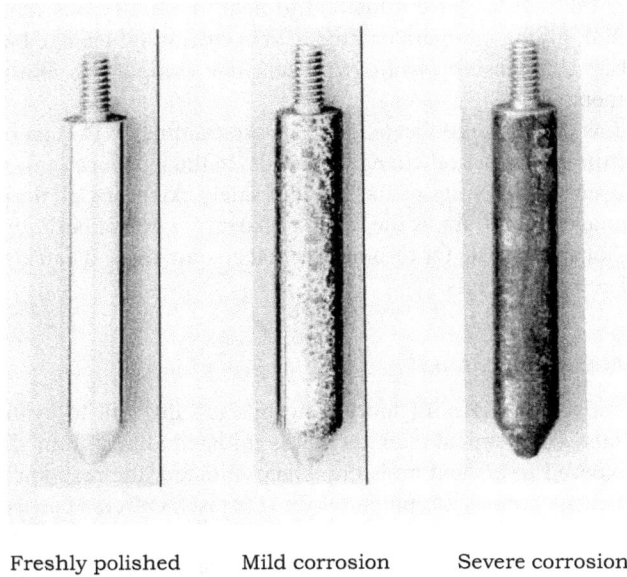

Freshly polished   Mild corrosion   Severe corrosion

FIGURE 6.2   ASTM D665/IP 135 steel corrosion specimens.

Source: Pathmaster Marketing Ltd.

The test uses a steel "lifter check" ball, which is placed in a test tube together with a measured quantity of test engine oil. The test tube is placed in a shaker, under shaking speed and temperature control. Air and a corrosive fluid are fed into the tube to simulate a corrosive environment. The corrosive fluid consists of representative samples of the types of chemicals that can be generated during the combustion process in short-trip winter driving. After 18 hours, the test ball is evaluated for the extent of rusting using a computerised scanning system with optical imaging. Excellent correlation between the BRT and API Sequence IID engine testing has been demonstrated for a wide variety, but not all, engine oil compositions.

The ASTM D5968 "Standard Test Method for Evaluation of Corrosiveness of Diesel Engine Oil at 121°C" test is intended to simulate the corrosion process of non-ferrous metals in heavy-duty diesel engine oils. The corrosion process under investigation is that believed to be induced primarily by inappropriate lubricant chemistry rather than lubricant degradation or contamination. This test method has been found to correlate with an extensive fleet database containing corrosion-induced cam and bearing failures.

The test method is used to test diesel engine lubricants to determine their tendency to corrode various metals, specifically alloys of lead and copper commonly used in cam followers and bearings. The test is based on four individual metal coupons of copper, lead, tin and phosphor-bronze, immersed in 100 ml of test oil at a temperature of 121°C. Air is blown through the oil at a rate of 5 l/hour for 168 hours. When the test has been completed, the coupons are examined to detect corrosion.

The increases in concentrations of copper, lead and tin in the test oil are also measured. Correlation with field experience has been established.

A variant of the test, originally referred to as ASTM D5968 Modified (The Cummins High-Temperature Corrosion Bench Test [CBT]), is run at a temperature of 135°C. This test, now designated ASTM D6594, was introduced for the API CH-4, API CI-4 and ACEA E5-99 engine oil specifications.

Another corrosion test is ASTM D7038 "Moisture Corrosion Resistance of Automotive Gear Lubricants", which simulates a type of severe field service in which corrosion-promoting moisture in the form of condensed water vapour accumulates in an axle assembly. This may happen as a result of volume expansion and contraction of the axle lubricant and the accompanied breathing in of moisture-laden air through the axle vent. The test screens gear oils for their ability to prevent the expected corrosion. The ASTM Test Monitoring Center (TMC) provides reference oils and an assessment of the test results obtained on those oils by a laboratory. The test method uses a bench-mounted hypoid differential housing assembly. It is commonly referred to as the L-33-1 test.

The test uses a Dana Corporation model 30 hypoid differential assembly, 4:10 ratio, standard differential with uncoated ring and pinion, without axle tubes. Before each test, the differential housing assembly is completely disassembled and cleaned. After thorough cleaning and coating all internal parts with test oil, the unit is reassembled. The unit is charged with 40 oz of test oil. After driving the unit at 2,500 rpm pinion speed, 1 oz of water is added to the oil. A pressure relief system is installed, and when the lubricant temperature reaches 180°F, the motoring phase is continued for 4 hours at this temperature. The motor is stopped at the completion of the motoring phase, and the test units are removed from the monitoring rig and placed in a storage box at 125°F for 162 hours. At the end of the storage phase, the test is complete. The differential assembly is drained, disassembled and rated for rust, stain and other deposits.

None of these lengthy tests is realistically suitable for an oil condition monitoring programme, although ASTM D665, IP 135 and ISO 7120 tests are used occasionally.

6.2.7 COPPER CORROSION

The most commonly used test is the copper strip corrosion test (ASTM D130, IP 154, ISO 2160, DIN 51759 and AFNOR M07-015) in which a polished copper strip is immersed in the sample and heated, and its colour and condition after the test are compared with a bank of standards. For lubricating oils, the test is usually performed at a temperature of 100°C for 3 hours.

At the end of the exposure period, the strip is removed, wiped clean and matched with coloured reproduction strips characteristic of the descriptions provided as follows:

1. Slight tarnish   a = Light orange, almost the same as the freshly polished strip
                       b = Dark orange
2. Moderate tarnish   a = Claret red
                       b = Lavender

c = Multicoloured with lavender blue or silver,
      or both, overlaid on claret red

d = Silvery

e = Brassy or gold

3. Dark tarnish   a = Magenta overcast on brassy strip

                      b = Multicoloured with red and green
                          (peacock), but no grey

4. Corrosion   a = Transparent black, dark grey or brown with
                          peacock green barely showing

                      b = Graphite or lusterless black

                      c = Glossy or jet black

Classification results are reported, together with the duration of the test and the test temperature. An illustration of the different degrees of copper corrosion is shown in Figure 6.3.

A more rapid test, probably of more value in a lubricant condition monitoring programme, is ASTM D7095 "Corrosiveness to Copper from Petroleum Products Using a Disposable Copper Foil Strip". The test is applicable to a wide range of petroleum products, including aviation gasoline, aviation turbine fuel, automotive gasoline, natural gasoline or other hydrocarbons having a vapour pressure not greater than 124 kPa, as well as to cleaners such as Stoddard solvent, kerosene, diesel fuel, distillate fuel oil, lubricating oil and other petroleum products. This test method is similar to ASTM D130, but it involves three major differences. It uses a single-use copper foil strip in place of the ASTM D130 multi-use copper strip, a different polishing technique is used in preparing the copper foil strip, and it involves a shorter analysis time of 45 minutes for all product types.

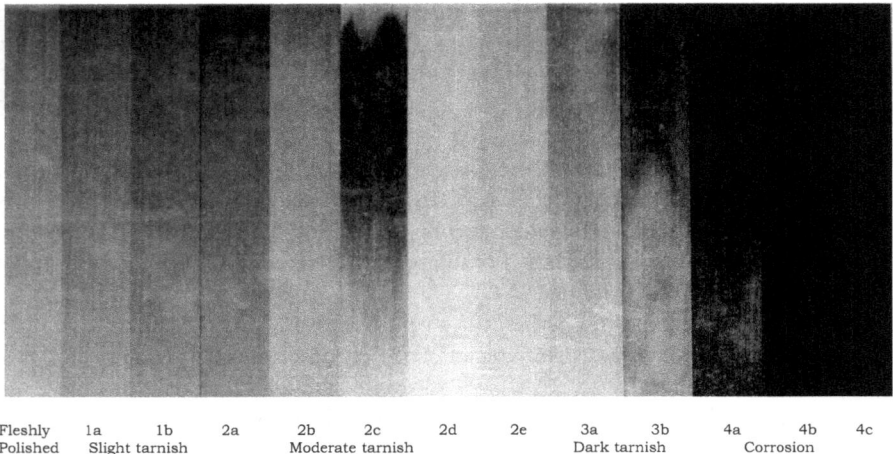

| Fleshly
Polished | 1a
Slight tarnish | 1b | 2a | 2b
Moderate tarnish | 2c | 2d | 2e | 3a
Dark tarnish | 3b | 4a
Corrosion | 4b | 4c |

FIGURE 6.3   ASTM D130/IP 154 copper corrosion specimens.

Source: Pathmaster Marketing Ltd.

A polished copper strip is immersed in a specific volume of the sample being tested and heated under conditions of temperature and time that are specific to the class of material being tested. At the end of the heating period, the copper foil strip is removed, washed, and the colour and tarnish level are assessed against the ASTM copper strip corrosion standard.

Over the years, manufacturers of large engines have required various bench corrosion tests to be performed on samples of their bearing metals before approving oils. Typical of such tests would be the silver corrosion test of EMD (formerly the Electro-Motive Division of General Motors and now the Progress Rail division of Caterpillar) and the Mirrlees corrosion test. These tests are not generally used in lubricant condition monitoring programmes.

6.2.8 EMULSION CHARACTERISTICS

Metalworking fluids used in the form of aqueous emulsions should remain stable in use. The IP 263 test is designed to assess the stability of water-mix metalworking fluids.

An emulsion of the metalworking fluid in synthetic hard water is prepared using a magnetic stirrer. The emulsion is poured into a special 150 ml flask with a 10 ml graduated neck and is allowed to stand for 24 hours at a temperature of 20°C ± 4°C. Any separated layer is measured. For obvious reasons, the test flask must be scrupulously clean. If less than 0.5 ml of oil has separated, the result is a pass and the water-mix emulsion is considered to be stable.

6.2.9 DEMULSIBILITY

The ASTM D1401 "Standard Test Method for Water Separability of Petroleum Oils and Synthetic Fluids" test method measures the demulsibility characteristics of oils. Test methods IP 412, ISO 6614 and AFNOR T60-125 are equivalent to ASTM D1401. Demulsibility is the ability of an oil to separate from water. The test measures how rapidly and completely an oil/water emulsion separates after mixing equal volumes of the oil and water for 5 minutes at an elevated temperature. Industrial oils are often contaminated with water. Water contamination can range from condensation from the atmosphere to exposure of condensed steam as in the case of turbine oils. For the effective removal of water, an oil must possess good demulsibility characteristics.

The test apparatus consists of a bath where 40 ml of the test oil and 40 ml of water are mixed in a 100 ml graduated cylinder for 5 minutes. The test is conducted at 54°C for oils having a viscosity less than 90 cSt at 40°C and at 82°C for oils having a viscosity greater than 90 cSt at 40°C. The separation of the oil/water emulsion is observed at 5-minute intervals. The time required for the emulsion to separate to 3 ml of emulsion or less is reported. The time limits for the test are 30 minutes at 54°C and 60 minutes at 82°C. For very high-viscosity oils where there is insufficient mixing of oil and water, test method ASTM D2711 is recommended.

The IP 19 test is different to ASTM D1401. The test gives a measure of the ability of an oil to separate from an emulsion. It is commonly applied to turbine oils, but it may be used for other lubricating oils. The test is commonly applied to used turbine oils, but since it is sensitive to aging and contamination of the oil, precision will be lower than stated in the test method. Oils containing inhibitors may give much higher results than the corresponding uninhibited oils and the precision may be less satisfactory

In IP 19, the demulsification number is defined as the number of seconds required for an oil to separate when it is emulsified and separated under specified conditions. In total, 20 ml of the oil is emulsified with steam at about 90°C. The emulsion is then placed in a bath at about 94°C, and the time for 20 ml of oil to separate is recorded.

The results obtained by the test method are readily affected by the presence of traces of impurities in the sample. Consequently, it is very important to avoid contamination of the sample and to exclude light from the sample until it can be tested.

ASTM D1401 uses a stirring technique and a heating bath at 54°C or 82°C; test results from IP 19 cannot be compared with those from ASTM D1401. ASTM D1935 "Steam Emulsion of Electrical Insulating Oils of Petroleum Origin" is technically similar to IP 19.

With ASTM D2711, a 405 ml sample of the test oil and 45 ml of distilled water are stirred for 5 minutes at 82°C in a graduated funnel. After a 5-hour settling period, a 50 ml sample is drawn from the top of the oil and is centrifuged to determine the percent of water in the oil. Then any free water is drawn off and the amount is determined.

After removing the free water, the sample is siphoned off until 100 ml remains in the funnel. The remaining sample is then centrifuged, and the volumes of the water and emulsion components are measured. The amount of water from this step is added to the amount of free water and is reported as "total free water". The following measurements are reported:

- Water in oil (%).
- Free water from funnel after centrifuging (ml).
- Total free water (ml).
- Emulsion (ml).

The average values from four determinations on each oil are reported. While the procedure remains the same, when EP gear oils are tested, the oil sample size is reduced to 360 ml and the amount of water used is increased to 90 ml.

6.2.10   HYDROLYTIC STABILITY

The ASTM D2619 "Standard Test Method for Hydrolytic Stability of Hydraulic Fluids (Beverage Bottle Method)" test is used to measure the hydrolytic stability of hydraulic oils and turbine oils. Hydrolytically unstable oils form acidic and insoluble contaminants which can cause system malfunctions due to corrosion, valve sticking or change in viscosity of the fluid. A 75 gm sample of the test oil, 25 gm of distilled water and a pre-weighed copper strip are sealed in a pressure-type beverage bottle. The bottle is rotated end for end at 5 rpm for 48 hours in an oven at 93°C. Then the liquid layers are separated and the following determinations are made:

- Viscosity change of the test oil.
- AN change of the test oil.

- Total acidity of the water.
- Weight of insoluble material that formed.
- Weight change of the copper strip.
- Appearance of the copper strip under 20× magnification.

Because the beverage bottles used in the test at first were "Coca-Cola" bottles, the test became known as the "Coke Bottle Test". Because of the length of the test, it is unsuitable for use in a lubricant condition monitoring programme and is only used to test new oils and with the development of improved formulations.

6.2.11 FOAMING PROPERTIES

The ASTM D892, IP 146, ISO 6247, DIN 51566 and AFNOR T60-129 "Standard Method of Test for Foaming Characteristics of Lubricating Oils" tests are intended to determine the ability or otherwise of lubricating oils to resist the tendency to foam at specified temperatures. The test is performed in three "Sequences".

The foaming test apparatus consists of a 1,000 ml graduated cylinder and an air-inlet tube, to the bottom of which is fastened a 25.4 mm (1 in) diameter spherical gas diffuser stone made of fused crystalline alumina grain. The cylinder has a diameter such that the distance from the inside bottom to the 1,000 ml graduated mark is 360 ± 25 mm. It is circular at the top and is fitted with a rubber stopper having one hole at the centre for the air-inlet tube and a second hole off-centre for an air-outlet tube. The air-inlet tube is adjusted so that, when the rubber stopper is fitted tightly into the cylinder, the diffuser stone just touches the bottom of the cylinder and is approximately at the centre of the circular cross section. Diffuser stones are specified in the test method. The graduated cylinders used in the tests are placed in heating baths, such that immersion of the cylinder reaches at least to the 900 ml mark (see Figure 6.4).

In Sequence I, 190 ml of test oil is placed in one cylinder in a heating bath maintained at 24°C. When the oil and diffuser stone has reached the test temperature, air is blown through the delivery tube and diffuser stone at a flow rate of 94 ± 5 ml/min for 5 minutes, timed from the first appearance of bubbles from the stone. At the end of 5 minutes, the air supply is turned off and the volume of foam recorded immediately. The cylinder is allowed to stand for 10 minutes and the volume of foam recorded again.

In Sequence II, a second 180 ml sample of test oil is placed in a second cylinder, in a heating bath maintained at 93.5°C. The same air blowing, foam volume recording, standing and foam volume recording procedure is used as in Sequence I.

In Sequence III, any foam remaining in the cylinder used in Sequence II is allowed to collapse by stirring and the cylinder is allowed to cool to below 43°C. A cleaned diffuser stone and air delivery tube is inserted into the cylinder, and it is then placed in the heating bath at 24°C and the air blowing procedure is repeated.

A later test, ASTM D6082 "Standard Test Method of High Temperature Foaming Characteristics of Lubricating Oils" uses the same equipment, but tests oils at 150°C. The test is designed to be used to assess the anti-foaming properties of engine oils, automotive gear oils and automatic transmission fluids. The test results are sometimes reported as Sequence IV.

FIGURE 6.4   Foaming test apparatus.

Source: Dott. Gianni Scavini & C., with permission.

A number of companies supply equipment for performing ASTM D892, ASTM D6082, IP 146, ISO 6247 DIN 51566 and AFNOR T60-129 tests. Some suppliers offer automated equipment.

6.2.12   CARBON RESIDUE

The carbon residue of an oil is the amount of deposit, in percentage by weight (%wt), left after evaporation and pyrolysis of the oil under prescribed conditions. In the test, oils from any given type of crude oil show lower values than those of similar viscosity containing residual stocks. Oils of naphthenic type usually show lower residues than those of similar viscosity made from paraffinic crude oils. The more severe the refining treatment (whether an oil is subjected to solvent processing or hydroprocessing), the lower the carbon residue value will be.

Originally the carbon residue test was developed to determine the carbon-forming tendency of steam cylinder oils. Subsequently, unsuccessful attempts were made to relate carbon residue values to the amount of carbon formed in the combustion chambers and on the pistons of internal combustion engines. Since such factors as fuel composition and engine operation and mechanical conditions, as well as other lubricating oil properties, are of equal or greater importance, carbon residue values alone have only limited significance. The carbon residue determination is now made mainly on base oils used for engine oil manufacture, straight mineral engine oils such as aircraft engine oils and some products of the cylinder oil type used for reciprocating air compressors. In these cases, the determination is an indication of the degree of refining to which the base oil has been subjected.

ASTM D4530 "Standard Test Method for Determination of Carbon Residue (Micro Method)" measures the "Conradson Carbon Residue" of a base oil or lubricant. The method offers advantages of better control of test conditions, smaller samples and less operator attention compared with test method ASTM D189 "Standard Test Method for Conradson Carbon Residue of Petroleum Products", to which it is equivalent. IP 13, ISO 6615, DIN 51551 and AFNOR T60-116 test methods are equivalent to ASTM D189. IP 398 and ISO 10370 test methods are equivalent to ASTM D4530.

The ASTM D4530 test covers the determination of the amount of carbon residue formed after evaporation and pyrolysis of petroleum materials under certain conditions and is intended to provide some indication of the relative coke-forming tendency of such materials. Up to 12 samples may be run simultaneously, including a control sample when a vial holder shown in the test method is used exclusively for sample analysis.

The procedure is a modification of the original method and apparatus for carbon residue of petroleum materials, where it has been demonstrated that thermogravimetry is another applicable technique. However, it is the responsibility of the operator to establish operating conditions to obtain equivalent results when using thermogravimetry.

The test is applicable to petroleum products that partially decompose on distillation at atmospheric pressure and was tested for carbon residue values of 0.10 to 30%wt. Samples expected to be below 0.10%wt residue should be distilled to remove 90%v/v of the flask charge. The 10% bottom remaining is then tested for carbon residue.

Ash-forming constituents, as defined by test method ASTM D482 (see later), or non-volatile additives present in the sample will add to the carbon residue value and be included as part of the total carbon residue value reported. As a consequence, ASTM D482 test results will not be the same as ASTM D189, IP 13 and ISO 6615 test results.

ASTM D524 "Standard Test Method for Ramsbottom Carbon Residue of Petroleum Products" is another method for determining the carbon residue of base oils and lubricants. Test methods IP 14, ISO 4262 and AFNOR T60-117 are equivalent to ASTM D524. The carbon residue value of motor oil, while at one time regarded as indicative of the amount of carbonaceous deposits a motor oil would form in the combustion chamber of an engine, is now considered to be of doubtful significance due to the presence of additives in many oils. For example, an ash-forming detergent additive can increase the carbon residue value of an oil yet will generally reduce its tendency to form deposits.

The carbon residue values of crude oil atmospheric residues, cylinder oils and bright stocks are useful in the manufacture of lubricants.

The test also covers the determination of the amount of carbon residue left after evaporation and pyrolysis of an oil, and it is intended to provide some indication of relative coke-forming propensity. The test is generally applicable to relatively non-volatile petroleum products which partially decompose on distillation at atmospheric pressure. This test method also covers the determination of carbon residue on 10%v/v distillation residues. As with ASTM D4530, petroleum products containing ash-forming constituents as determined by ASTM D482 will have an erroneously high carbon residue, depending upon the amount of ash formed.

The term carbon residue is used throughout the test to designate the carbonaceous residue formed during evaporation and pyrolysis of a petroleum product. The residue is not composed entirely of carbon, but is a coke which can be further changed by pyrolysis. The term carbon residue is continued in this test method only in deference to its wide common usage.

The values obtained by the test are not numerically the same as those obtained by ASTM D189 or ASTM D4530. Approximate correlations have been derived, but they need not apply to all materials which can be tested because the carbon residue test is applicable to a wide variety of petroleum products. The Ramsbottom Carbon Residue test method is limited to those samples that are mobile below 90°C.

6.2.13 Ash and Sulphated Ash

The ASTM D482, IP 4, ISO 6245 and AFNOR M07-045 "Standard Method of Test for Ash from Petroleum Products" test methods describe a procedure for determining the ash from distillate and residual fuel oils, crude oils, lubricating oils, waxes and other petroleum products, in which any ash-forming materials present are normally considered to be undesirable impurities or contaminants. The method is limited to petroleum products which are free from added ash-forming additives, including certain phosphorous compounds.

The sample of oil, contained in a suitable vessel, is ignited and allowed to burn until only ash and carbon remain. The carbonaceous residue is then reduced to an ash by heating in a muffle furnace at 775°C, cooled and weighed.

The method is not suitable for the analysis of new or used lubricating oils that contain additives, so cannot be used in an oil condition monitoring programme. For this, ASTM D874 "Standard Method of Test for Sulphated Ash from Lubricating Oils and Additives" is used. IP 163, ISO 3987, DIN 51575 and AFNOR T60-143 test methods are identical to ASTM D874. The test is also used to determine the sulphated ash from additive concentrates used in blending and compounding. These additives usually contain one or more compounds of calcium, magnesium, zinc, potassium, sodium, tin, sulphur, phosphorus and chlorine. Previously, some additives contained compounds of barium, but these are not widely used now.

Application of the test to sulphated ash levels below 0.02% is restricted to oils containing ashless additives. The lower limit of the method is 0.005% sulphated ash. There is evidence that magnesium does not react the same as other alkali metals in the test. If magnesium additives are present, the data should be interpreted with caution.

In the test, the sample is ignited and burned until only ash and carbon remain. After cooling, the residue is treated with sulphuric acid and heated at 775°C until oxidation of carbon is complete. The ash is then cooled, re-treated with sulphuric acid and heated at 775°C to constant weight.

The sulphated ash may be used to indicate the concentration of known metal-containing additives in new oils. When phosphorus is absent, calcium, magnesium, sodium and potassium are converted to their sulphates and tin (stannic) and zinc to their oxides. Sulphur and chlorine do not interfere, but when phosphorus is present with metals, it remains partially or wholly in the sulphated ash as metal phosphates.

6.2.14 INSOLUBLES CONTENTS

ASTM D893 "Standard Test Method for Insolubles in Used Lubricating Oils" is used to determine the level and composition of insoluble contaminants in a lubricant. The level of contamination is derived from measuring material that can be mechanically separated from the fluid by a centrifuge. The composition and origin of the material are obtained by identifying the contaminant's solubility in pentane and toluene.

Pentane insolubles can include oil-insoluble materials and some oil-insoluble resinous matter originating from oil or additive degradation or both. Toluene insoluble materials can come from external contamination, from fuel carbon and highly carbonised materials from degradation of fuel, oil and additives or from engine wear and corrosion materials. A significant change in pentane insolubles, toluene insolubles (with or without coagulant) and insoluble resins indicates a change in the oil which could lead to lubrication system problems. Insolubles measured can also assist in evaluating the performance characteristics of a used oil or in determining the cause of equipment failure.

The test method has two procedures. Procedure A covers the determination of insolubles without the use of coagulant in the pentane. It provides an indication of the materials that can readily be separated from the oil–solvent mixtures by centrifuging. Procedure B covers the determination of insolubles in oils containing detergents and employs a coagulant for both the pentane and toluene insolubles. In addition to the materials separated by using Procedure A, this coagulation procedure separates some finely divided materials that may be suspended in the oil. Results obtained by Procedures A and B should not be compared since they usually give different values. The same procedure should be employed when comparing values obtained periodically on an oil in use or when comparing results determined by two or more laboratories.

In the test, the amount of material deposited in the tip of the tube represents oil-insoluble, resinous matter that may originate from fluid and/or additive degradation (soft contaminants). This material is measured and identified as the pentane insolubles. In the second portion of the method, following the determination of pentane insolubles, the remaining deposit is mixed with a toluene-based solution, recentrifuged and the fluid is decanted. The remaining deposit is composed of material associated with external sources of contamination, wear debris or densely carbonised matter generated from high-temperature thermal events (hard contaminants). Variations of the test method include membrane filtration with toluene rinse.

A more useful test for an oil condition monitoring programme is ASTM D7317 "Coagulated Pentane Insolubles in Used Lubricating Oils by Paper Filtration (LMOA Method)". Coagulated pentane insolubles can include oil-insoluble materials, some oil-insoluble resinous matter originating from oil or additive degradation, soot from incomplete diesel fuel combustion or a combination of all three. A significant change in the amount of coagulated pentane insolubles indicates a change in the oil, which could lead to lubrication system problems. Coagulated pentane insolubles measurements can assist in evaluating the performance characteristics of a used oil or in determining the cause of equipment failure. This test method was originally developed by the Locomotive Maintenance Officers Association (LMOA). This test method, in general, does not correlate with ASTM D893 on insolubles in lubricating oils, because it uses separation by centrifugation and a more concentrated solution of anticoagulant. The correlation between this test method and enhanced thermal gravimetric analysis procedure in Test Method D5967 has not been investigated.

A sample of used lubricating oil is mixed with pentane–coagulant solution and filtered under vacuum. The filter is washed with pentane, dried and weighed to give coagulated pentane insolubles.

6.2.15 DISSOLVED GAS ANALYSIS

The ASTM D3612 "Standard Test Method for Analysis of Gases Dissolved in Electrical Insulating Oil by Gas Chromatography" is an analytical test that can be used for lubricants other than insulating oils. As noted earlier, some degradation mechanisms cause an oil's hydrocarbon molecules to crack, which can generate gases that are dissolved in the oil. The type and distribution of these gases can indicate which degradation mechanism was responsible.

According to the test description, oil and oil-immersed electrical insulation materials may decompose under the influence of thermal and electrical stresses, and in doing so generate gaseous decomposition products of varying composition which dissolve in the oil. The nature and amount of the individual component gases that may be recovered and analysed may be indicative of the type and degree of the abnormality responsible for the gas generation. The rate of gas generation and changes in concentration of specific gases over time are also used to evaluate the condition of the electric apparatus. Guidelines for the interpretation of gas-in-oil data are given in IEEE C57.104.

The test method covers three procedures for extraction and measurement of gases dissolved in electrical insulating oil having a viscosity of 20 cSt or less at 40°C and the identification and determination of the individual component gases extracted. Other methods have been used to perform this analysis.

Gases that are useful to examine when applying this test to fluid degradation include hydrocarbons (methane, ethane, ethylene and acetylene), carbon oxides (carbon monoxide and carbon dioxide) and hydrogen. Researchers have attempted to identify the specific gases produced once the hydrocarbon molecule is cracked (temperatures in excess of 300°C). Acetylene, for example, is created at temperatures greater than 1,000°C.

6.2.16 WATER CONTENT

Testing for water in hydraulic, turbine, gear or air compressor oils is important to minimise the risk of possible undetected oxidation and corrosion. Rust formation often leads to particle formation that can cause abrasive wear in bearings. Excessive water can also increase or decrease an oil's viscosity, depending on conditions. Water in oils in warm reservoirs can promote microbial growth that could foul system filters, small diameter pipes or transducer line extensions. When a turbine, hydraulic or air compressor oil has cooled to room temperature, dissolved water may come out of solution as free water, so care must be taken to minimise an oil's water content to protect both the equipment and the oil.

A simple way to check for the presence of water droplets in an oil is to use the "crackle test". The end of a heated rod is dipped into the oil, and if the oil is heard to "crackle", free water is present. (This can often be determined simply by the hazy appearance of the oil.)

A more accurate method used to determine the water content of lubricants is ASTM D6304 "Standard Test Method for Determination of Water in Petroleum Products, Lubricating Oils, and Additives by Coulometric Karl Fischer Titration". The method uses standard automatic Karl Fischer titration equipment and reagents which allow the determination of water contents in the range of 10 to 25,000 mg/kg (ppm) entrained water. The method also covers the indirect analysis of water thermally removed from samples and swept with dry inert gas into the Karl Fischer titration cell. The presence of sulphur, sulphides (including H_2S) and mercaptans is known to interfere with the test result.

ASTM D7546 "Standard Test Method for Determination of Moisture in New and In-Service Lubricating Oils and Additives by Relative Humidity Sensor" covers the quantitative determination of water in new and in-service lubricating oils and additives in the range of 10 to 100,000 mg/kg (0.001 to 10%wt./wt.) using a relative humidity (RH) sensor. Methanol, acetonitrile and other compounds are known to interfere with this test method. However, this test is very suitable for use in an oil condition monitoring programme for selected oils.

The ASTM D4377, IP 356 and ISO 10336 test method "Standard Test Method for Water in Crude Oils by Potentiometric Karl Fischer Titration" was withdrawn in January 2020. A replacement test had not been published at the time of writing. The test method covered the determination of water in the range from 0.02% to 2% in crude oils. Mercaptan and sulphur compounds were known to interfere with the test. It was a modification of the Karl Fischer method, aiming to increase sensitivity by using a potential difference electrode system to indicate the end point and by using a special titration cell which could be totally sealed and rapidly conditioned to the end point before addition of the sample to be analysed. The test method was withdrawn in accordance with Section 10.6.3 of the Regulations Governing ASTM Technical Committees, which requires that standards shall be updated by the end of the eighth year since the last approval date.

An alternative is method IP 358, ASTM D4006 or ISO 9029 "Standard Test Method for Water in Crude Oil by Distillation". Although the method was originally developed for determining water in crude oil, it may be applied to other petroleum hydrocarbons. The sample is heated under reflux with a water-immiscible solvent which co-distils with the water in the sample. Condensed solvent and water are continuously separated in a trap, the water settling in the graduated section and the solvent returning to the still. The method is similar to IP 74, ASTM D95, ISO 3733, DIN 51582 and AFNOR T60-113.

6.2.17 SULPHUR, NITROGEN AND PHOSPHOROUS CONTENTS

ASTM D5291 "Standard Test Methods for Instrumental Determination of Carbon, Hydrogen, and Nitrogen in Petroleum Products and Lubricants" is the first ASTM standard covering the simultaneous determination of carbon, hydrogen and nitrogen in petroleum products and lubricants. Carbon, hydrogen and particularly nitrogen analyses are useful in determining the complex nature of petroleum products covered by the test.

The concentration of nitrogen is a measure of the presence of nitrogen containing additives. Knowledge of its concentration can be used to predict performance. Some petroleum products also contain naturally occurring nitrogen. Knowledge of hydrogen content in samples is helpful in addressing their performance characteristics. Hydrogen-to-carbon ratio is useful to assess the performance of base oil refinery upgrading processes.

The test method covers the instrumental determination of carbon, hydrogen and nitrogen in laboratory samples of petroleum products and lubricants. Values obtained represent the total carbon, the total hydrogen and the total nitrogen. These test methods were tested in the concentration range of at least 75 to 87 mass % for carbon, at least 9 to 16 mass % for hydrogen and <0.1 to 2 mass % for nitrogen.

The nitrogen test method is not applicable to light materials or those containing <0.75 mass % nitrogen, or both, such as gasoline, jet fuel, naphtha, diesel fuel or chemical solvents. However, using Test Method D, levels of 0.1 mass % nitrogen in lubricants could be determined. These test methods are not recommended for the analysis of volatile materials such as gasoline, gasoline-oxygenate blends or gasoline-type aviation turbine fuels.

The test instrument works on basis of a combustion method to convert the sample elements to simple gases (CO_2, H_2O and N_2), at an elevated temperature (975°C) and in a pure oxygen environment.

Elements such as halogens and sulphur are removed by scrubbing reagents in the combustion zone. The combustion product gas stream is passed over a heated reduction zone, which contains copper to remove excess oxygen and reduce NOx to N_2 gas. The gases are then homogenised and controlled, in a mixing chamber, to exact conditions of pressure, temperature and volume. The homogenised gases are allowed to de-pressurise through a column where they are separated in a stepwise steady-state manner and detected as a function of their conductivities.

ASTM D4294, IP 336, ISO 8754 and AFNOR M07-053 "Standard Test Method for Sulphur in Petroleum and Petroleum Products by Energy Dispersive X-ray Fluorescence Spectrometry" provide rapid and precise measurement of total sulphur in petroleum and petroleum products with a minimum of sample preparation. A typical analysis time is 1 to 5 minutes per sample.

The quality of many petroleum products is related to the amount of sulphur present. Knowledge of sulphur concentration is necessary for processing purposes. There are also regulations promulgated in national and international standards that restrict the amount of sulphur present in some fuels.

The test covers the determination of total sulphur in petroleum and petroleum products that are single-phase and either liquid at ambient conditions, liquefiable with moderate heat or soluble in hydrocarbon solvents. These materials can include diesel fuel, jet fuel, kerosene, other distillate oil, naphtha, residual oil, lubricating base oil, hydraulic oil, crude oil, unleaded gasoline, gasoline–ethanol blends, biodiesel and similar petroleum products.

The concentrations of substances in the test method were determined by the calculation of the sum of the mass absorption coefficients times mass fraction of each element present. This calculation was made for dilutions of representative samples containing approximately 3%wt of interfering substances and 0.5%wt sulphur. For samples with high oxygen contents (>3%wt), sample dilution as described in the test method or matrix matching must be performed to assure accurate results.

Inter-laboratory studies on precision revealed the scope to be 17 mg/kg to 4.6 mass %. An estimate of this test method's pooled limit of quantitation (PLOQ) is 16.0 mg/kg as calculated by the procedures in Practice ASTM D6259. However, because instrumentation covered by this test method can vary in sensitivity, the applicability of the test method at sulphur concentrations below approximately 20 mg/kg must be determined on an individual basis. An estimate of the limit of detection is three times the reproducibility standard deviation, and an estimate of the limit of quantitation is ten times the reproducibility standard deviation.

Samples containing more than 4.6 mass % sulphur can be diluted to bring the sulphur concentration of the diluted material within the scope of the test method. Samples that are diluted can have higher errors than indicated in Section 16 of the test method than non-diluted samples.

A fundamental assumption in the test method is that the standard and sample matrices are well matched, or that the matrix differences are accounted for. Matrix mismatch can be caused by C/H ratio differences between samples and standards or by the presence of other heteroatoms.

6.2.18 METALS CONTENTS

The metal contents of new lubricants (and the additives they contain) and lubricants in use are best measured individually by methods such as atomic absorption, ICP spectroscopy or X-ray fluorescence. The old techniques of wet chemical precipitation, filtering and weighing have been superseded except for reference purposes by modern analytical equipment which uses comparative instrumental techniques. These relate the test sample readings to those given by standard calibration material. Some methods can use oil as received, in others it needs to be diluted with a solvent, while some require more complex sample preparation techniques.

At one time flame photometry (ASTM D3340) was a general-purpose technique used for many elemental analyses in chemical laboratories, but now its use is largely confined to older laboratories and for the rapid measurement of sodium and lithium contents found in greases and some lubricating oils. The sample can be diluted with solvent, or for greater sensitivity can be ashed and dissolved in water. In either case the liquid is aspirated into a flame. Characteristic wavelengths of light are emitted in the flame depending on the elements present, the light is dispersed by a spectrometer, and the relevant wavelengths are measured electronically. The intensity of emission from the sample is compared with standards of known elemental concentration and the concentrations in the sample thereby estimated.

Atomic absorption (AA) spectroscopy (ASTM D4628) has largely replaced flame photometry for routine analysis in many petroleum laboratories. The IP 308 and DIN 51391T1 test methods are equivalent to ASTM D4628. A monochromatic beam of light, characteristic of the element to be measured, is produced by a hollow-cathode lamp containing the element. This light is passed through an intense flame, such as from an acetylene/nitrous oxide burner, and the sample in solvent solution is aspirated into the flame. Atoms of the element being measured are produced in the flame and absorb the characteristic radiation, and the drop in its intensity can be measured and related to the elemental content of the sample using calibration standards. Elements routinely measured by atomic absorption include barium, calcium, magnesium and zinc.

Flame photometry is a form of emission spectroscopy, but the term is usually applied to more complex apparatus employing an electric arc or a plasma to excite the spectral lines from the sample, rather than a flame. The higher-energy excitation source gives higher sensitivity and produces more useful spectral lines for the purpose of analysis, which can help to overcome problems of interferences between elements.

Inductively coupled plasma emission (ICP) spectroscopy (ASTM D4951) is a more recent technique of emission spectroscopy, tending to be less sensitive than arc systems for some elements but normally giving greater accuracy. A stream of ionisable gas such as argon is excited by a powerful radio-frequency coil and is converted into a plasma. The sample in a solvent is sprayed into the plasma and characteristic spectral lines are emitted. The method has moderate sensitivity for barium and phosphorus but more for calcium, magnesium, sulphur and zinc. It can also be used for trace metals in used oil analysis, when the ASTM D5185 test is used.

X-Ray fluorescence (XRF) spectroscopy (ASTM D4927, IP 407 and DIN51391T2 tests) has become a very widely used analytical technique of high speed and good precision. Sample preparation is not normally needed, the oil being placed in a small cell with a thin plastic film window at the bottom. This is attached to a vacuum chamber and the sample is irradiated through the window with powerful X-rays. The sample emits secondary X-rays whose wavelengths are characteristic of the elements present. These are analysed with a crystal diffraction spectrometer and the intensity at selected wavelengths is measured with an X-ray counter (for example, a Geiger counter). The method is sensitive for most elements, but it is unsuitable for those lighter than silicon and therefore excludes sodium and magnesium. The precision of the method is relatively good and can sometimes be further improved by extending the counting time (the time taken to measure the emitted X-rays). There are considerable inter-element interference effects, and correct calibration is essential for accurate analysis.

ASTM D5185 "Standard Test Method for Multi-Element Determination of Used and Unused Lubricating Oils and Base Oils by Inductively Coupled Plasma Atomic Emission Spectrometry (ICP-AES)" covers the rapid determination of 22 elements in used and unused lubricating oils and base oils. It provides rapid screening of used oils for indications of wear. Test times approximate a few minutes per test specimen and detectability for most elements is in the low mg/kg range. In addition, the test covers a wide variety of metals in virgin and re-refined base oils.

When the predominant source of additive elements in used lubricating oils is the additive package, significant differences between the concentrations of the additive elements and their respective specifications can indicate that the incorrect oil is being used. The concentrations of wear metals can be indicative of abnormal wear if there are baseline concentration data for comparison. A marked increase in boron, sodium or potassium levels can be indicative of contamination as a result of coolant leakage in the equipment. The test can be used to monitor equipment condition and define when corrective actions are needed. It can also be used for unused oils to provide more complete elemental composition data than ASTM D4628, ASTM D4927 or ASTM D4951.

A weighed portion of a thoroughly homogenised used oil is diluted ten-fold by weight with mixed xylenes or other suitable solvent. Standards are prepared in the same manner. An optional internal standard can he added to the solutions to compensate for variations in sample introduction efficiency. The solutions are introduced to the ICP instrument by free aspiration or an optional peristaltic pump. By comparing emission intensities of elements in the specimen with emission intensities measured with the standards, the concentrations of elements in the specimen are calculable.

The elements covered by the test are aluminium (Al), barium (Ba), boron (B), calcium (Ca), chromium (Cr), copper (Cu), iron (Fe), lead, (Pb), magnesium (Mg), manganese (Mn), molybdenum (Mo), nickel (Ni), phosphorous (P), potassium (K), silicon (Si), silver (Ag), sodium (Na), sulphur (S), tin (Sn), titanium (Ti), vanadium (V) and zinc (Zn).

ASTM D6595 "Wear Metals and Contaminants in Used Oils or Hydraulic Fluids by Rotating Disc Electrode Atomic Emission Spectrometry" is used to determine debris in used lubricating oils, a key diagnostic method practiced in machine condition monitoring programmes. The presence or increase in concentration of specific wear metals can be indicative of the early stages of wear if there are baseline concentration data for comparison. A marked increase in contaminant element can be indicative of foreign materials in the lubricant, such as anti-freeze or sand, which may lead to wear or lubricant degradation. The test method identifies the metal and their concentration so that trends relative to time or distance can be established and corrective action can be taken before more serious or catastrophic failure occurs. This test method uses oil-soluble metals for calibration and does not purport to relate quantitatively the values determined as insoluble particles to the dissolved metals. Analytical results are particle size-dependent and low results may be obtained for those elements present in used oil samples as large particles.

Wear metals and contaminants in a used oil test specimen are evaporated and excited by a controlled arc discharge using the rotating disk technique. The radiant energies of selected analytical lines and one or more references are collected and stored by way of photomultiplier tubes, charge coupled devices or other suitable detectors. A comparison is made of the emitted intensities of the elements in the used oil test specimen against those measured with calibration standards. The concentrations of the elements present in the oil test specimen are calculated, displayed or entered into a database for processing.

Lubrication engineers will find it difficult to appreciate the comparatively low levels of precision found in chemical analysis of petroleum products compared with physical measurements within their own disciplines. There are many reasons for this, but the most important are:

- Precise basic standards can be defined for most measurements of length or distance, for example, a bar of metal or the wavelength of light. Concentration standards are, however, much more difficult to set up.
- Where concentration standards can be produced, for example, by adding weighed amounts of a pure substance to a diluent, such samples often have limited application.
- This is because in most cases the standard samples need to be quite similar to the unknown samples for analysis. In spectrographic analytical methods, the different elements interfere mutually with each other's output signals, and the base matrix can also have a considerable influence.
- For classical wet chemical analysis, precipitated material can be lost, or false precipitates of other interfering elements can be unknowingly produced.
- Spectroscopic devices have considerable "noise" in both the excitation and detection areas which is significant in relation to the magnitude of the signals being measured and limits the sensitivity at low concentration levels.
- Petroleum samples are normally liquids which are not completely stable either chemically or physically.

Repeatability of spectrographic methods for lubricating oil analysis lies in the range of 2% to 5% of the mean value, while estimates of reproducibility from co-operative test programmes show values ranging from around 4% to 20%. This does not mean that two laboratories working together, such as a lubricant supplier and a lubricant customer, cannot come to considerably closer agreement for a specific analysis. The problems of establishing meaningful specifications for new lubricants depend on close co-operation between the supplier and the user. Accuracy depends on the quality of the reference or calibration standards used.

6.2.19 FOURIER TRANSFORM INFRARED SPECTROSCOPY (FTIR)

FTIR spectroscopy is a technique which is used to obtain an infrared spectrum of absorption or emission of a solid, liquid or gas. An FTIR spectrometer simultaneously collects high spectral resolution data over a wide spectral range. This confers a significant advantage over a dispersive spectrometer, which measures intensity over a narrow range of wavelengths at a time. To convert the raw spectral data into an actual spectrum, a Fourier transform (a mathematical process) is required. An example of a FTIR spectrum is shown in Figure 6.5.

The aim of any absorption spectroscopy is to measure how well a sample absorbs light at each wavelength. The most straightforward way to do this, the dispersive spectroscopy method, is to shine a monochromatic light beam at a sample, measure how much of the light is absorbed and repeat for each different wavelength.

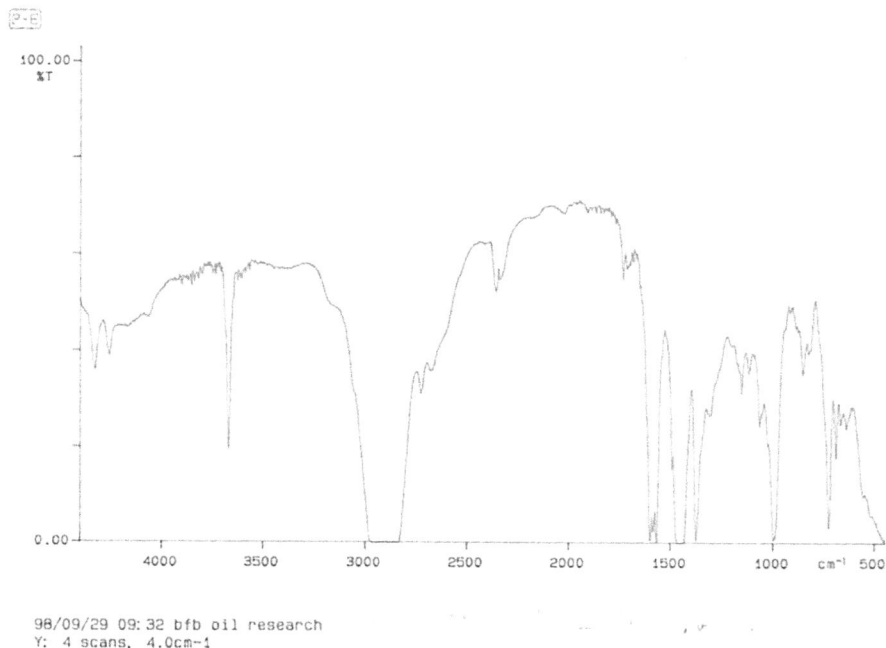

98/09/29 09:32 bfb oil research
Y: 4 scans, 4.0cm-1

FIGURE 6.5   Example infrared spectrum.

Source: Pathmaster Marketing Ltd.

FTIR is a less intuitive way to obtain the same information. Rather than shining a monochromatic beam of light at the sample, this technique shines a beam containing many frequencies of light at once and measures how much of that beam is absorbed by the sample. Next, the beam is modified to contain a different combination of frequencies, giving a second data point. This process is repeated many times. Then, a computer program works backward from all the data points to infer what the absorption is at each wavelength.

The beam is generated by a broadband light source that contains the full spectrum of wavelengths to be measured. The light shines into a Michelson interferometer, a specific configuration of mirrors, one of which is moved by a motor. As this mirror moves, each wavelength of light in the beam is periodically blocked, transmitted, blocked, transmitted, by the interferometer, due to wave interference. Different wavelengths are modulated at different rates so that at each moment the beam coming out of the interferometer has a different spectrum.

A Fourier transform algorithm (calculation) converts the raw data (light absorption for each mirror position) into the desired result (light absorption for each wavelength). The raw data is sometimes called an interferogram.

With hydrocarbon liquids, whether containing additives or not, (base oils, lubricants and additives), FTIR is one way of determining the presence or absence of organic compounds in a sample.

One FTIR test method that is used for oils in use is ASTM D7686 "Field-Based Condition Monitoring of Soot in In-Service Lubricants Using a Fixed-Filter Infrared (IR) Instrument". The test provides a simple field-based technique for condition monitoring of soot in lubricants associated with combustion engines and industrial and military equipment. (It should be noted that critical applications should use laboratory-based test methods such as thermal gravimetric analysis used in ASTM D5967, Annex A4.) ASTM D7686 can be used to monitor soot build-up in lubricants and can indicate whether soot has accumulated to an extent which could significantly degrade the oil's performance or cause filter and oil passage blockage. The test method is applicable for soot levels of up to 12% and is intended as a field test only.

The test method describes the acquisition of spectral data with the use of a fixed-filter IR instrument. A well-homogenised oil sample is applied to a Horizontal Attenuated Total Reflectance (HATR) crystal. Infrared light is totally internally reflected through the HATR crystal and then passed through a broadband infrared filter centred on 3.9 μm. The amount of infrared light reaching a detector is converted to a soot concentration via a calibration curve generated from the instrument response on various soot standards. Some types of dirt or wear debris may cause a positive bias. In most cases the bias will be small and will likely also affect thermal gravimetric analysis as described in ASTM D5967.

A further FTIR test for oils in use is ASTMD7889 "Field Determination of In-Service Fluid Properties Using IR Spectroscopy", which provides a means for obtaining useful in-service fluid analysis properties in the field. Each of monitored properties has been shown over time to indicate contamination in the oil system or a particular breakdown modality of the oil, which is critical information to assess the health of the oil as well as the machinery. The method uses a grating spectrometer to analyse properties of an in-service fluid sample, such as oxidation, nitration, sulphation, soot and anti-wear additives. Applicable oils include all mineral hydrocarbons in API (Groups I to IV) used in machinery lubricants, including reciprocating engine oils, turbine oils, hydraulic oils and gear oils.

The test method uses a self-contained field apparatus to provide detailed information concerning the condition status of in-service fluids. An absorbance spectrum of the sample under test conditions is obtained. For the background spectrum, the cell background is used. Test methods have been developed for FTIR devices using absorbance spectra obtained using Standard Practice ASTM D7418. The essence of this test method is to capture the underlying chemical trends associated with each property for in-service fluid analysis using a self-contained field apparatus and a coupled wipe-clean transmission cell.

6.3 CHEMICAL TESTS FOR GREASES

6.3.1 CORROSION RESISTANCE

Since the early 1960s, one of the most popular grease rust tests in the United States has been ASTM D1743. This method determines the corrosion preventive properties of greases using tapered roller bearings which are stored under static conditions in the presence of distilled water.

Tapered roller bearings are packed with grease and run under a light load to distribute the grease evenly. The bearings are exposed to distilled water, then stored at 52°C and 100% relative humidity for 48 hours. After cleaning, the bearing races are examined for corrosion. Since 1987, the ASTM D1743 procedure has specified a pass or fail rating on the basis of a single corrosion spot of 1.0 mm or larger in the longest dimension on two of three bearings tested simultaneously.

The ASTM D5969 test method is the synthetic sea water version of ASTM D1743 which specifies the use of only distilled water. The test methods are identical except that the tapered roller bearings are exposed to desired concentrations of synthetic sea water (prepared as in test method ASTM D665B) diluted with distilled water, and the test time is reduced from 48 to 24 hours. The pass or fail criteria are the same as for ASTM D1743, on the basis of a single corrosion spot of 1.0 mm or larger in the longest dimension on two of three bearings tested simultaneously.

The IP 220 "Emcor" method is used widely in Europe for greases. This procedure employs actual ball bearings, which are visually rated for rust on the outer races at the end of the test. This is a dynamic test that can be run with a given amount of water flowing into the test bearing housings. Severity may be increased with the use of salt water or acid water in place of the standard distilled water.

Ten grams of grease is tested in a ball bearing running at 80 rpm under no applied load in the presence of water (distilled, salt or synthetic sea water). The bearing is run for 30 minutes immediately after assembly, to distribute the grease evenly. Ten millilitres of water is then introduced into the bearing, into each side of the plummer block. The bearing is then run in the Emcor rig, first for 8 hours followed by a 16-hour stop, for another 8 hours followed by a 16-hour stop and finally for 8 hours followed by a 108-hour stop. The grease is evaluated by examining the outer race of the bearing, after test, for freedom from rust.

The ASTM D6138 test method is commonly referred to as the "Emcor Test" and is an adopted version of the IP 220 test method. This test is a dynamic procedure used to determine the corrosion protection of a grease in the presence of water approximating typical service conditions.

Double-row self-aligning ball bearings are packed with grease and run to distribute the grease evenly. The bearings are exposed to either distilled water, synthetic sea water or sodium chloride solution, and the test rig is operated under alternating running and standing conditions for one week. After cleaning, the bearing races are examined for corrosion and rated from "0" to "5" as follows:

0   No evidence of corrosion.
1   No more than three spots of a size just sufficient to be visible.
2   Up to 1% surface corrosion.
3   Between 1% and 5% surface corrosion.
4   Between 5% and 10% surface corrosion.
5   More than 10% surface corrosion.

The test rig will simultaneously run eight bearings. Duplicate determinations are required, so the rig is normally used to evaluate four different greases at a time.

6.3.2 COPPER CORROSION

The ASTM D4048 test method is similar to the ASTM D130, IP 154 and ISO 2160 test used for industrial oils but is designed to evaluate the copper corrosion prevention properties of greases. A polished and cleaned 3-inch copper strip as prescribed in ASTM D130 is placed in a jar in which the copper strip is totally immersed in the test grease. The jar is capped and heated to a specified temperature for a defined period of time. Commonly used conditions are 100°C for 24 hours. At the end of the test the strip is removed, wiped clean and matched with coloured reproduction strips characteristic of the descriptions in the copper corrosion test for oils (ASTM D130, IP 154), as described in Section 6.2.7.

6.3.3 OXIDATION RESISTANCE

ASTM D942, IP 142 and DIN 51808 "Standard Test Method for Oxidation Stability of Lubricating Greases by the Oxygen Pressure Vessel Method" (the Norma Hoffman Grease Oxidation Test) is the test most usually used for determining the oxidation stability of greases.

The test measures the net change in pressure resulting from consumption of oxygen by oxidation and gain in pressure due to formation of volatile oxidation by-products. The test may be used for quality control to indicate batch-to-batch uniformity. It predicts neither the stability of greases under dynamic service conditions, nor the stability of greases stored in containers for long periods nor the stability of films of greases on bearings and motor-parts. It should not be used to estimate the relative oxidation resistance of different grease types.

The test defines the degree of oxidation after a given period of time by the corresponding decrease in pressure. This method should be used to compare similar greases and should not be used to estimate the relative oxidation resistance of different grease types.

In the test, five glass dishes are filled with 4 grams (g) of test grease and placed in the pressure vessel. The vessel is sealed and pressurised to 100 psi with oxygen and then placed in a bath held at 99°C. The pressure in the vessel is recorded at prescribed intervals throughout the test. At the end of the specified test time, usually 100, 200 or 500 hours, the pressure drop is calculated and reported. Obviously, this test takes too long to be of much practical value for a lubricant condition monitoring programme.

The more suitable test is ASTM D7527 "Antioxidant Content in Lubricating Greases by Linear Sweep Voltametry". This test measures the amount(s) of oxidation inhibitor(s) added to the base oil as protection against oxidation of the grease. With greases in-service, the test can be used to measure the amount(s) of antioxidant(s) that remain after oxidation has/have reduced its/their concentration. The test method is applicable to mineral oil-based and synthetic oil-based greases, containing all types of thickeners. It is applicable to greases containing at least one type of antioxidant and the presence of other types of additives, such as corrosion inhibitors or metal deactivators, will not interfere with the test method.

A measured quantity of sample is weighed into a vial contained a measured quantity of acetone-based electrolyte and containing a layer of sand. When the vial is shaken, the dissolved antioxidant(s) and other solution-soluble components in the sample are extracted into the solution. The remaining droplets suspended in the solution are agglomerated by the sand. The sand–droplet suspension is allowed to settle out, and the antioxidants dissolved in the solution are quantified by voltammetric analysis.

Another test to assess the oxidation stability of greases is ASTM D8206 "Oxidation Stability of Lubricating Greases – Rapid Small-Scale Oxidation Test (RSSOT)", which measures the net changes in pressure resulting from the consumption of oxygen by oxidation and gain in pressure caused by the formation of volatile oxidation by-products. It determines the resistance of lubricating greases to oxidation when stored statically in an oxygen atmosphere in a sealed system at an elevated temperature under the conditions of the test. It covers the quantitative determination of the oxidation stability of lubricating greases with a dropping point above the test temperature. The test method may be used for quality control to indicate batch-to-batch uniformity. It does not predict either the stability of greases stored in containers for long periods or the stability of films of greases on bearings or machinery components.

A 4.00 ± 0.01 gram (g) of test grease is weighed in a glass dish at a temperature of $20°C \pm 5°C$ and then introduced into a pressure vessel, which is subsequently charged with oxygen to $700 \pm 5 kPa$. The test is initiated by heating the pressure vessel to a temperature of $140°C \pm 0.5°C$ or $160°C \pm 0.5°C$, depending on the selected test. The pressure is recorded continuously until the break point is reached. The induction period as determined under the test conditions can be used as an indication of oxidation stability. However, no correlation has been determined between the results of this test method and service performance, and the test method is probably too long to be of much value in a lubricant condition monitoring programme.

6.3.4 METALS CONTENTS

ASTM D7303 "Metals in Lubricating Greases by ICP-AES" is used specifically to analyse greases. Lubricating greases are composed of about 90% additised oil and soap or other thickening agent. There are over a dozen metallic elements present in grease, either blended as additives for performance enhancements or as thickeners, or in used grease present as contaminants and wear metals. Determination of their concentration can be an important aspect of grease manufacture. The metal content can also indicate the amount of thickeners in the grease.

This is the first industry standard available for simultaneous multi-element analysis of lubricating greases. The metals determined by this method include Al, Sb, Ba, Ca, Fe, Li, Mg, Mo, P, Si, Na, S and Zn. It may also be possible to determine other elements such as Bi, B, Cd, Cr, Cu, Pb, Mn, K, Ti and so on by this test method.

A weighed portion of the grease sample subjected to alternate means of sample dissolution which may include sulphated ashing in a muffle furnace or by closed vessel microwave digestion in acid. Ultimately, these diluted acid solutions are analysed using ICP-AES. Aqueous calibration standards are used. The solutions are introduced to the ICP instrument by free aspiration or an optional peristaltic pump.

Spectral interferences can usually avoided by careful choice of analytical wavelengths. If the grease sample contains refractory additives such as Si or Mo, it is possible that some of these elements may remain undissolved in the residue and may result in lower recoveries. If hydrofluoric acid (HF) is used for dissolution of grease residues, elements such as silicon may be lost as SiF_6. Residual HF can also attack the ICP sample introduction system. HF can be passivated by adding dilute boric acid to the acid solution. If the dry ashing step in sample preparation is used, elements such as sulphur will be volatilised during combustion.

6.4   FUTURE CHEMICAL TESTS FOR OILS AND GREASES

New types of machines, equipment and systems that use lubricants are being developed and introduced all the time. For example, the increasing numbers of electric and hybrid vehicles of all types, including cars, vans, trucks, buses and other vehicles, have required the development of new types of lubricants and cooling fluids. In the future, civil and military aircraft are likely to have to be powered by fuels that have less damaging environmental emissions, such as hydrogen. An increasing number of ships may be powered by natural gas.

New tests have been required to evaluate the properties and performance of the new lubricants and fluids required for the new applications and the impacts they have on the new designs and arrangements of engines, motors, gearboxes, bearings and other components.

Most of these tests are still in development and are not yet standard methods. Some involve chemical properties and performance. Many have been and will be used to develop new and improved lubricants and fluids. Many are likely to involve lengthy testing and evaluation times. As a consequence, they will not be suitable for use in a lubricant condition monitoring programme.

It would therefore seem likely that many of the tests that will be required in the future to monitor the condition of these new types of oils and greases in the new applications are those that have already been described in this chapter. Some new lubricant condition monitoring tests may be required in the future. When they have gained at least some utility, credibility and acceptance, they will be included in future editions of this book.

6.5   SUMMARY

A very large number of tests can be used to assess the chemical properties and performance of lubricating oils and greases. However, not all of these tests are suitable for use in a lubricant condition monitoring programme, either because they take too long to be of practical benefit or they require large volumes of test sample.

While the ideal tests for both new and in-service lubricants provide results relatively quickly (generally within 24 hours) and are comparatively low cost, some of the equipment required to perform the tests can be very expensive. In recent years, a number of the test methods have been automated, thereby reducing test times and manpower costs.

Some of these tests are applicable for lubricants used in a wide range of machines or items of equipment. Other tests are more applicable to specific types of machines or systems. Some tests should only be used to assess new lubricants. Other tests can be used for both new and in-service lubricants.

NOTES

1 Deficiencies of membrane patch colorimetry (MPC) test The Varnish Potential Test, *Machinery Lubrication*, 6, 2014.
2 A, Sasaki, CPA Newsletter, October 2015.
3 See note 1.

7 Physical Tests for Lubricants

7.1 INTRODUCTION

As in the previous chapter on chemical tests for oils and greases, there are many tests for various physical properties of oils and greases. These are presented and discussed below.

However, not all of these tests are suitable for use in a lubricant condition monitoring programme. This chapter will describe and discuss those physical tests that could or should be used to monitor the condition of oils and greases. Chemical tests were the subject of Chapter 6, and later chapters will look at mechanical tests for lubricant condition monitoring.

The international and national organisations that develop, publish, update and monitor these tests were summarised in Chapter 4. A number of original equipment manufacturers (OEMs) also develop, publish and update test methods, some of which can be useful in a lubricant condition monitoring programme. Those physical tests that should (or sometimes could) be used to monitor each type of lubricant application will be discussed in later chapters.

The tests described in this chapter cover only those that can or should be used to monitor the condition of oils and greases in operating machinery. Other physical tests not described in this chapter are used during the formulation and development of new or improved lubricants and/or during the blending of lubricating oils or the manufacture of greases.

7.2 PHYSICAL TESTS FOR OILS

7.2.1 COLOUR

The colour of base oils and lubricating oils, as observed by light transmitted through them, varies from practically clear or transparent to opaque or very dark brown. The methods of measuring colour are based on a visual comparison of the amount of light transmitted through a specified depth of oil with the amount of light transmitted through one of a series of coloured glasses. The colour is then given as a number corresponding to the number of the coloured glass.

Colour variations in base oils result from differences in crude oils, viscosity and method and degree of treatment during refining. During processing, colour is a useful guide to a refiner to indicate whether processes are operating properly. In finished lubricants, colour has little significance except in the case of medicinal and industrial white oils, which are often compounded into, or applied to, products in which staining or discoloration would be undesirable.

DOI: 10.1201/9781003245254-7

The test method of choice is ASTM D1500, IP 196, ISO 2049, DIN 51578 or AFNOR T60-104 "Standard Method of Test for ASTM Colour of Petroleum Products". Using a standard light source, a liquid sample is placed in the test container and compared with coloured glass disks ranging in value from 0.5 to 8.0. When an exact match is not found and the sample colour falls between two standard colours, the higher of the two colours is reported.

The colour of an oil in service may not be a particularly useful test for monitoring its condition. Some lightly coloured oils when new are likely to darken in use, because that is what they are supposed to do. Examples include gasoline and diesel engine oils. These are supposed to keep the insides of engines clean, for optimum energy efficiency, by keeping soot in suspension and removing varnish and oil degradation products form internal surfaces. Other oils, particularly air compressor oils, are likely to darken in service as they gradually degrade thermally. When trend analysis of other chemical or physical tests indicates that a darker colour might confirm unacceptable amounts of thermal degradation, it is probably time to change the oil.

Conversely, a gradual darkening in the colour of a turbine, hydraulic or gear oil may be indicative of accelerating oxidative and/or thermal degradation. In these cases, oil colour could help to add to the information obtained from other chemical and physical tests.

7.2.2 DENSITY

The density of a substance is the mass of a unit volume of it at a standard temperature. The specific gravity (relative density) is the ratio of the mass of a given volume of a material at a standard temperature to the mass of an equal volume of water at the same temperature.

Several methods are used to measure the density of petroleum liquids. One of the first, IP 59, used a "density balance" with which to measure density. The method required buoyancy corrections and temperature corrections, and has been superseded by more accurate test methods.

IP 190 "Density and Relative Density by Capillary-Stoppered Pyknometer Method" gives more accurate results than IP 59. The weights of equal volumes of the test oil and water are compared. Equal volumes are ensured by placing the capillary-stoppered pyknometers in a bath at the test temperature until equilibrium is reached. The standard test temperature is 15°C. Several types of pycnometer can be used. They should be left in the bath for at least 20 minutes or until the liquid level in the pycnometer has stabilised. They are then stoppered, removed from the bath, wiped clean of any water on the outside and weighed.

ASTM D1217 "Standard Test Method for Density and Relative Density (Specific Gravity) of Liquids by Bingham Pycnometer" is widely used. The liquid sample is introduced into a pycnometer equilibrated to the desired temperature and then weighed. The weight of water required to fill the pycnometer at the same temperature is also measured. Both weights are corrected for the buoyancy of air before calculating density and relative density. IP 249, "Density of Liquids by Bingham Pyknometer Method" is technically equivalent to ASTM D1217. Standard test temperatures are 15°C, 20°C or 25°C.

The more recent test method is ASTM D4052 "Standard Test Method for Density, Relative Density, and API Gravity of Liquids by Digital Density Meter". A small volume (approximately 0.7 millilitre (ml)) of liquid sample is introduced into an oscillating sample tube and the change in oscillating frequency caused by the change in the mass of the tube is used in conjunction with calibration data to determine the density of the sample. The IP 365 "Density and Relative Density of Liquids by Digital Density Meter" test is equivalent to ASTM D4052, as are ISO 12185, DIN 51757D and AFNOR T60-172.

The test covers the determination of the density or relative density of petroleum distillates and viscous oils that can be handled in a normal fashion as liquids at test temperatures between 15°C and 35°C. Its application is restricted to liquids with vapour pressures below 600 mm Hg (80 kPa) and viscosities below about 15,000 cSt (mm²/s) at the temperature of test. The test method should not be applied to samples so dark in colour that the absence of air bubbles in the sample cell cannot be established with certainty.

A portable version of this test has been developed. ASTM D7777 "Density, Relative Density, or API Gravity of Liquid Petroleum by Portable Digital Density Meter" is easily calibrated and primarily suitable for field applications. The density, relative density and API gravity of petroleum products are important quality indicators and are used in quantity calculations or to satisfy application, transportation, storage and regulatory requirements. ASTM D7777 should not be used to determine density for custody transfer quantity calculations, particularly in cases in which mass or the weight is the unit of quantity measurement. ASTM D4052 is more appropriate for such applications.

The test method is used at temperatures between 0°C and 40°C. Its applications are restricted to samples with a dry vapour pressure equivalent up to 80 kPa and a viscosity below 100 mm²/s at the test temperature. The test method is suitable for determining the density to nearest 1 kg/m³. The test sample is brought to a temperature close to that of the density meter. If present, the density meter's software program for calculating the density at the reference temperature is selected, and the oscillating U-tube of the density meter is filled with the test portion. After stabilising, the density meter's built-in software calculates the density at the reference temperature. If the density meter does not have such software, the density can be calculated from observed density at the test temperature using the Petroleum Measurement Tables. The IP 559 test method is equivalent to ASTM D7777. Although the portable apparatus was developed for a wide range of petroleum products, it is very suitable for a lubricant condition monitoring programme.

For many years, the US oil industry used API gravity rather than density or relative density. API gravity is a special function of specific gravity (SG), which is related to it by the following equation:

$$\text{API gravity} = \frac{141.5}{\text{SG } 60 / 60°\text{F}} - 131.5$$

The API gravity value, therefore, increases as the specific gravity decreases. Since both density and gravity change with temperature, determinations are made at a controlled temperature and then corrected to a standard temperature by use of special tables.

ASTM D287 "Standard Test Method for API Gravity of Crude Petroleum and Petroleum Products (Hydrometer Method)" is most suitable for determining the API gravity of low-viscosity transparent liquids. The test method can also be used for viscous liquids by allowing sufficient time for the hydrometer to reach temperature equilibrium and for opaque liquids by employing a suitable meniscus correction. Additionally, for both transparent and opaque fluids, the readings need to be corrected for the thermal glass expansion effect before correcting to the reference temperature.

Density determinations are made quickly and easily. Because products of a given crude oil, having definite boiling ranges and viscosities, will fall into definite ranges, this property is widely used for control in refinery operations. It is also useful for identifying oils, provided the distillation range or viscosity of the oils is known. The primary use of API gravity, however, is to convert weighed quantities to volume and measured volumes to weight.

The density of an in-service oil is not a particularly useful guide in a condition monitoring programme, simply because trend analyses of other test results are likely to indicate oil deterioration before changes in density show it.

7.2.3 KINEMATIC AND DYNAMIC VISCOSITY

Probably the most important single property of a lubricating oil is its viscosity. A factor in the formation of lubricating films under both thick and thin film conditions, viscosity affects heat generation in bearings, cylinders and gears. It also determines the ease with which machines may be started under cold conditions, and it governs the sealing effect of the oil and the rate of consumption or loss. For any piece of equipment, the first essential for satisfactory results is to use an oil of proper viscosity to meet the operating conditions.

The basic concept of viscosity can be illustrated by moving a plate at a uniform speed over a film of oil. The oil adheres to both the moving surface and the stationary surface. Oil in contact with the moving surface travels with the same velocity (U) as that surface, while oil in contact with the stationary surface has zero velocity. In between, the oil film may be visualised as being made up of many layers, each drawn by the layer above it at a fraction of velocity U that is proportional to its distance above the stationary plate. A force F must be applied to the moving plate to overcome the friction between the fluid layers. Since this friction is the result of viscosity, the force is proportional to viscosity. Viscosity can be determined by measuring the force required to overcome fluid friction in a film of known dimensions. Viscosity determined in this way is called dynamic or absolute viscosity.

Dynamic viscosities are usually reported in poise (P) or centipoise (cP; 1 cP = 0.01 P), or in SI units in pascal-seconds (Pa s; 1 Pa s = 10 P). Dynamic viscosity, which is a function only of the internal friction of a fluid, is the quantity used most frequently in bearing design and oil flow calculations. Because it is more convenient to measure viscosity in a manner such that the measurement is affected by the density of the oil, kinematic viscosities normally are used to characterise lubricants.

The kinematic viscosity of a fluid is the quotient of its dynamic viscosity divided by its density, both measured at the same temperature and in consistent units. The most common units for reporting kinematic viscosities are stokes (St) or centistokes (cSt; 1 cSt = 0.01 St), or in SI units, square millimetres per second (mm²/s; 1 mm²/s = 1 cSt). Dynamic viscosities, in centipoise, can be converted to kinematic viscosities, in centistokes, by dividing by the density in grams per cubic centimetre (g/cm³) at the same temperature. Kinematic viscosities, in centistokes, can be converted to dynamic viscosities, in centipoise, by multiplying by the density in grams per cubic centimetre. Kinematic viscosities, in square millimetres per second (mm²/s), can be converted to dynamic viscosities, in pascal-seconds, by multiplying by the density, in grams per cubic centimetre, and dividing the result by 1,000.

Other viscosity systems, including those of Saybolt, Redwood and Engler, are in use and will probably continue to serve for many years because of their familiarity to some people. However, the instruments developed to measure viscosities in these systems are rarely used. Most actual viscosity determinations are made in centistokes and converted to values in the other systems by means of published SI conversion tables.

The viscosity of any fluid changes with temperature, decreasing as the temperature is increased and increasing as the temperature is decreased. It is therefore necessary to have some method of determining the viscosities of lubricating oils at temperatures other than those at which they are measured. This is usually accomplished by measuring the viscosity at two temperatures, then plotting these points on special viscosity–temperature charts developed by ASTM. A straight line can then be drawn through the points and viscosities at other temperatures read from it with reasonable accuracy. The line should not be extended below the pour point or above approximately 150°C (for most lubricating oils) since in these regions it may no longer be straight.

ASTM D445 "Standard Test Method for Kinematic Viscosity of Transparent and Opaque Liquids (and Calculation of Dynamic Viscosity)" and IP 71, ISO 3104, DIN 51562 and AFNOR T60-100 are equivalent test methods.

The time is measured in seconds for a fixed volume of liquid to flow under gravity through the capillary of a calibrated viscometer under a reproducible driving head and at a closely controlled temperature. The kinematic viscosity is the product of the measured flow time and the calibration constant of the viscometer. The two temperatures most used for measuring and reporting kinematic viscosities are 40°C (104°F) and 100°C (212°F).

Dynamic viscosity can be calculated from kinematic viscosity using the equation:

$$\eta = \rho v$$

where η is dynamic viscosity in centipoises, ρ is density in gm/cc at the same temperature used for measuring the flow time and v is the kinematic viscosity in centistokes.

Dynamic viscosity can also be measured using ASTM D7483 "Standard Test Method for Determination of Dynamic Viscosity and Derived Kinematic Viscosity of Liquids by Oscillating Piston Viscometer". This viscometer, sometimes referred to as an electromagnetic viscometer, or EMV viscometer, was invented at Cambridge

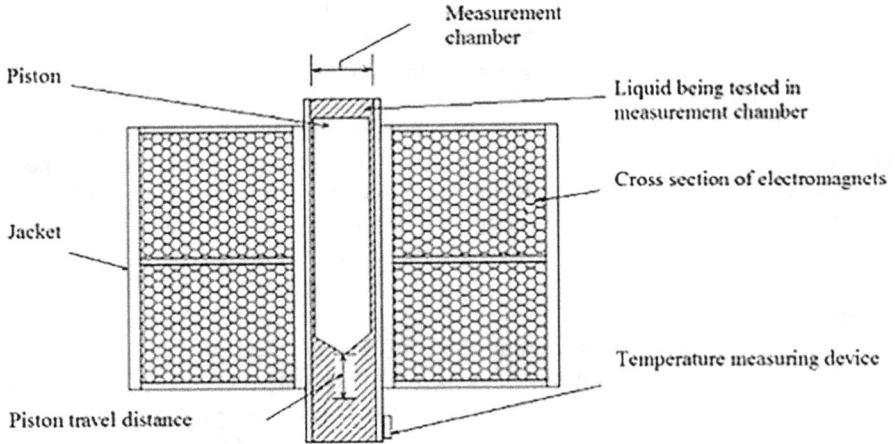

FIGURE 7.1   Oscillating piston viscometer.

Source: Pathmaster Marketing Ltd.

Viscosity (Formally Cambridge Applied Systems) in 1986. A schematic diagram of the viscometer is shown in Figure 7.1. The sensor comprises a measurement chamber and magnetically driven piston. A small sample is first introduced into the thermally controlled measurement chamber where the piston resides. Electronics drive the piston into oscillatory motion within the measurement chamber with a controlled magnetic field. A shear stress is imposed on the liquid (or gas) due to the piston travel, and the viscosity is determined by measuring the travel time of the piston. The construction parameters for the annular spacing between the piston and measurement chamber, the strength of the electromagnetic field and the travel distance of the piston are used to calculate the dynamic viscosity according to Newton's Law of Viscosity.

Oscillating piston viscometers allow viscosity measurement of a broad range of materials including transparent, translucent and opaque liquids. The measurement principle and stainless steel construction makes the oscillating piston viscometer resistant to damage and suitable for portable operations. The measurement itself is automatic and does not require an operator to time the oscillation of the piston. The electromagnetically driven piston mixes the sample while under test. The instrument requires a sample volume of less than 5 ml and typical solvent volume of less than 10 ml, which minimises clean-up effort and waste.

The test is applicable to Newtonian and non-Newtonian liquids. The range of dynamic viscosity covered by the test is from 0.2 to 20,000 mPa s (which is approximately the kinematic viscosity range of 0.2 to 22,000 mm²/s for new oils) in the temperature range between −40°C and 190°C. However, the precision has been determined only for new and used oils in the range of 34 to 1,150 mPa s at 40°C, 5.7 to 131 mPa s at 100°C and 46.5 to 436 mm²/s at 40°C.

The oscillating piston viscometer technology has been adapted for small sample viscosity and micro-sample viscosity testing in laboratory applications. It has also been adapted to measure high pressure viscosity and high-temperature viscosity measurements in both laboratory and process environments. The viscosity sensors have been scaled for a wide range of industrial applications such as small-size viscometers for use in compressors and engines, flow-through viscometers for dip coating processes, in-line viscometers for use in refineries and hundreds of other applications.

ASTM D8092 "Field Determination of Kinematic Viscosity Using A Microchannel Viscometer" provides a reliable field test method to determine kinematic viscosity of transparent and opaque liquids (such as new and in-service lubricating oils) at 40°C without requiring solvents or chemicals for cleaning. The analysis requires only 60 µl of sample for testing. It uses a miniature microchannel viscometer and has a range of 12.9 to 174 mm²/s viscosity.

The liquid sample is placed into a loading funnel of the miniature capillary viscometer and kinematic viscosity at 40°C is determined by measuring the time in seconds it takes the liquid to travel between the beam produced by two light-emitting diodes (LEDs), LED #1 and LED #3. These times associated with the liquid passing each LED are determined by monitoring the voltage of the corresponding photodiodes. The liquid sample thermalises rapidly to 40°C as the entire unbounded microchannel capillary is constructed of aluminium, which is stabilised at 40°C. Consequently, when the oil sample is placed into the microchannel capillary, it immediately comes into contact with the aluminium and the liquid flowing out of the loading funnel thermalises to 40°C within tenths of a second. The presence of large particles in the oil that would tend to get stuck in the capillary and impede the sample flow may interfere with the measurement. Ways to overcome this interference are suggested in the standard.

Another test applicable to a lubricant condition monitoring programme is ASTM D7279 "Kinematic Viscosity of Transparent and Opaque Liquids by Automated Houillon Viscometer", which is based on the standard laboratory test for kinematic viscosity, ASTM D445. The Houillon viscometer tube method offers automated determination of kinematic viscosity. A typically a sample volume of less than 1 ml is required for the analysis. The method is applicable to both fresh and used lubricating oils. The range of kinematic viscosity covered by the test method is from 0.2 to 1,000 mm²/s in the temperature range between 20°C and 150°C.

The kinematic viscosity is determined by measuring the time taken for a sample to fill a calibrated volume at a given temperature. The test oil is injected into the apparatus and then flows into the viscometer tube that is equipped with two detection cells. The oil reaches the test temperature of the viscometer bath and when the leading edge of the specimen passes in front of the first detection cell, the automated instrument starts the timing sequence. When the leading edge of the oil passes in front of the second detection cell, the instrument stops timing the flow. The measured time interval allows the calculation of the kinematic viscosity using a viscometer constant determined earlier by calibration with certified viscosity reference standards.

ASTM provides guidelines in ASTM D8185 "Standard Guide for In-Service Lubricant Viscosity Measurement", to assist developers and users of lubricant condition monitoring programmes in selecting the most appropriate tests to measure the viscosities of oils and obtain information for in-service oil analysis. The guide explains basic viscosity measurements and their significance and discusses each of the currently available ASTM tests for kinematic viscosity, dynamic viscosity, apparent viscosity, high-temperature viscosity and low-temperature viscosity.

7.2.4 Apparent Viscosity

Apparent viscosity is the shear stress applied to a fluid divided by the shear rate. For a Newtonian fluid, the apparent viscosity is constant, and equal to the Newtonian viscosity of the fluid, but for non-Newtonian fluids, the apparent viscosity depends on the shear rate. Apparent viscosity has the SI derived unit Pa s (Pascal-second), but the centipoise (cP) is frequently used in practice.

A single viscosity measurement at a constant speed in a typical viscometer is a measurement of the apparent viscosity of a fluid. In the case of non-Newtonian fluids, measurement of apparent viscosity without knowledge of the shear rate is of limited value: the measurement cannot be compared with other measurements if the speed and geometry of the two instruments are not identical. An apparent viscosity that is reported without the shear rate or information about the instrument and settings, for example, speed and spindle type for a rotational viscometer, is meaningless.

Multiple measurements of apparent viscosity at different, well-defined shear rates can give useful information about the non-Newtonian behaviour of a fluid and allow it to be modelled.

Two methods are used to assess the apparent viscosity of lubricating oils. ASTM D4683 "Standard Test Method for Measuring Viscosity of New and Used Engine Oils at High Shear Rate and High Temperature by Tapered Bearing Simulator Viscometer at 150°C" covers the laboratory determination of the viscosity of engine oils at 150°C and $1 \times 10^6 \mathrm{s}^{-1}$ shear rate using a tapered bearing simulator (TBS) viscometer equipped with a refined thermoregulator system.

The Newtonian calibration oils used to establish this test method cover the range from approximately 1.5 to 5.6 cP (mPa·s) at 150°C. The non-Newtonian reference oil used to establish this test method has a viscosity of approximately 3.5 cP (mPa·s) at 150°C and a shear rate of $1 \times 10^6 \mathrm{s}^{-1}$. Applicability to petroleum products other than engine oils was not determined in preparing the test method.

A motor drives a tapered rotor that is closely fitted inside a matched stator. The rotor exhibits a reactive torque response when it encounters a viscous resistance from an oil that fills the gap between the rotor and stator and the value of this torque response is used to determine the apparent viscosity of the test oil. Two oils, a calibration oil and non-Newtonian reference oil, are used to determine the gap distance between the rotor and stator so that a shear rate of $1 \times 10^6 \mathrm{s}^{-1}$ is maintained. Additional calibration oils are used to establish the viscosity/torque relationship that is required for the determination of the apparent viscosity of test oils at 150°C.

The same method is used in ASTM D6616 "Viscosity at High Shear Rate by Tapered Bearing Simulator Viscometer At 100°C". Viscosity at the shear rate and temperature of this test is thought to be particularly representative of bearing conditions in large, medium-speed reciprocating engines and automotive and heavy-duty engines operating in this temperature regime. The importance of viscosity under these conditions has been stressed in railroad specifications. The test method covers the laboratory determination of the viscosity of engine oils at 100°C and $1 \times 10^6 s^{-1}$ using a tapered bearing simulator (TBS) viscometer. The Newtonian calibration oils used to establish this test method range from approximately 5 to 12 mPa s at 100°C and either the manual or automated protocol was used to develop the test precision. The viscosity range of the test method at this temperature is from 1 mPa s to more than 25 mPa s, depending on the model of TBS used. The non-Newtonian reference oil used to establish the shear rate of $1 \times 10^6 s^{-1}$ for the test method has a viscosity of approximately 10 mPa s at 100°C.

ASTM D4741 "Standard Test Method for Measuring Viscosity at High Temperature and High Shear Rate by Tapered-Plug Viscometer" covers the laboratory determination of the viscosity of oils at 150°C and $1 \times 10^6 s^{-1}$ using Ravenfield high shear rate tapered-plug viscometer, models BE or BS. The test method may readily be adapted to other conditions if required.

The test lubricant fills the annular space between the tapered plug and matching stator. The rotor is spun at a fixed speed and the lubricant's dynamic viscosity determined from a measurement of the applied torque and a calibration line prepared using Newtonian calibration oils. The shear rate is adjusted by varying the penetration (and hence the clearance) of the rotor within the stator.

Newtonian calibration oils are used to adjust the working gap and for calibration of the apparatus. These calibration oils cover a range from approximately 1.8 to 5.9 cP (mPa·s) at 150°C. The test should not be used for extrapolation to higher viscosities than those of the Newtonian calibration oils used to calibrate the apparatus. IP 370 and CEC L-36-90 test methods are technically equivalent to ASTM D4741.

ASTM D7042 "Dynamic Viscosity and Density of Liquids by Stabinger Viscometer (and the Calculation of Kinematic Viscosity)" is another test that might be used for lubricating oils. Density and viscosity are important properties of petroleum and non-petroleum liquids. The test method provides concurrent measurements of both dynamic viscosity and density of liquid petroleum products and crude oils. The kinematic viscosity can be obtained by dividing the dynamic viscosity by density obtained at the same temperature. This test method is intended for Newtonian liquids. Although precision only in a specific range has been determined, the method should be applicable to a wider range of products.

The test liquid is introduced into the measuring cells, which are at a closely controlled and known temperature. The measuring cell consists of a pair of rotating concentric cylinders and an oscillating U-tube. The dynamic viscosity is determined from the equilibrium rotational speed of the inner cylinder under the influence of the shear stress of the test liquid and an eddy current brake in conjunction with adjustment data. The density is determined by oscillation frequency of the U-tube in conjunction with adjustment data. The kinematic viscosity is calculated by dividing the dynamic viscosity by the density. Unfortunately, because the test can only be used for liquids that exhibit Newtonian behaviour, it cannot be used for either new or in-service lubricants that contain polymeric viscosity index (VI) improving polymers, such as multigrade engine oils and many high-viscosity-index hydraulic and gear oils.

7.2.5 Low-Temperature Viscosity

Rotational viscometers measure dynamic viscosity, using the viscous drag on a rotor immersed in oil and measuring the torque required to rotate it at a given speed or the speed achieved for a given torque. Important examples for base oils and automotive engine oils include the Brookfield Viscometer, the Cold-Cranking Simulator (CCS), the Mini-Rotary Viscometer (MRV), all for low-temperature use. (The TBS is used for measuring high shear dynamic viscosity at 150°C, applicable only to engine oils.)

The Brookfield Viscometer is a rapid, direct-reading instrument that uses an assortment of rotary spindles and variable speed settings (very low to medium) to provide a wide range of measurements. The appropriate spindle is immersed in the test liquid, the speed of rotation is selected and the torque reading is a measure of the apparent viscosity.

ASTM D2983, IP 267, ISO 9262 and AFNOR T42-011 "Standard Test Method for Low-Temperature Viscosity of Automatic Transmission Fluids, Hydraulic Fluids, and Lubricants using a Rotational Viscometer" describes the use of the Brookfield viscometer for the determination of the low-shear-rate viscosity of automotive fluid lubricants in the temperature range from −5°C to −40°C. The viscosity range is 500 to 1,000,000 cP (mPa·s). Brookfield viscosity is expressed in centipoises (1 cP = 1 mPa·s). Its value may vary with the spindle speed (shear rate) of the Brookfield viscometer because many automotive fluid lubricants are non-Newtonian at low temperatures.

The test fluid sample is cooled in an air bath at test temperature for 16 hours. It is carried in an insulated container to a nearby Brookfield viscometer where its Brookfield viscosity is measured at the specified test temperature.

Between 1957 and 1967, a viscometer with a very specific use was developed. This was the cold-cranking simulator (CCS) whose purpose was to define the winter grades of motor oil in a way that correlated with engine crankability, rather than by use of the VI system. It uses a non-cylindrical rotor enclosed closely by a stator and was originally operated at a single temperature, namely 0°F or −18°C.

Recent modifications to the viscosity classification system now require test temperatures ranging from −5°C for 25W-XX oils to −30°C for 0W-XX viscosity oils. The method is given in ASTM D5293 "Standard Test Method for Apparent Viscosity of Engine Oils and Base Stocks Between −10°C and −35°C Using a Cold-Cranking Simulator". An electric motor drives a rotor that is closely fitted inside a stator. The space between the rotor and stator is filled with oil. Test temperature is measured near the stator inner wall and maintained by regulated flow of refrigerated coolant through the stator. The speed of the rotor is calibrated as a function of viscosity. Test oil viscosity is determined from this calibration and the measured rotor speed.

The test covers the laboratory determination of apparent viscosity of engine oils at temperatures between −5°C and −35°C, at shear stresses of approximately 50,000 to 100,000 Pa and shear rates of approximately 10^5 to $10^4 s^{-1}$ and viscosities of approximately 500 to 10,000 mPa·s. The results are related to engine-cranking characteristics of engine oils.

In the 1980s, problems of oil pumpability in cases where cranking was satisfactory led to further modifications of the viscosity classification and the adoption of new low-temperature viscometers. Low-temperature, low-shear viscosity is important for predicting the possibility of "air binding" in motor oils after vehicles have stood at low temperatures for a considerable period. The non-Newtonian motor oil can gel to a semi-solid and fail to flow to the oil pump inlet when the engine is started. The oil pump then pumps air instead of oil to the engine and both the pump and other engine parts can be rapidly damaged. Even if "air binding" does not take place, an oil can be so viscous after standing at low temperatures that the rate of pumping oil to sensitive bearings and rockers may be inadequate, and again engine damage can result. The Scanning Brookfield method ASTM D5133 "Standard Test Method for Low Temperature, Low Shear Rate, Viscosity/Temperature Dependence of Lubricating Oils Using a Temperature-Scanning Technique" is believed to correlate with these problems, and it is recommended that this test is performed on new oil formulations. The low-temperature, low-shear viscometric behaviour of an engine oil determines whether the oil will flow to the sump inlet screen, then to the oil pump, then to the sites in the engine requiring lubrication in sufficient quantity to prevent engine damage immediately or ultimately after cold temperature starting. Two forms of flow problems have been identified, flow-limited and air-binding behaviour. The first form of flow restriction, flow-limited behaviour, is associated with the oil's viscosity; the second, air-binding behaviour, is associated with gelation. The temperature-scanning technique employed by the test method was designed to determine the susceptibility of the engine oil to flow-limited and air-binding response to slow cooling conditions by providing continuous information on the rheological condition of the oil over the temperature range of use. In this way, both viscometric and gelation responses are obtained in one test. However, the test is time-consuming and does not readily permit tests on large numbers of samples, so is less suited in an oil condition monitoring programme than other tests.

For base oils, low-temperature flow properties are a better guide as to their suitability for use in automotive engine oils, automatic transmission fluids and some gear oils and hydraulic oils than pour point.

ASTM D3829 "Standard Test Method for Predicting the Borderline Pumping Temperature of Engine Oil" measures the lowest temperature at which an engine oil can be continuously and adequately supplied to the oil pump inlet of an automotive engine. The test method covers the prediction of the borderline pumping temperature (BPT) of engine oils through the use of a 16 hour cooling cycle over the temperature range from 0°C to −40°C. Applicability to petroleum products other than engine oils has not been determined.

An engine oil sample is cooled from 80°C to the desired test temperature at a nonlinear programmed cooling rate over a 10 hour period and held at the test temperature for the remainder of a 16 hour period. After completion of the soak period, two standard torques of increasing severity are applied to the rotor shaft and the speed of rotation in each case is measured. From the results at three or more temperatures, the borderline pumping temperature is determined.

Alternatively, for some specification or classification purposes, it may be sufficient to determine that the BPT is less than a certain specified temperature.

In 1999, the ASTM formed a task force to address potential test method and measurement issues for low-temperature rheological determinations of used engine oils. With anticipated soot loadings of 5% to 10% in some of these used oils, there was concern that standard ASTM test methods developed for fresh oil pumpability might not be suitable for these sooted oils. The task force conducted some preliminary work on a used Mack T8E test sample of approximately 5% soot loading. These data indicated that variation in preheating conditions could have significant influence on the low-temperature properties measured by standard procedures. Additional rheometric studies showed that quiescent heating was probably building structure in the oil due to soot agglomerisation. To address these concerns, a modified MRV procedure was developed in which the sample is heated externally, then agitated to remove any soot-agglomerised structure and then cooled via the "TP-1" cooling portion of the standard ASTM D4684 test procedure.

The result was the development of ASTM D7110 "Viscosity-Temperature Relationship of Used and Soot-Containing Engine Oils at Low Temperatures". The low-temperature, low-shear viscometric behaviour of an engine oil, whether new, used or sooted, determines whether the oil will flow to the sump inlet screen, then to the oil pump and then to the sites in the engine requiring lubrication in sufficient quantity to prevent engine damage immediately or ultimately after cold temperature starting. Two forms of flow problems were identified. Flow-limited behaviour is associated with the oil's viscosity, and air-binding behaviour is associated with gelation.

A specially made glass stator/metal rotor cell is attached to the viscometer and subjected to a programmed temperature change for both calibration and sample analysis. Following calibration of the rotor/stator set, an approximately 20 ml test sample of a lubricating test oil is poured into the stator and preheated for 1.5 to 2.0 hours in an oven or water bath. Shortly after completing the preheating step, the room temperature rotor is put into the stator containing the heated oil and coupled to a torque sensing viscometer head using an adapter to automatically centre the rotor in the stator during test. A programmable low-temperature bath is used to cool the cell at a specified rate of 3°C per hour from −5°C to the temperature at which the maximum torque recordable is exceeded when using a speed of 0.3 rpm for the rotor. After the desired information has been collected, the computer program generates the desired viscometric and rheological values from the recorded data.

In the ASTM D4684 "Standard Test Method for Determination of Yield Stress and Apparent Viscosity of Engine Oils at Low Temperature", a fresh engine oil is slowly cooled through a temperature range where wax crystallisation is known to occur, followed by relatively rapid cooling to the final test temperature. Laboratory test results have predicted as failures the known engine oils that have failed in the field because of lack of oil pumpability. The documented field failing oils all consisted of oils normally tested at −25°C. These field failures are believed to be the result of the oil forming a gel structure that results in either excessive yield stress or viscosity of the engine oil or both.

An engine oil sample is held at 80°C and then cooled at a programmed cooling rate to a final test temperature. A low torque is applied to the rotor shaft of the mini-rotary viscometer (MRV) test equipment to measure the yield stress. A higher torque is then applied to determine the apparent viscosity of the sample.

For oils to be tested at −20°C or colder, the cooling profile described in the test method is based on the viscosity properties of the ASTM Pumpability Reference Oils (PRO). This series of oils includes oils with normal low-temperature flow properties and oils that have been associated with low-temperature pumpability problems. Significance for the −35°C and −40°C temperature profiles is based on the data collected from the "Cold Starting and Pumpability Studies in Modern Engines" conducted by the ASTM.

For oils to be tested at −15°C or −10°C, a different cooling regime is used. No significance has been determined for this temperature profile because of the absence of appropriate reference oils. Similarly, precision of the test method using this profile for the −10°C test temperature is unknown.

ASTM D4684 contains two procedures. Procedure A incorporates several equipment and procedural modifications from the earlier version of ASTM D4684 issued in 2002 that have shown to improve the precision of the test. Procedure B is unchanged from ASTM D4684-02. Additionally, Procedure A applies to those instruments that use thermoelectric cooling technology or direct refrigeration technology of recent manufacture for instrument temperature control. Procedure B can use the same instruments used in Procedure A or those cooled by circulating methanol.

The test method is applicable for unused oils, sometimes referred to as fresh oils, designed for both light-duty and heavy-duty engine applications. It has also been shown to be suitable for in-service diesel and gasoline engine oils, so can be used in an oil condition monitoring programme. The applicability to petroleum products other than engine oils has not been determined.

A more suitable low-temperature test designed for in-service diesel engine oils is the newer ASTM D6896 "Yield Stress and Apparent Viscosity of Used Engine Oils at Low Temperature". When an engine oil is cooled, the rate and duration of cooling can affect its yield stress and viscosity. In ASTM D6896, used engine oil is slowly cooled through a temperature range at which wax crystallisation is known to occur, followed by relatively rapid cooling to the final test temperature. As in other low-temperature rheological tests, such as ASTM D3829, D4684 and D5133, a preheating condition is required so that all residual waxes are solubilised in the oil before cooling. This removes any thermal memory. However, it has also been found that diesel engine oils which contain high levels of soot can experience a soot agglomerisation phenomenon when heated under quiescent conditions. ASTM D6896 uses a separate preheat and agitation step to break up any soot agglomerisation that may have occurred before cooling. The viscosity of highly sooted diesel engine oils as measured in the test method has been correlated to pressurisation times in a motored engine test. This test method covers the measurement of the yield stress and viscosity of engine oils after cooling at controlled rates over a period of 43 or 45 hours to a final test temperature of −20°C or −25°C. The viscosity measurements are made at a shear stress of 525 Pa over a shear rate of 0.4 to $15\,s^{-1}$. The test method is suitable for measurement of viscosities ranging from 400 to >400,000 mPa·s and is suitable for yield stress measurements of 7 to >350 Pa.

In the test, a used engine oil sample is heated at 80°C and then is vigorously agitated. The sample is then cooled at a programmed cooling rate to a final test temperature. A low torque is applied to the rotor shaft to measure the yield stress. A higher torque is then applied to determine the apparent viscosity of the sample. However, the test may be too long to be of much value in a lubricant condition monitoring programme if quick results are required in specific circumstances.

7.2.6   VISCOSITY INDEX

Different oils have different rates of change of viscosity with temperature. For example, a distillate oil from a naphthenic crude oil would show a greater rate of change of viscosity with temperature than would a distillate oil from a paraffinic crude oil. The viscosity index (VI) is a method of applying a numerical value to this rate of change, based on comparison with the relative rates of change of two arbitrarily selected types of oil that differ widely in this characteristic. A high VI indicates a relatively low rate of change of viscosity with temperature; a low VI indicates a relatively high rate of change of viscosity with temperature. For example, consider a high VI oil and a low VI oil having the same viscosity at, say, room temperature: as the temperature increased, the high VI oil would thin out less and, therefore, would have a higher viscosity than the low VI oil at higher temperatures.

The VI of an oil is calculated from kinematic viscosities determined at two temperatures by means of tables published by the ASTM, IP, ISO and AFNOR. Tables based on kinematic viscosities determined at both 40°C and 100°C and 100°F and 212°F are available.

Mineral oil base oils refined through special hydroprocessing techniques can have VIs well above 100. Some synthetic lubricating oils have VIs both below and above this range.

VI calculations are done using ASTM D2270, IP 226, ISO 2909 or AFNOR T60-136, all of which are equivalent. An illustrative graph of oils having three different viscosity indices is shown in Figure 7.2.

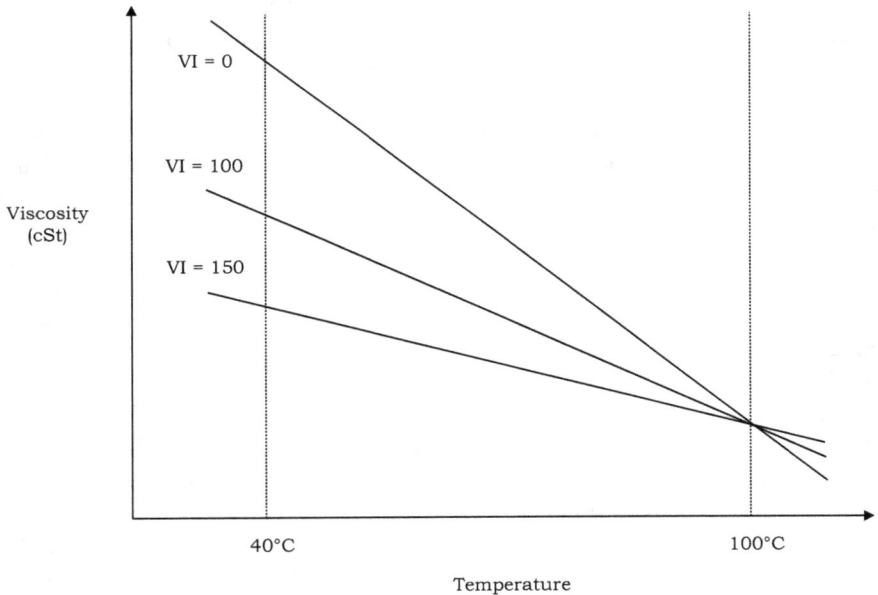

FIGURE 7.2   Illustrative graph of viscosity index.

Source: Pathmaster Marketing Ltd.

7.2.7   POUR POINT

The pour point of a lubricating oil is the lowest temperature at which it will pour or flow when it is chilled, without disturbance, under prescribed conditions. Most base oils contain some dissolved wax, and as an oil is chilled, this wax begins to separate as crystals that interlock to form a rigid structure that traps the oil in small pockets in the structure. When this wax crystal structure becomes sufficiently complete, the oil will no longer flow under the conditions of the test. Since, however, mechanical agitation can break up the wax structure, it is possible to have an oil flow at temperatures considerably below its pour point. Cooling rates also affect wax crystallisation; it is possible to cool an oil rapidly to a temperature below its pour point and still have it flow.

While the pour point of most oils is related to the crystallisation of wax, certain oils, which are essentially wax free, have viscosity-limited pour points. In these oils the viscosity becomes progressively higher as the temperature is lowered until at some temperature no flow can be observed. The pour points of such oils cannot be lowered with pour point depressants, since these agents act by interfering with the growth and interlocking of the wax crystal structure.

The first standard test was ASTM D97 "Standard Test Method for Pour Point of Petroleum Products", now adopted as IP 15, ISO 3016, DIN 51597 and AFNOR T60-105. After preliminary heating, the sample is cooled at a specified rate and examined at intervals of 3°C for flow characteristics. The lowest temperature at which the movement of the oil is observed is recorded as the pour point.

Three new pour point test methods are now available for Herzog, IT Phase Technology and ATPEM automatic machines. The three methods show somewhat different precisions, but show no real bias against ASTM D97, which is to be considered the referee method in case of a dispute.

With ASTM D5949 "Standard Test Method for Pour Point of Petroleum Products (Automatic Pressure Pulsing Method)", after inserting the test oil into the automatic pour point apparatus, and initiation of the test program, the oil is heated and then cooled by a Peltier device at a rate of 1.5 ± 0.1°C/min. At temperature intervals of 1°C, 2°C or 3°C, depending on the selection made by the user, a moving force in the form of a pressurised pulse of nitrogen gas is imparted onto the surface of the oil. Multiple optical detectors are used in conjunction with a light source to monitor movement of the surface of the oil. The lowest temperature at which movement of the oil surface is observed upon application of a pulse of nitrogen gas is recorded as the pour point by ASTM D5949.

In the ASTM D5895 "Standard Test Method for Pour Point of Petroleum Products (Rotational Method)" test, after inserting the test oil into the automatic pour point apparatus, and initiation of the program, the oil is heated and then cooled by maintaining a constant temperature differential between a cooling block and the sample. The oil is continuously tested for flow characteristics by rotating the test cup at approximately 0.1 rpm against a stationary, counterbalanced, sphere-shaped pendulum. The temperature of the oil at which a crystal structure or a viscosity increases, or both, within the oil causes the displacement of the pendulum and is recorded with a resolution of 0.1°C. The oil is then heated to the original starting temperature.

With ASTM D6749 "Standard Test Method for Pour Point of Petroleum Products (Automatic Air Pressure Method)", after inserting the test jar containing the test oil into the automatic pour point apparatus and initiating the test program, the oil is automatically heated to the designated temperature and then cooled at a controlled rate. At temperature intervals of 1 or 3°C, depending on the selection made by the user prior to the test, a slightly positive air pressure is gently applied onto the surface of the oil which is contained in an air-tight test jar equipped with a communicating tube. Since one end of the communicating tube is inserted into the oil, while the other end is maintained at atmospheric pressure, a small amount of downward movement or deformation of the oil surface, as a result of the application of air pressure, is observed by means of upward movement of the oil in the communicating tube. This upward movement of the oil is detected by a pressure sensor which is installed at the atmospheric end of the communicating tube. The lowest temperature at which deformation of the oil is observed upon application of air pressure is recorded as the pour point.

In the ASTM D6892 "Standard Test Method for Pour Point of Petroleum Products (Robotic Tilt Method)" test, after insertion of the oil into the automatic pour point apparatus and initiation of the testing program, the oil is heated and then cooled according to a prescribed profile. The oil surface is examined periodically for movement using an optical camera system mounted on top of the test jar, while tilting the test jar. The test jar is removed from the jacketed cooling chamber prior to each examination. The lowest temperature, when movement of the surface of the oil is detected, is recorded as the pour point.

When testing at 3°C intervals, relative bias among certain samples was observed versus results from ASTM D97. The relative bias was not a fixed value but appeared to be a linear function of the pour point value. When measuring at 1°C intervals, a bias of 1.1°C on the average was observed against ASTM D97.

7.2.8 FLASH POINT

The flash point of an oil is the temperature at which the oil releases enough vapour at its surface to ignite when an open flame is applied. For example, if a base oil is heated in an open container, ignitable vapours are released in increasing quantities as the temperature rises. When the concentration of vapours at the surface becomes great enough, exposure to an open flame will result in a brief flash as the vapours ignite. When a test of this type is conducted under certain specified conditions, as in the Cleveland Open Cup (COC) method, the bulk oil temperature at which this happens is reported as the flash point.

The release of vapours at this temperature is not sufficiently rapid to sustain combustion, so the flame immediately dies out. However, if heating is continued, a temperature will be reached at which vapours are released rapidly enough to support combustion. This temperature is called the fire point.

The flash point of new oils varies with viscosity; higher-viscosity oils have high flash points. Flash points are also affected by the type of crude oil and by the refining process. For example, naphthenic oils generally have lower flash points than paraffinic oils with similar viscosity. Flash point tests are of value to refiners for control purposes and are significant to consumers under certain circumstances for safety considerations.

ASTM D92 "Standard Test Method for Flash and Fire Points by Cleveland Open Cup Tester" covers the determination of the flash and fire points of all petroleum products except fuel oils and those having an open-cup flash point above 79°C and below 400°C.

The test cup is filled to a specified level with sample. The temperature is rapidly increased at first and then at a slow constant rate as the flash point is approached. At specified intervals a small test flame is passed across the cup. The lowest temperature at which the vapours above the surface of the liquid ignite is taken as the flash point. The test is continued until the application of the test flame causes the oil to ignite and burn for at least 5 seconds. This temperature is the fire point.

IP 36, ISO 2592, DIN 51376 and AFNOR T60-118 are equivalent test methods to ASTM D92.

ASTM D93 "Standard Test Methods for Flash Point by Pensky-Martens Closed Cup Tester" is used for the analysis of fuel oils, lubricating oils, suspension of solids, liquids that tend to form a surface film under test conditions and other liquids.

The sample is heated at a slow, constant rate with continual stirring. A small flame is directed into the test cup (which, unlike the test cup used in ASTM D92, has a cover) at regular intervals with simultaneous interruption of stirring. The flash point is the lowest temperature at which the vapour above the sample ignites.

Three procedures are allowed, which differ in their heating rate and the stirrer speed of the sample. These tests cover the determination of the flash point of petroleum products in the temperature range from 40°C to 370°C by a manual Pensky-Martens closed cup (PMC) apparatus or an automated PMC apparatus, and the determination of the flash point of biodiesel in the temperature range of 60°C to 190°C by an automated PMC apparatus.

Procedure A is applicable to distillate fuels (diesel, biodiesel blends, kerosine, heating oil and turbine fuels), new and in-use lubricating oils and other homogeneous petroleum liquids not included in the scope of Procedure B or C.

Procedure B is applicable to residual fuel oils, cutback residua, used lubricating oils, mixtures of petroleum liquids with solids, petroleum liquids that tend to form a surface film under test conditions or are petroleum liquids of such kinematic viscosity that they are not uniformly heated under the stirring and heating conditions of Procedure A.

Procedure C is applicable to biodiesel (B100). Since a flash point of residual alcohol in biodiesel is difficult to observe by manual flash point techniques, automated apparatus with electronic flash point detection has been found to be suitable.

Flash point determinations above 250°C can be performed; however, the precision has not been determined above this temperature. For residual fuels, precision has not been determined for flash points above 100°C. The precision of in-use lubricating oils has not been determined. Some specifications state an ASTM D93 minimum flash point below 40°C, but the precision has not been determined below this temperature.

IP 34, ISO 2719, DIN 51758 and AFNOR M07-019 are equivalent test methods to ASTM D93.

ASTM D6450 "Standard Test Method for Flash Point by Continuously Closed Cup (CCCFP) Tester" covers the determination of the flash point of fuel oils, lubricating oils, solvents and other liquids by a continuously closed cup tester. The measurement is made on a test specimen of 1 ml. The test uses a closed but unsealed cup with air injected into the test chamber. It is suitable for testing samples with a flash point from 10C°C to 250°C. Flash points below 10°C and above 250°C can be performed; however, the precision has not been determined at these temperatures.

The lid of the test chamber is regulated to a temperature of at least 18°C below the expected flash point. A 1 ml test liquid is introduced into the sample cup, ensuring that both specimen and cup are at a temperature of at least 18°C below expected flash point, cooling if necessary. The cup is then raised and pressed onto the lid of specified dimensions to form the continuously closed but unsealed test chamber with an overall volume of 4 ml. The lid is heated at a prescribed, constant rate. An arc of defined energy is discharged inside the test chamber at regular intervals. After each ignition, 1.5 ml of air is introduced into the test chamber to provide the necessary oxygen for the next flash test. After each arc, the instantaneous pressure increase above the ambient barometric pressure inside the test chamber is monitored. When the pressure increase exceeds a defined threshold, the temperature at that point is recorded as the uncorrected flash point.

The above tests can be used to measure and describe the properties of materials, products or assemblies in response to heat and a test flame under controlled laboratory conditions. They should not be used to describe or appraise the fire hazard or fire risk of materials, products or assemblies under actual fire conditions. However, results of this test method may be used as elements of a fire risk assessment that takes into account all of the factors that are pertinent to an assessment of the fire hazard of a particular end use. The fire point is one measure of the tendency of the test specimen to support combustion.

Two other tests are sometimes suggested for determining flash points of oils. ASTM D56 "Standard Test Method for Flash Point by Tag Closed Cup Tester" is used to measure the flash point of liquids with a viscosity below 5.5 cSt (mm²/sec) at 40°C or below 9.5 mm²/s (cSt) at 25°C and a flash point below 93°C. ASTM D1310 "Standard Test Method for Flash Point and Fire Point of Liquids by Tag Open-Cup Apparatus" covers the determination of the flash point and fire point of liquids having flash points between −18°C and 165°C and fire points up to 165°C. When applied to paints and resin solutions that tend to skin over or that are very viscous gives less reproducible results than when applied to solvents. Neither test is really suitable for most lubricating oils.

7.2.9 VOLATILITY

There is a direct association with an oil's volatility characteristics and oil consumption rates. Although volatility is not the sole reason for oil consumption in any given engine, it provides a measure of the oil's ability to resist vaporisation at high temperatures. Typically, distillations were run to determine volatility characteristics of base oils used to formulate engine oils. This is still true currently. The objective of further defining an oil's volatility led to the introduction of several new non-engine bench tests. The determination of engine oil volatility at 371°C is a requirement in some lubricant specifications.

The most common of these tests are the NOACK Volatility and the GCD (Gas Chromatography Distillation) Volatility (ASTM D6417). Simulated Distillation, or "sim-dis" (ASTM D2887), was used previously. All these tests measure the amount of oil in percent that is lost upon exposure to high temperatures and therefore, serve as a measure of the oil's relative potential for increased or decreased oil consumption during severe service.

With the introduction of a maximum volatility limit in European engine oil specifications, there was a requirement for a volatility test with acceptable precision limits, since volatility limits may have a significant impact on the formulation and ultimate production costs of engine oils.

DIN 51581 was developed to attempt to assess the volatility properties of base oils and formulated engine oils. The test has now been adopted as CEC L-40-93, IP 421 and ASTM D5800 "Standard Test Method for Evaporation Loss of Lubricating Oils by the NOACK Method".

The test is based on the principle of loss of mass at constant temperature under a constant stream of air. Standardisation of the equipment is necessary to maintain the required precision and is obtained by checking the value with a datum oil of known value. The test method described relates to one set of operating conditions but may be readily adapted to other conditions if required. The test has three alternative procedures.

Procedure A is the original DIN and CEC method which uses Wood's metal as a heating medium. Procedure B is the automated ISL method which uses a non-Wood's metal heating method. Procedure C called "Selby Noack" method also does not use the Wood's metal. It can also collect the evolved gases for separate analysis if desired. NOACK results determined using Procedures A and B show consistent differences. Procedure A gives slightly lower results compared with Procedure B on formulated engine oils, while Procedure A gives higher results versus procedure B on base oils. Inter-laboratory tests have shown that Procedures A, B and C yield essentially equivalent results with a correlation coefficient R2 of 0.996.

The evaporation loss is defined as that mass of oil lost when the sample is heated in a test crucible, through which a constant flow of air is drawn. A measured quantity of sample is placed in an evaporation crucible which is then heated to 250°C with a constant flow of air drawn through it for 60 minutes. The loss in mass of the oil is determined. The reproducibility of this test method is dependent on a mandatory use of standard apparatus and procedure.

Attempts have been made to correlate NOACK results with thermogravimetry, high-temperature GCD and MCRT.

A great deal of concern has been expressed both in ASTM and IP regarding the use and handling of hazardous material used in the NOACK test. The safety concerns are two-fold:

- Volatilisation of heated oil.
- Release of lead, tin, bismuth and cadmium from the Wood's metal heating block at 250°C during testing. These are highly toxic, chronic poisons.

An incident in Europe showed contamination of laboratory benches around the NOACK analysers with these metals. The analysers **must** be operated in well-ventilated hoods. Further specific precautions may be necessary to prevent exposure to these toxic metal fumes and oil vapours. Savant Laboratory and ISL are now selling commercial instruments which do not use the Wood's metal. This is a safe and fast alternative for the determination of NOACK evaporation loss of a lubricant. The test method (Procedure C) is applicable to base oils and fully formulated engine oils having a NOACK evaporative loss from 0 to 30 mass %. The procedure requires a much smaller sample size.

ASTM D6417 "Standard Test Method for Estimation of Engine Oil Volatility by Capillary Gas Chromatography" is intended as an alternative to ASTM D 5800 and ASTM D5480. The data obtained by the test method are not directly equivalent to that obtained by ASTM D5800. The results by the test method can be biased by the presence of additives (polymeric materials) or by heavier base oils, which may not completely elute from the gas chromatographic column. The results of the test method may also not correlate with other oil volatility methods for non-hydrocarbon synthetic oils. The test method can be used on lubricant products not within the scope of other test methods using simulated distillation methodologies, such as ASTM D2887.

The test applicability is limited to samples having an initial boiling point greater than 126°C. This test method may be applied to both lubricant oil base stocks and finished lubricants containing additive packages. However, because of the non-elution of heavier components of these additive packages, the results by this test method are biased low compared with those by the ASTM D5480, which uses an internal standard to compensate for non-eluted material.

A sample aliquot diluted with a viscosity reducing solvent is introduced into the gas chromatographic system which uses a non-polar open tubular capillary gas chromatographic column for eluting the hydrocarbon components of the sample in the order of increasing boiling point. The column oven temperature is raised at a reproducible linear rate to effect separation of the hydrocarbons. The quantitation is achieved with a flame ionisation detector. The sample retention times are compared with those of known hydrocarbon mixtures, and the cumulative corrected area of the sample determined to the 371°C retention time is used to calculate the percentage of oil volatilised at 371°C.

Reduced volatility limits are placing more restraints on base oil processing and selection and the additive levels to achieve the required volatility levels. Volatility is one of the properties of a base oil that cannot be corrected by the use of additives, so having a low-volatility base oil as the basis for a low-volatility automotive engine oil is critically important.

7.2.10 AIR RELEASE

ASTM D3427, IP 313, ISO 2120, DIN 51381 and AFNOR E48-614 "Standard Test Method for Air Release Properties of Hydrocarbon Based Oils" are all equivalent methods for assessing the air release properties of lubricating oils.

Agitation of a lubricating oil with air in equipment such as bearings, couplings, gears, pumps and oil return lines may produce a dispersion of finely divided air bubbles in the oil. If the residence time in the reservoir is too short to allow the air bubbles to rise to the oil surface, a mixture of air and oil will circulate through the lubricating oil system. This may result in an inability to maintain oil pressure (particularly with centrifugal pumps), incomplete oil films in bearings and gears and poor hydraulic system performance or failure.

The test measures the time for the entrained air content to fall to the relatively low value of 0.2% volume under a standardised set of test conditions and hence permits the comparison of the ability of oils to separate entrained air under conditions where a separation time is available. The significance of this test has not been fully established. However, sponginess and lack of sensitivity of the control systems of some turbines may be related to the air release properties of the oil. System design and system pressure are other variables. Currently the applicability of this test method appears to be directed towards turbines manufactured outside the United States. It may not be suitable for ranking oils in applications where residence times are short and gas contents are high.

The test apparatus consists of a jacketed sample tube fitted with an air-inlet capillary, baffle plate, air outlet tube and a tapered sinker density balance. Compressed gas is blown through the test oil which has been heated to a temperature of 25°C, 50°C or 75°C for 7 minutes. After the gas flow is stopped, the time required for the gas entrained in the oil to reduce in volume to 0.2% is recorded as the gas bubble separation time. Gas bubble separation time is defined as the number of minutes needed for gas entrained in the oil to reduce in volume to 0.2% under the conditions of the test and at the specified temperature. The maximum time for the test is 30 minutes.

7.2.11 ELECTRICAL PROPERTIES

Tests for the electrical properties of fluids are mainly used for electrical transformer, switchgear and cable oils and insulating greases. However, with the rapid introduction of hybrid and electrical vehicles in many countries, these tests are likely to be needed to evaluate the properties and performance of fluids used in these cars, vans, motorcycles, trucks and buses. Unfortunately, there are a number of electrical properties that are measured currently, some or all of which may need to be used for lubricants or cooling fluids used in electric and hybrid vehicles.

ASTM D1169 "Standard Test Method for Specific Resistance (Resistivity) of Electrical Insulating Liquids" measure a liquid's electrical insulating properties under conditions comparable to those of the test. High resistivity reflects low content of free ions and ion-forming particles and normally indicates a low concentration of conductive contaminants. The test method covers the determination of specific resistance (resistivity) applied to new electrical insulating liquids, as well as to liquids in service or subsequent to service, in cables, transformers, circuit breakers and other electrical apparatus. It covers a procedure for making referee tests with direct current (DC) potential. When it is desired to make routine determinations requiring less accuracy, certain modifications to the test method are permitted as described in Sections 19 to 26. Whether the test might be applicable to the cooling fluids used in electric or hybrid vehicles has not yet been determined.

ASTM D1816 "Standard Test Method for Dielectric Breakdown Voltage of Insulating Liquids Using VDE Electrodes" is used to determine a liquid's ability to withstand electric stress without failure. The dielectric breakdown voltage serves to indicate the presence of contaminating agents such as water, dirt, cellulosic fibres or conducting particles in the liquid, one or more of which may be present in significant concentrations when low breakdown voltages are measured. However, a high-dielectric breakdown voltage does not necessarily indicate the absence of all contaminants. It may simply indicate that the concentrations of contaminants in the liquid between the electrodes are not large enough to significantly affect the average breakdown voltage of the liquid when tested by the method. The method can be used in laboratory or field tests. For field breakdown results to be comparable to laboratory results, all criteria including room temperature (20°C to 30°C) must be met.

The test method covers the determination of the dielectric breakdown voltage of insulating liquids, including mineral oils, silicone fluids, high fire-point electrical insulating oils, synthetic ester fluids and natural ester fluids. This test method is applicable to insulating liquids commonly used in cables, transformers, oil circuit breakers and similar apparatus as an insulating and cooling medium. The test method is sensitive to the deleterious effects of moisture in solution especially when cellulosic fibres are present in the liquid. It has been found to be especially useful in diagnostic and laboratory investigations of the dielectric breakdown strength of insulating liquid in insulating systems.

The IEC 60156, IP 295 and DIN 60156 tests "Insulating liquids – Determination of the breakdown voltage at power frequency – Test method" (also called "Electric Strength of Insulating Oils") specify a method for determining the dielectric breakdown voltage of insulating liquids at power frequency. The test procedure is performed in a specific apparatus, where the oil sample is subjected to an increasing alternating current (AC) electrical field until breakdown occurs. The method applies to all types of insulating liquids of nominal viscosity up to 350 mm²/s at 40°C. It is appropriate both for acceptance testing on unused liquids at the time of their delivery and for establishing the condition of samples taken in monitoring and maintenance of equipment. The third edition of IEC 60156 was published in 2018 to replace the second edition published in 1995. It is a technical revision and mainly confirms the content of the previous edition, and although some improvements are included, the test method has not been changed for practical reason, due to the very large number of test instruments spread around the world.

DIN 51412 "Testing of petroleum products – Determination of the electrical conductivity" has two parts, both of which specify a test method for determining the electrical conductivity of petroleum products and other organic liquids. Part 1, DIN 51412-1, specifies the test method to be used in a laboratory, while Part 2, DIN 51412-2, specifies the test method to be used in the field. Mineral oil products and other organic liquids can become electrostatically charged during filling processes, so that if an ignitable mixture is present, electrical discharges can cause it to ignite. Knowledge of the electrical conductivity is necessary in order to be able to assess the potentially resulting hazardous situations. The field method can be used to measure the electrical conductivities in the range of 5 to 1,999 picosiemens/metre (pS/m) in jet fuels and other mineral oil products to which conductivity improvers have been added. The conductivities measured according to the method are the quiescent conductivity. It is possible that either test method might be applicable to low-viscosity cooling fluids used in electric and hybrid vehicles.

ASTM D4308 "Standard Test Method for Electrical Conductivity of Liquid Hydrocarbons by Precision Meter" covers the determination of the "rest" electrical conductivity of aviation fuels and other similar low-conductivity hydrocarbon liquids in the range from 1 to 2,000 pS/m. The generation and dissipation of electrostatic charge in fuel due to handling depend largely on the ionic species present which may be characterised by the rest or equilibrium electrical conductivity. The time for static charge to dissipate is inversely related to conductivity. The test method can supplement ASTM D2624 test method, which is limited to fuels containing a static dissipator additive.

ASTM D2624 and IP 274 test methods "Standard Method of Test for Electrical Conductivity of Aviation Fuels Containing a Static Dissipator Additive" cover the determination of the electrical conductivity of aviation fuels containing a static dissipator additive. The method normally gives a measurement of the conductivity when the fuel is uncharged, that is, electrically at rest (known as rest conductivity). The method is primarily intended as a field test to measure electrical conductivities in situ in tanks. It may also be used as a laboratory or field test on sampled fuels, but care must be taken to avoid contamination. After pumping operations, it may be necessary to wait for at least 30 minutes before taking measurements to allow the fuel to become electrically at rest and reach an equilibrium conductivity. An alternative technique is provided for the measurement of electrical conductivity (equivalent to rest conductivity) in a flowing stream.

A voltage is applied across two electrodes immersed in the fuel and the resulting current expressed as a conductivity value. With portable meters, the current measurement is made almost instantaneously upon application of the voltage to avoid errors arising from ion depletion. Ion depletion or polarisation is eliminated in dynamic monitoring systems by continuous replacement of the sample in the measuring cell. The procedure, with the correct selection of electrode size and current measurement apparatus, can be used to measure conductivities from 1 pS/m or greater. The field equipment referred to in the method is designed to cover a conductivity range of 50–1,000 pS/m with good precision.

ASTM D4308 can be used in the laboratory or in the field. The test method may be applicable to some of the lower-viscosity hydrocarbon liquids used as cooling fluids for electric motors in electric or hybrid vehicles. Its utility in this regard is still being evaluated.

The IEC 60247 and DIN 60247 test methods "Insulating liquids – Measurement of relative permittivity, dielectric dissipation factor (tan d) and d.c. resistivity" describe methods for the determination of the dielectric dissipation factor, relative permittivity and direct current (DC) resistivity of any insulating liquid material at the test temperature. The methods are primarily intended for making reference tests on unused liquids. They can also be applied to liquids in service in transformers, cables and other electrical apparatus. However, the method is only applicable to single-phase liquids. When it is desired to make routine determinations, simplified procedures, as described in Annex C, may be adopted. With insulating liquids other than hydrocarbons, alternative cleaning procedures may be required.

7.2.12 ULTRASOUND

Ultrasound is any sound pressure wave with a repetition frequency higher than 20 kHz. This characteristic, particularly between 30 and 40 kHz, is very useful for maintenance engineers listening for symptoms of component, machine, equipment or system deterioration or impending failure. Although it is not immediately obvious that ultrasound can be used to monitor lubricants, this can sometimes be the case. Maintenance engineers who use ultrasound are aware of its favourable ease-of-use to cost–benefit ratio. The current typical cost of an ultrasound system ranges from US$2,500 to US$20,000. It has been reported to be not uncommon for an engineer or supervisor to recover the cost the first time the system is used. Although ultrasound has been considered to be a companion to routine predictive tools, such as vibration analysis and infrared imaging, stand-alone ultrasound programmes are becoming more common.

In an increasing number of situations, the first signs of change in a component or machine's condition can be picked up in ultrasound frequencies. Because of this, in many applications, ultrasound is often the best way to find defects that can lead to asset failure, allowing for remediation before any further deterioration occurs.

Operating environments can be very noisy, but ultrasound detectors are tuned to hear higher frequencies exclusively and to block out competing sounds, enabling monitoring to be performed at any time of day, including times of peak equipment or system use.

Ultrasound equipment detects and converts ultrasound to audible sound that can be monitored with headphones. Almost all ultrasound equipment can create a bar graph display or a decibel measurement. Some ultrasound devices can capture a time-specified wave signal for software analysis. Many additional functionalities are also available. The devices have high-frequency piezo-ceramic crystal sensors that detect sources of low-frequency ultrasound that are then converted to corresponding audible signals. During this heterodyning conversion, the qualities and characteristics of the original ultrasound are preserved, giving humans the ability to hear sounds far beyond their normal range. The energy in the signal is defined by the root mean square (RMS) or effective value.

The RMS is a reliable metric since it is both stable and repeatable. The equipment provides both dynamic signal analysis, for which continuous data is acquired for a user-defined time frame, and four static condition indicators. These are:

- Overall RMS: This displays the RMS average dBμV (decibels-microvolt) value for the entire acquisition time, providing guidance about the equipment's lubrication condition as well as a comparative trend against previous measurements.
- Maximum RMS: This breaks down the overall RMS into quarter second samples. If the acquisition time is set to 3 seconds, there will be 12 blocks of quarter of a second samples. The loudest of the 12 blocks is taken as the maximum RMS and is a good indication of signal stability for the duration of the acquisition time.
- Peak RMS: This is the single loudest measured value for the acquisition time. It is the loudest of the 24,000 possible measurement samples for the

3-second measurement duration. High peak RMS values indicate impacting as the probable cause of high overall dBμV values leading to the conclusion that there are defects present in the component, machine, equipment or system.

- Crest Factor: This is a linear ratio between peak RMS and overall RMS, and is designed to return a comparison that tracks impacting relative to friction. When overall RMS increases, but peak RMS does not, this is indicative of friction. When peak RMS increases without overall RMS, this is indicative of impacting. When they increase together, it is indicative of early-stage failure. When overall RMS continues to increase and peak RMS decreases, then the failure defects are critical.

Ultrasound is being promoted as very versatile. With machine condition monitoring, under-lubrication or over-lubrication of bearings, for example, can lead to increased friction and eventually to impacting. These faults can be detected with ultrasound before the bearing fails and while corrective maintenance can still be performed. Detecting internal leaks in pumps, motors or servo-valves in hydraulic systems are more detectable with ultrasound than are external leakages. Installing an ultrasound contact sensor above, on and below the direction of flow will detect turbulence. Ultrasonic testing is able to identify functioning and malfunctioning valves easily, so they can be tagged and repaired quickly to avoid any loss of production.

Detecting the sounds of impacting or higher than normal friction in a noisy operating environment can be difficult, so ultrasonic monitoring is potentially very useful for mining, off-highway and marine equipment or systems. It has the potential to detect over-lubrication, leading to lower costs, and under-lubrication, leading to improved maintenance scheduling or preventative maintenance.

When compressed gas flows through a leak, it produces turbulence and a characteristic high-frequency hissing. While the hissing is audible in isolation, it can be very difficult to hear in a noisy environment. Ultrasound is easier and more effective at detecting gas leaks than other traditional tests. Ultrasound is effective for both open and enclosed electrical equipment across a range of voltages. Electrical apparatuses such as switchgear, transformers, insulators or disconnects and splices can all fail, but before they do, faults can be detected. Unlike electrical infrared thermography monitoring, ultrasound devices can detect electrical discharges in their earliest stages, long before they begin to generate heat.

7.2.13 Particle Counts

In almost all machines and items of equipment, it is very important to have clean oil. Filters will remove particles from oil, but it is also important to know how clean an oil needs to be for a specific machine or item of equipment. Particle count data is an invaluable part of any proactive condition monitoring programme, from ensuring that abrasive bearing wear is minimised to determining if a hydraulic fluid is clean enough for reliable operation.

Particle counting was introduced during the 1970s, but it has only been the last 15 to 20 years that industry, determined to work smarter, has started to fully realise the importance of fluid cleanliness. Examples of particles in oil are shown in Figure 7.3.

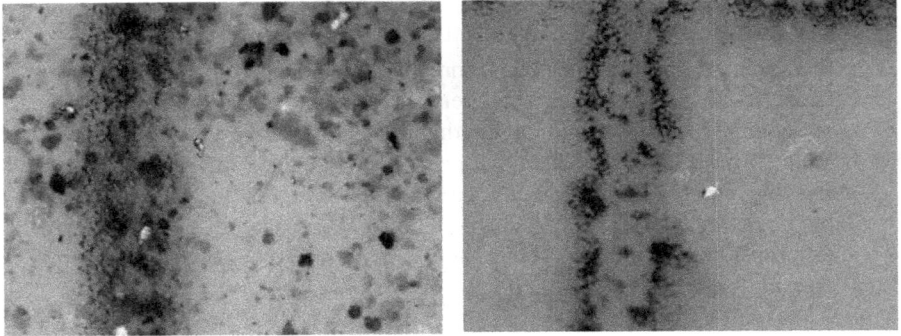

FIGURE 7.3 Examples of particles in oil.

Source: Pathmaster Marketing Ltd.

The most common way of reporting fluid cleanliness is the ISO Code System, covered under ISO standard 4406:99. In this standard, the number of particles in three different size categories, >4, >6 and >14 µm are determined in one millilitre of sample. ISO 4406:99 states that the number of particles in each size category should be counted with the absolute count converted to an ISO code, using the ISO range code chart. This chart is reproduced in Table 7.1.

Other standards for measuring and reporting fluid cleanliness include NAS 1638 and MIL-STD 1246C as well as outdated standards such as the SAE fluid cleanliness rating system. Whichever method of reporting is selected, the first step is to count the number of particles in a volume of fluid. Three basic methods can be used to determine the absolute number of particles in any given sample.

Optical microscopy (ISO 4407), the original method for determining fluid cleanliness levels, examines a representative portion of the sample under an optical microscope. In this procedure, the particles are counted manually, which can then be used to determine the fluid cleanliness of the bulk sample. While this method may seem outdated, slow and cumbersome, it is still in use and is considered by many to be the most reliable and accurate method of particle counting, unaffected by some of the limitations of more modern, automated methods.

Automatic Optical Particle Counting (ISO 11500) is perhaps the most widely used method for determining fluid cleanliness currently, using an automatic optical particle counter. A variety of instruments are available commercially, from portable units for on-site use that cost as little as $15,000, to large, sophisticated laboratory-based instruments that may cost in excess of $40,000. A low-cost, online optical particle counter is available for under $1,000. However, all instruments use one of two methods, either a white-light source or more commonly today, a laser.

In a white light instrument, particles pass through the capillary detection zone and create a shadow on a photocell detector. The drop in voltage produced by the photocell is directly proportional to the size of the shadow and hence the size of the particle passing through.

In a laser-based instrument, due to the near-parallel nature of the laser beam, light scattering from the unimpeded laser beam is minimal because it is focused into a

TABLE 7.1

ISO 4406 Particle Range Numbers

ISO	Number of Particles per Millilitre	
Range Number	Greater Than	Less Than
24	80,000	160,000
23	40,000	80,000
22	20,000	40,000
21	10,000	20,000
20	5,000	10,000
19	2,500	5,000
18	1,300	2,500
17	640	1,300
16	320	640
15	160	320
14	80	160
13	40	80
12	20	40
11	10	20
10	5	10
9	2.5	5
8	1.3	2.5
7	0.64	1.3
6	0.32	0.64
5	0.16	0.32
4	0.08	0.16
3	0.04	0.08
2	0.02	0.04
1	0.01	0.02

Source: ISO.

Note that for Range Number 1, the number of particles is between 1 and 2 per 100 ml and so on for Range Numbers up to 9. That means that the number of particles per 100 ml for Range Number 24 is between 8 and 16 million.

beam stop, until a particle passes through the instrument. As the laser strikes the particle, light scatters and hits the photocell. As with a white-light instrument, the change in voltage across the photocell is directly related to the size of the particle. An increased signal is compared with what should (in theory) be a zero background signal. Laser optical particle counters are generally considered to be slightly more accurate and sensitive than white-light instruments.

Several subtleties must be considered with automatic optical particle counters. First, in general, particles from used oil samples are not perfectly spherical. This can create problems for optical counters. Because ISO 4406 is one-dimensional, a particle that is $5\,\mu m$ across the minor axis and $40\,\mu m$ across the major axis would be difficult to categorise. To resolve this issue, developers of automatic optical particle counters have devised a compromise known as the equivalent spherical diameter.

With the equivalent spherical diameter method, a particle is counted in the size range under which the shadow or scattering effect observed would have appeared, if the particle had been a perfect sphere. This allows the average fluid cleanliness to be estimated, permitting the ISO code to be trended over subsequent samples.

Another concern is the effect of false positives. For example, both air bubbles and free and emulsified water appear as if they are a particle using the optical particle counting method. While the effect of air and water can be negated using an ultrasonic bath and vacuum de-gassing to remove air, and solvent extraction to dissolve free and emulsified water, other false positives are possible with multiple particle coincidences and additive floc. For this reason, care and attention to procedural details must be exercised when performing optical particle counts.

Pore Blockage Particle Counting (BS 3406) is a widely used method of obtaining an automatic particle count. In this method, a volume of fluid is passed through a mesh screen with a clearly defined pore size, commonly $10\,\mu m$. Two instrument types use this method.

One instrument measures the flow decay across the membrane as it becomes plugged while pressure is held constant, first with particles greater than $10\,\mu m$ and later by smaller particles as the larger particles plug the screen. The second measures the rise in differential pressure across the screen, while the flow rate is held constant as it becomes plugged with particles. Both instruments are tied to a software algorithm, which turns the time-dependent flow decay or pressure rise into an ISO cleanliness rating according to ISO 4406.

While pore block particle counters do not suffer the same problems as optical particle counters with respect to false positives caused by air, water, dark fluid and so on, they do not have the same dynamic range as an optical particle counter. Also, because the particle size distribution is roughly estimated, they are dependent on the accuracy of the algorithm to accurately report ISO fluid cleanliness codes according to ISO 4406. Nevertheless, they accurately report the aggregate concentration of particulates in the oil, and in certain situations, particularly dark fluids such as diesel engine oils and other heavily contaminated oils, pore block particle counting does offer advantages.

To try to overcome the problems of interference from water droplets and soft particles with particle counting, ASTM has published a set of techniques in ASTM D7647 "Automatic Particle Counting of Lubricating and Hydraulic Fluids Using Dilution Techniques to Eliminate the Contribution of Water and Interfering Soft Particles by Light Extinction". The method is intended for use in analytical laboratories, including on-site in-service oil analysis laboratories. It covers the determination of particle concentration and particle size in new and in-service oils used for lubrication and hydraulic applications. Particles considered are in the range from 4 to 200 microns (μm) with the upper limit being dependent on the specific automatic particle counter being used. Lubricants that can be analysed by this test method are categorised as petroleum products or synthetic-based products, such as polyalphaolefins, polyalkylene glycols or phosphate esters. The applicable viscosity range is up to 1,000 mm²/s at 40°C. Samples containing visible particles or that are opaque after dilution may not be suitable for analysis using the test method. The test sample is agitated before obtaining an aliquot from a homogeneous sample. It is diluted with appropriate diluent for the sample type. The diluted sample is agitated and the sample is degassed. The testing is begun within 90 seconds and the particle count is obtained in triplicate. The data are analysed after validity checks are conducted.

Converting particle counts into an ISO cleanliness rating is straightforward, as shown in Table 7.2. The example fluid in Table 7.2 has an ISO cleanliness rating of 18/14/12.

Discussion of ISO cleanliness targets for different types of oils in specific machines and items of equipment will be included in later chapters.

A complimentary test method is ASTM D8127 "Coupled Particulate and Elemental Analysis Using X-Ray Fluorescence (XRF) for In-Service Lubricants". Analysts have found that separating information regarding small or dissolved elemental materials in a lubricant from suspended particulate is crucial. In many cases, only an overall elemental analysis is provided, which may not capture significant wear or even machinery failure events. Such events often are accompanied by a sudden increase in production of large particulates, which are suspended in and can be detected in the machinery's lubricant. This test method specifically targets such particulate, which historically has been difficult to quantify. Users of the test include military organisations and maintenance engineers of wind turbines, nuclear power facilities and off-shore oil and gas rigs. This test method is applicable to all known in-service mineral oil lubricants (API Groups I to IV) at any stage of degradation.

The test method uses a combination of pore blockage particle counting and energy dispersive X-ray fluorescence (ED-XRF) spectrometry for the quantitative determination of solid particle counts larger than 4 mm and elemental content of suspended particulates of iron greater than 4 µm in the range of 6 to 223 mg/kg and copper greater than 4 µm in the range of 3.5 to 92.4 mg/kg in test lubricants. Total particle counts greater than 4 µm are determined in the range of 11,495 particles/ml greater than 4 µm to 2,169,500 particles/ml greater than 4 µm in the lubricant. An ED-XRF

TABLE 7.2
Example Particle Counts and Corresponding ISO Code

Particle Size, µm	Particles per ml	ISO Code Range	ISO Code
4	1,648	1,300–2,500	18
6	153	80–160	14
10	64		
14	32	20–40	12
21	10		
38	3		

Source: Pathmaster Marketing
The ISO Code is, therefore, 18/14/12.

spectrometer provides the fluorescence spectrum, from which the elemental concentrations of iron and copper are calculated using their respective fundamental $K\alpha$ lines by the way of established calibration that includes inter-element corrections. Compton backscattering corrections also may be applied. A number of factors may interfere with the analysis, and these are explained in the standard.

7.2.14 FERROGRAPHY

Ferrography is a method of studying wear particles in samples of oils taken from machines and items of equipment, to try to predict and diagnose potential or actual problems with the machinery. Since the advent of ferrography in the 1970s, it has been used in many industries as an aid to condition monitoring and preventative maintenance.

Ferrography was pioneered in the 1970s by Vernon C. Westcott to help the US military diagnose problems related to bearing failure before the issues became deadly. At that time, the best method of analysis could not detect small particles, so by the time the military found problems it was too late to identify solutions. To solve this problem, Mr. Westcott developed the first ferrograph. The ferrograph first saw major use by the United Kingdom to detect helicopter failures during the Falklands War in 1982.

Ferrography is unique because it can deliver information about enclosed parts of a machine or system while lubricating oil circulates through these areas and is still accessible. Rinsing vital components with particle free lubricant and analysing the output can offer a detailed report of machine wear without disassembling anything. Wear particle analysis is a powerful technique for non-intrusive examination of oil-wetted parts of a machine. The particles contained in the lubricant carry detailed and important information about the condition of the machine.

This information may be deduced from particle shape, composition, size distribution and concentration. The particle characteristics are sufficiently specific so that the operating wear modes within the machine may be determined, allowing prediction of the imminent behaviour of the machine. Ferrographic analysis helps to prevent catastrophic equipment failure through timely and accurate prediction of abnormal or critical machine wear.

Direct-reading ferrography magnetically separates wear particles and optically measures the quantity of large and small particles present in the oil sample. Using results from direct-reading ferrography that indicate the rate, intensity and severity of wear with these measurements, machine wear baselines can be established and trends in wear conditions can be monitored. If there is a significant increase in the wear trend levels, a detailed analytical ferrography should be performed.

When direct-reading ferrographs and/or other analysis indicate abnormal wear, analytical ferrography can further pinpoint its source and the specific type of wear. Skilled analysts will extract, classify and visually analyse wear particles and solid contaminants. Particles are examined under a powerful optical microscope to determine the size, concentration, colour, shape and particle composition. Results provide timely corrective maintenance, based on a machine's actual condition.

Analytical ferrography is among the most powerful diagnostic tools in oil analysis. The test is qualitative, because it relies on the skill and knowledge of the ferrographic analyst. While this can have definite advantages, the interpretation is somewhat sub-jective and requires detailed knowledge of wear debris failure modes. The test proce-dure is also lengthy and requires the skill of a trained ferrographic analyst.

A sample of the machine's lubricating oil is taken and diluted, then run across a glass slide. This glass slide is then placed on a magnetic cylinder that attracts the contaminants. Non-magnetic contaminants remain distributed across the slide from the wash. These contaminants are then washed, to remove excess oil, heated to 315°C (600°F) for 2 minutes and the slide is analysed under a microscope. After analysis, the particles will be ranked according to size. Particles larger than 30 μm in size are considered "abnormal" and indicate severe wear.

Particles are divided into six categories, with an additional five subcategories under ferrous wear:

- Copper.
- White non-ferrous: Usually aluminium or chromium.
- Babbitt: Particles containing tin and lead.
- Contaminants that do not change appearance after heating, usually dirt.
- Fibres: Typically from filters.
- Ferrous wear: Magnetic particles that are attracted to the magnetic cylinder.
- High alloys: Rarely found on ferrograms.
- Low alloys.
- Cast iron.
- Dark metallic oxides: Darkness indicates oxidation.
- Red oxides.

Being able to identify different particles can prove to be invaluable, because the prom-inence of certain particles can point to specific locations of wear. Additionally, the presence of particles that do not make contact with the lubricating oil can uncover con-tamination. This kind of analysis requires a trained professional and can be prohibi-tively expensive for smaller operations. An example ferrogram is shown in Figure 7.4.

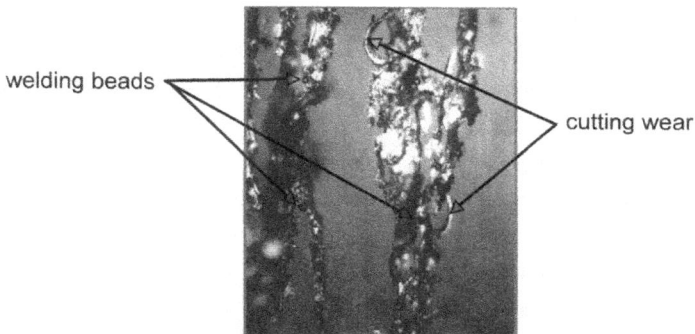

FIGURE 7.4   Example ferrogram.

Source: Pathmaster Marketing Ltd.

ASTM D7690 practice "Microscopic Characterization of Particles from In-Service Lubricants by Analytical Ferrography" covers the identification by optical microscopy of wear and contaminant particles commonly found in used lubricant samples that have been deposited on ferrograms. The practice relates to the identification of particles, but not to methods of determining the particle concentration. This practice interfaces with but generally excludes particles generated in the absence of lubrication, such as those that could result from corrosion, impaction, gouging or polishing. In usual practice of a routine condition monitoring programme, a ferrogram is not prepared for every sample taken, but may be prepared when routine tests, such as spectrochemical analysis, particle counting or ferrous debris monitoring, indicate abnormal results. ASTM D7690 provides guidelines for identifying particles arising from different types of wear or from contaminants.

7.3   PHYSICAL TESTS FOR GREASES

7.3.1   PENETRATION

The consistency of a grease can determine its effectiveness in providing proper lubrication. If it is too soft, the grease will not stay in place, resulting in poor lubrication. Conversely, if it is too hard, the grease may not be properly distributed, also resulting in poor lubrication.

ASTM D217, IP 50, ISO 2137 and AFNOR T60-132 "Standard Test Methods for Cone Penetration of Lubricating Grease" covers four procedures for measuring the consistency of lubricating greases by the penetration of a cone of specified dimensions, mass and finish. Penetration numbers are useful in classifying greases for particular applications or service requirements as well as for measuring the consistency of a given grease from batch to batch. The consistency normally changes somewhat when a grease is worked (as in actual use). The National Lubricating Grease Institute (NLGI) and the European Lubricating Grease Institute (ELGI) classify greases according to their consistency as measured by worked penetration. These classifications are shown in Table 7.3.

TABLE 7.3
NLGI and ELGI Grease Grades

Worked Penetration Range	NLGI and ELGI Grade
445–475	000
400–430	00
355–385	0
310–340	1
265–295	2
220–250	3
175–205	4
130–160	5
85–115	6

Sources: NLGI and ELGI.

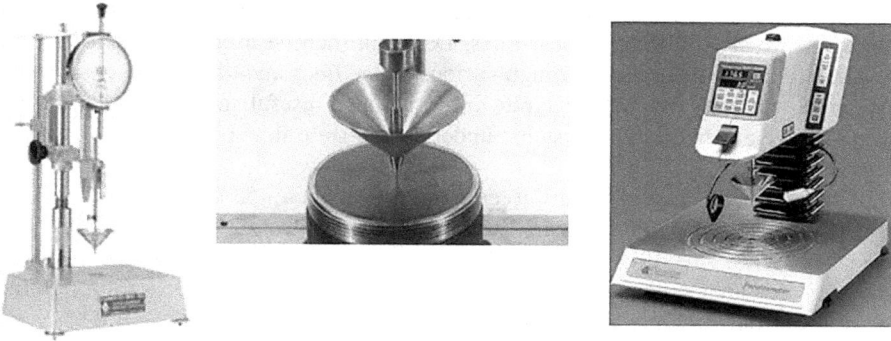

FIGURE 7.5   Grease penetration test equipment.

Source: Pathmaster Marketing Ltd.

To measure the penetration, the test grease is packed into a worker cup and brought to 25°C. The cone assembly is placed directly over the surface of the grease. The cone is then released and allowed to drop freely into the grease for 5 seconds. Photographs of the type of equipment used to measure grease penetration are shown in Figure 7.5. The penetration of the cone into the grease is measured in tenths of a millimetre. Three determinations are made and are averaged to give the reported result. The four test variations consist of measuring the penetrations of unworked and worked grease samples. The test variations are:

* Unworked Penetration.
* Worked Penetration: 60 double strokes in the grease worker.
* Prolonged Penetration: Typically 10,000+ double strokes in the grease worker.
* Blocked Penetration: Unworked penetration of a hard cube of grease.

ASTM D1403 "Standard Test Methods for Cone Penetration of Lubricating Grease Using One-Quarter and One-Half Scale Cone Equipment" is intended for use only when the size of the sample is limited. Precision is better in the full-scale method.

The test methods include procedures for the measurement of unworked and worked penetrations. Unworked penetrations do not generally represent the consistency of greases in use as effectively as do worked penetrations. The latter are usually preferred for inspecting lubricating greases. There are two test procedures, one for one-fourth scale cone and the other for the half scale cone. The penetration is determined at 25°C (77°F) by releasing the cone assembly from the penetrometer and allowing the cone to drop freely into the grease for 5 seconds.

7.3.2   APPARENT VISCOSITY

ASTM D1092 "Standard Test Method for Measuring Apparent Viscosity of Lubricating Greases" can be used to determine the apparent viscosity of grease at varying shear rates. The results provide a better indication of pressure drops in grease distribution systems under steady-state flow at constant temperatures. The test determines apparent viscosity at 16 different shear rates. Data can then be used to predict the flow characteristics of the grease through various pipes, lines and dispensing equipment. Apparent viscosity versus shear rate information is useful in predicting pressure drops in grease distribution systems under steady-state flow conditions at constant temperature.

The test covers measurements at temperature between $-53°C$ and $37.8°C$ ($-65°F$ and $100°F$). The measurements are limited to the range from 25 to 100,000 poises at $0.1 s^{-1}$ and from 1 to 100 at poises at $15,000 s^{-1}$. At very low temperatures, the shear rate may be reduced because of the great force required to force grease through the smaller capillaries. Precision has not been determined below $10 s^{-1}$.

The sample is forced through a capillary by means of a floating piston actuated by the hydraulic system. From the predetermined flow rate and the force developed in the system, the apparent viscosity is calculated by means of Poiseuille's equation. A series of eight capillaries and two pump speeds are used to determine the apparent viscosity at 16 shear rates. The results are expressed as a log-log plot of apparent viscosity versus shear rate.

Unfortunately, the ASTM D1092 test requires several kilograms of grease. It also can be very time-consuming, especially if apparent viscosity determinations are required at several different temperatures. It is, therefore, unsuitable for use in a grease condition monitoring programme.

The US Steel Grease Mobility test, similar to ASTM D1092, was developed to measure grease flow resistance at various temperatures and pressures. It helps predict the pumpability characteristics of grease at low temperatures. By varying temperature and pressure, grease can be evaluated for its performance in known supply systems. However, the method is not standardised, so its repeatability and reproducibility are unknown.

The Lincoln Ventmeter test was developed as a rapid alternative to ASTM D1092. Results can be used to approximate the maximum or minimum size of supply line needed for a particular grease at a given temperature. Pumpability can be predicted by comparing the Ventmeter viscosity reading with supply line charts supplied with the instrument. One of the known drawbacks of the test is that its repeatability and reproducibility are doubtful.

7.3.3   PUMPABILITY AND FLOW PROPERTIES

Grease must be able to be pumped to some applications at temperatures below ambient, then be able to flow and provide adequate lubrication at the contact. The ASTM D1478 test evaluates the extent to which a grease retards the rotation of a slow-speed ball bearing by measuring the starting and running torques of a bearing packed with the test grease at temperatures below $-20°C$.

ASTM D1478 "Standard Test Method for Low-Temperature Torque of Ball Bearing Grease" was developed using greases having very low torque characteristics at $-54°C$ ($-65°F$). Specifications for greases of this type commonly require testing at this temperature. Specifications for greases of other types can require testing at temperatures from $-75°C$ to $-20°C$.

The test has proved helpful in the selection of greases for low-powered mechanisms, such as instrument bearings used in aerospace applications. The suitability of this test method for other applications requiring different greases, speeds and temperatures should be determined on an individual basis.

ASTM D4693 "Standard Test Method for Low-Temperature Torque of Grease-Lubricated Wheel Bearings" may be better suited for applications using larger bearings or greater loads. However, greases having such characteristics that permit torque evaluations by either ASTM D1478 or ASTM D4693 will not give the same values in the two test methods (even when converted to the same torque units) because the apparatus and test bearings are different.

A Number 6204 open ball bearing is packed completely full of grease and cleaned off flush with the sides. The bearing remains stationary while ambient temperature is lowered to the test temperature and held there for 2 hours. At the end of this time, the inner ring of the ball bearing is rotated at 1 ± 0.05 rpm while the restraining force on the outer ring is measured. Torque is measured by multiplying the restraining force by the radius of the bearing housing. Both starting torque and after 60 minutes of rotation are determined.

The method is helpful for applications requiring greases with low yield stress and that can maintain suitable consistency at low temperatures and speeds. However, it requires a substantial amount of time and large samples to determine data at several temperatures.

The ASTM D4693 test differentiates among greases having distinctly different low-temperature characteristics. The test is used for specification purposes and correlates with its precursor which has been used to predict the performance of greases in automotive wheel bearings in low-temperature service. It is the responsibility of the user to determine the correlation with other types of service.

The test covers the determination of the extent to which a test grease retards the rotation of a specially manufactured, spring-loaded, automotive-type wheel bearing assembly when subjected to low temperatures. Torque values, calculated from restraining-force determinations, are a measure of the viscous resistance of the grease. The test was developed with greases giving torques of less than 35 N·m at $-40°C$.

A freshly stirred and worked sample of grease is packed into the bearings of a specially manufactured, automotive-type spindle bearings hub assembly. The assembly is heated and then cold soaked at $-40°C$. The spindle is rotated at 1 rpm and the torque required to prevent rotation of the hub is measured at 60 seconds.

Rheological testing to determine the deformation and flow of matter has been used in many other industries for years. Controlled stress/strain rheometers can evaluate the consistency, flow properties, pumpability, yield stress, thixotropic performance, temperature limits and even tackiness of grease at various shear rates and temperatures. The small amount of grease required for this evaluation is also ideal for evaluating used grease samples.

Using a controlled stress/strain rheometer can be more complicated than using other established test methods. However, it has been noted that the flexibility in test parameters and the data generated make rheology testing an ideal way to evaluate consistency, flow and pumpability of greases, so research into using rheometers to evaluate greases is ongoing.

Unfortunately, neither of these tests is at all suitable for a grease condition monitoring programme, as the volume of grease required to perform the tests is too great.

7.3.4 DROPPING POINT

The dropping point is the temperature at which a grease passes from a semi-solid to a liquid state. This change in state is typical of greases containing soaps of conventional types added as thickeners. These tests are useful in identifying the types of greases and for establishing and maintaining benchmarks for quality control.

There are two test methods for this purpose. In general, dropping points obtained by ASTM D566 and ASTM D2265 are in agreement up to 260°C. In the case where results differ, there is no known significance.

ASTM D566, IP 132, ISO 2176 and AFNOR T60-102 "Standard Test Method for Dropping Point of Lubricating Grease" is not recommended for temperatures above 288°C. A sample of test grease contained in a cup suspended in a test tube is heated in an oil bath at a prescribed rate. The temperature at which the material falls from the hole in the bottom of the cup is averaged with the temperature of the oil bath and recorded as the dropping point.

ASTM D2265 and ISO 6299 (there is no equivalent IP test method) "Standard Test Method for Dropping Point of Lubricating Grease over Wide Temperature Range" is used for greases with a dropping point above 288°C. A grease sample in a test cup is supported in a test tube placed in an aluminium block oven at a preset constant temperature. A thermometer is positioned in the cup so that it measures the temperature without coming in contact with the grease. As the temperature increases, a drop of material will fall from the test cup to the bottom of the test tube. Temperatures of both the thermometer and that of the block oven are recorded at this point. One-third of the difference between the two values is the correction factor which is added to the observed value and reported as the dropping point of the grease.

Both of these tests are well suited to a grease condition monitoring programme, as the sample of grease require is very small and the time required to complete the tests is comparatively short.

7.3.5 OIL SEPARATION

A few standard methods are available for testing oil separation from grease. ASTM D1742 and ASTM D6184 determine the amount of oil bleed from a grease under static conditions. These methods help evaluate the amount of oil bleed that could be detrimental to grease consistency at high temperatures. ASTM D1742 aims to evaluate the oil separation from a grease during storage.

ASTM D6184 "Standard Test Method for Oil Separation from Lubricating Grease (Conical Sieve Method)" covers the determination of the tendency of lubricating grease to separate oil at an elevated temperature. The test is conducted at 100°C for 30 hours, unless other conditions are required by a grease specification. The test is suitable for use with most greases, but the test precision was established using greases having a worked penetration (ASTM D217) greater than 220 (NLGI No. 3

grade) and less than 340 (NLGI No. 1 grade). However, the test duration is too long for it to be of much use in a grease condition monitoring programme.

The conical sieve method for measuring leakage from lubricating grease (commonly known as the cone bleed test) is virtually the same as Federal Test Method 791C Method 321.3. A weighed sample is placed in a cone-shaped, wire-cloth sieve, suspended in a beaker, then heated under static conditions for the specified time and temperature. At the end of the test, the separated oil is weighed and reported as a percentage of the mass of the starting test sample.

ASTM D4425 "Standard Test Method for Oil Separation from Lubricating Grease by Centrifuging (Koppers Method)" is useful in evaluating the degree to which a grease would separate into fluid and solid components when subjected to high centrifugal forces. Flexible shaft couplings, universal joints and rolling element thrust bearings are examples of machinery which subject lubricating greases to large and prolonged centrifugal forces. The test has been found to give results that correlate well with results from actual service. The test may be run at other conditions by agreement between supplier and user, but the precision noted in the test will no longer apply.

Pairs of centrifuge tubes are charged with grease samples and are placed in the centrifuge. The grease samples are subjected to a centrifugal force equivalent to a G value of 36,000 at 50°C ± 1°C, for a specific period of time. The resistance of the grease to separate the oil is then defined as a ratio of the percent oil separated to the total number of hours of testing.

The ASTM D6184 method includes the disclaimer, "Test results obtained with this procedure are not intended to predict oil separation tendencies of grease under dynamic service conditions." The dynamic test procedure ASTM D4425 may provide a better indication of grease bleed in applications producing high centrifugal forces such as flexible shaft couplings, universal joints and rolling element thrust bearings.

7.3.6 TACKINESS

Another important attribute to consider when selecting a grease is tackiness. Tackiness is defined as the ability of the grease to form threads when pulled to enable easy transfer of grease to the contacting areas. Tackiness can be a desired attribute for some applications, while it may be detrimental to others.

The amount of tack can be controlled to a large extent during formulation. Unfortunately, there are no established methodologies that efficiently and accurately evaluate the adhesion and tackiness of grease currently. However, current research is making strides in developing effective methods and equipment.

7.4 FUTURE PHYSICAL TESTS FOR OILS AND GREASES

As in Chapter 6, new types of machines, equipment and systems that use lubricants are being developed and introduced all the time. For example, the increasing numbers of electric and hybrid vehicles of all types, including cars, vans, trucks, buses and other vehicles, have required the development of new types of lubricants and cooling fluids. In the future, civil and military aircraft are likely to have to be powered by fuels that have less damaging environmental emissions, such as hydrogen. An increasing number of ships may be powered by natural gas.

New tests have been required to evaluate the properties and performance of these new lubricants and fluids and the impacts they have on the new designs and arrangements of engines, motors, gearboxes, bearing and other components.

Most of these tests are still in development and are not yet standard methods. Some involve physical properties and performance. Many have been and will be used to develop new and improved lubricants and fluids. Many are likely to involve lengthy testing and evaluation times. As a consequence, they will not be suitable for use in a lubricant condition monitoring programme.

It would therefore seem likely that many of the tests that will be required in the future to monitor the condition of these new types of oils and greases in the new applications are those that have already been described in this chapter. Some new lubricant condition monitoring tests may be required in the future. When they have gained at least some utility, credibility and acceptance, they will be included in future editions of this book.

7.5 SUMMARY

A very large number of tests can be used to assess the physical properties and performance of lubricating oils and greases. However, not all of these tests are suitable for use in a lubricant condition monitoring programme, either because they take too long to be of practical benefit or they require large volumes of test sample.

While the ideal tests for both new and in-service lubricants provide results relatively quickly (generally within 24 hours) and are comparatively low cost, some of the equipment required to perform the tests can be very expensive. In recent years a number of the test methods have been automated, thereby reducing test times and manpower costs.

Some of these tests are applicable for lubricants used in a wide range of machines or items of equipment. Other tests are more applicable to specific types of machines or systems. Some tests should only be used to assess new lubricants. Other tests can be used for both new and in-service lubricants.

8 Mechanical Rig Tests for Lubricants

8.1 INTRODUCTION

One of the main functions of a lubricant is to reduce mechanical wear. Closely related to wear reduction is the ability of lubricants of the extreme-pressure (EP) type to prevent scuffing, scoring and seizure as applied loads are increased. As a result, a considerable number of machines and procedures have been developed to try to evaluate anti-wear and EP properties. In a number of cases, the same machines are used for both purposes, although different operating conditions may be used.

Wear can be divided into four classifications based on the cause; abrasive wear, corrosive (chemical) wear, adhesive wear and fatigue wear.

Abrasive wear is caused by abrasive particles, either contaminants carried in from outside or wear particles formed as a result of adhesive wear. In either case, oil properties do not have much direct influence on the amount of abrasive wear that occurs, except through their ability to carry particles to filtering systems that remove them from the circulating oils.

Corrosive or chemical wear results from chemical action on the metal surfaces combined with rubbing action that removes corroded metal. A typical example is the wear that may occur on cylinder walls and piston rings of diesel engines burning high sulphur fuels. The strong acids formed by combustion of the sulphur can attack the metal surfaces, forming compounds that can be fairly readily removed as the rings rub against the cylinder walls. Direct measurement of this wear requires many hours of test unit operation, which is often done as the final stage of testing new formulations. However, useful indications have been obtained in relatively short periods of time by means of sophisticated electronic techniques.

Adhesive wear in lubricated systems occurs when, owing to load, speed or temperature conditions, the lubricating film becomes so thin that opposing surface asperities can make contact. If adequate extreme pressure additives are not present, scuffing and scoring can result, and eventually seizure may occur. Adhesive wear can also occur with EP lubricants when the reaction kinetics of the additives with the surfaces are such that metal-to-metal contact is not fully controlled.

In a mechanical engineering sense, seizure is what happens when adhesive wear becomes so catastrophic that sudden and acute localised welding of moving parts occurs. When this happens, the moving parts will suddenly stop moving and the machinery will either jam or parts will break. Broken parts are highly likely to cause further catastrophic damage to the machine.

Pitting occurs in two forms, pitting fatigue and pitting corrosion. Pitting fatigue usually happens during slow speed, shock loading conditions when the critical

DOI: 10.1201/9781003245254-8

contact (Hertzian) pressure allowable for a specific metal is exceeded. Pitting corrosion is a form of extremely localised corrosion that leads to the creation of small holes in metal surfaces.

It is generally accepted that alternating stresses lead to the fatigue of the metal, followed by pitting. The development of pits on metal surfaces is influenced by the metal, its surface treatment and finishing, the surface profile, operating conditions, the type of lubricant and the lubricating regime. Controlling pitting fatigue due to cracks initiated beneath the metal's surface appears to be possible only by reducing the Hertzian pressure, thereby reducing the alternating stresses.

Pitting corrosion results from the depassivation of a small area, which becomes anodic (oxidation reaction) while an unknown but potentially large area becomes cathodic (reduction reaction), leading to very localised galvanic corrosion. The corrosion penetrates the mass of the metal, with a limited diffusion of ions. The more conventional explanation for pitting corrosion is that it is an autocatalytic process. Metal oxidation results in localised acidity that is maintained by the spatial separation of the cathodic and anodic half-reactions, which creates a potential gradient and electromigration of aggressive anions into the pit.

8.2 REQUIREMENTS FOR MECHANICAL RIG TESTS FOR LUBRICANTS

A number of machines, such as the Almen, Alpha LFW-1, Falex, Four-Ball, SAE, Timken, Pin-on-Disk and Optimol SRV, are used to determine the loading conditions under which seizure, welding or drastic surface damage to test specimens would be permitted. Of these, the Alpha LFW-1, Four-Ball, Falex, Timken and Optimol SRV are also used to measure wear at loads below the failure load. The results obtained with these machines do not necessarily correlate with field performance, but in some cases, the results from certain machines have been reported to provide useful information for specific applications.

Either actual machines or scale model machines are used in testing the anti-wear properties of lubricants. Anti-wear hydraulic fluids are tested for anti-wear properties in pump test rigs using commercial hydraulic pumps of the piston-and-vane type. Frequently, industrial gear lubricants are tested in the FZG Spur Gear Tester, a scale model machine that reasonably approximates commercial practice with respect to gear tooth loading and sliding speeds. Correlation of the FZG test with field performance has been good.

Fatigue wear occurs under certain conditions when the lubricating oil film is intact and metallic contact of opposing asperities is either nil or relatively small. Cyclic stressing of the surfaces causes fatigue cracks to form in the metal, leading to fatigue spalling or pitting. Fatigue wear occurs in rolling element bearings and gears when there is a high degree of rolling, and adhesive wear, associated with sliding, is negligible. A variation of the fatigue pitting wear is the micropitting wear mechanism that results in small pits in the surfaces of some gears and bearings. Asperity interaction and high loads as well as metallurgy are factors influencing micropitting. An upgraded version of the standard FZG Spur Gear Tester is used to evaluate gear oil micropitting performance according to a German Research Institute test method called FVA (Forschungsvereinigung Antriebstechnik) Method 54.

The understanding of the influence of lubricant composition and characteristics is an evolving technology. Some products are currently available that address fatigue-induced wear in anti-friction bearings and gears caused by the presence of small amounts of water in the Hertzian load zones. In these instances, the fatigue is called water-induced fatigue, and oil chemistry can reduce the negative effects of water in the load zones of antifriction bearings and gears. Two tests have been introduced by Institute of Petroleum for studies of this type. In the IP 300 test, "Pitting Failure Tests for Oils in a Modified Four-Ball Machine", oils are tested under conditions that cause cyclic stressing of steel bearing balls. To shorten the test time, the load and stress are set higher than is usual for the ball bearings. In another test, IP 305, "The Assessment of Lubricants by Measurement of Their Effect on the Rolling Fatigue Resistance of Bearing Steel Using the Unisteel Machine", a ball thrust bearing with a flat thrust ring is used. The flat thrust ring is the test specimen. With a flat thrust ring, stresses are higher than would be normally encountered if a raceway providing better conformity were machined in the ring. Both tests are run in replicate. Results to date indicate that both tests provide useful information on the effect of lubricant physical properties and chemical composition on fatigue life under cyclic stressing conditions.

Mechanical rig tests that might be suitable in a lubricant condition monitoring programme should be relatively quick to complete, ideally in 2 hours or less. An example of a very useful test that does not meet this criterion is ASTM D8165 "Load-Carrying Capacity of Lubricants Used in Hypoid Final-Drive Axles Operated Under Low-Speed and High-Torque Conditions". Parts of the test method are written for use by laboratories that make use of ASTM Test Monitoring Center (TMC) services. The test, commonly referred to as the L-37-1 test, describes a procedure for evaluating the load-carrying capacity, wear performance and EP properties of a gear lubricant in a hypoid axle under conditions of low-speed, high-torque operation. The test measures a lubricant's ability to protect hypoid final-drive axles from abrasive wear, adhesive wear, plastic deformation and surface fatigue when subjected to low-speed, high-torque conditions. Lack of protection can lead to premature gear or bearing failure or both.

An axle ring and pinion gearset are mounted in an axle housing, which is installed on a test stand equipped with the appropriate controls for speed, torque, lubricant temperature, axle cooling and various other operating parameters. The axle assembly is driven by an electric motor. Before each test run, the axle assembly is built, cleaned and inspected and build specifications are measured and recorded, and the gear is conditioned under specified operating conditions. The test method consists of running the axle unit for 24 hours at 80 wheel rpm and 2,359 Nm wheel torque. Two test methods are described in the standard. The standard test has a lubricant temperature in the axle of 135°C. The Canadian test has a lubricant temperature of 93°C. The ring gear and pinion gear are removed and rated for various forms of distress.

Although the test is very useful for evaluating the performance of a new gear oil during the development of an improved formulation, it takes 24 hours to complete, making it too long for inclusion in a lubricant condition monitoring programme.

8.3 MECHANICAL RIG TESTS FOR OILS AND GREASES

8.3.1 Four-Ball Wear and Weld Load Tests

ASTM D4172 "Standard Test Method for Wear Preventive Characteristics of Lubricating Fluids (Four-Ball Method)" can be used to determine the relative wear preventive properties of lubricating fluids in sliding contact under the prescribed test conditions. No attempt has been made to correlate the test with balls in rolling contact. ASTM notes that "The user of this test method should determine to his or her own satisfaction whether results of this test procedure correlate with field performance or other bench test machines".

The oil is tested in a four-ball system where a rotating ball slides over three stationary balls. Three 1/2 inch diameter steel balls are clamped together in a test cup and covered with the test oil. The oil is heated to 75°C, and then a fourth identical ball (referred to as the top ball) is pressed into the cavity formed by the three clamped, stationary balls with a force of 15 or 40 kgf. The three lower balls are pressed against the top ball by means of a load lever, as shown in Figure 8.1. This arrangement forms a three-point contact. The top ball is then rotated at either 1,200 rpm (in the United States) or 1,500 rpm (in Europe) for 60 minutes. The average scar diameter (mm) of the three stationary balls and the load used in the test are reported.

The ASTM D2266 test procedure is used to determine the relative anti-wear properties of lubricating greases. The test method and procedure are identical to ASTM D4172. The average scar diameter (mm) of the three stationary balls and the load used in the test are reported.

ASTM D2783 "Standard Test Method for Measurement of Extreme-Pressure Properties of Lubricating Fluids (Four-Ball Method)" covers the determination of the load-carrying properties of lubricating fluids. The test is technically similar to

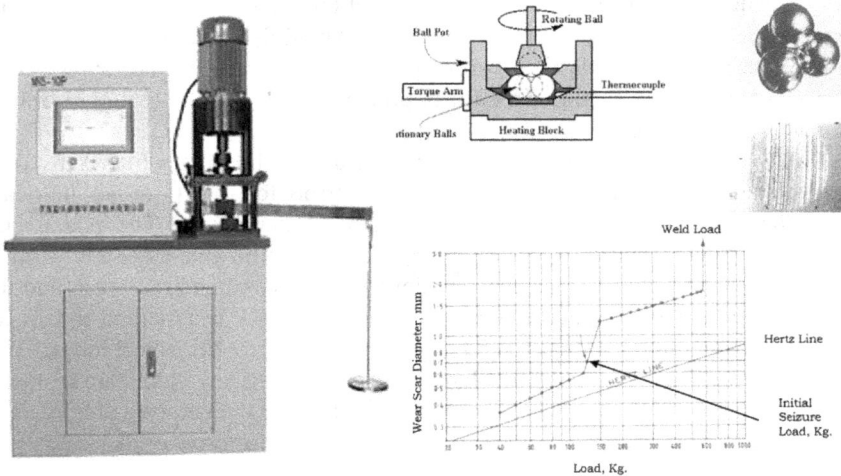

FIGURE 8.1 Four-ball test machine and balls.

Source: Pathmaster Marketing Ltd.

IP 239, although with modifications to speed and time, noted below. The following two determinations are made:

- Load-wear index (formerly Mean-Hertz load).
- Weld point (Weld load).

The oil is tested in the four-ball system where a rotating ball slides over three stationary balls. The test procedure differs from the four-ball wear test in that it is run at higher loads to the point where the test balls are "welded" together. Three 1/2 inch diameter steel balls are clamped together in the test cup and covered with the test oil. The oil is brought to between 65°F and 95°F (18°C and 35°C), and then a fourth identical ball (referred to as the top ball) is pressed downward into the cavity formed by the three clamped, stationary balls. This arrangement forms a three-point contact. In the United States, the top ball is rotated at 1,800 rpm, and a series of tests of 10 seconds duration are made at increasing loads until welding of the balls occurs. In Europe, the rotation of speed of the top ball is usually 1,500 rpm and the series of tests are run for 1-minute durations. The weld point (kgf) and the load-wear index (kgf) are reported.

The ASTM D2596 test procedure is used to determine the relative load-carrying properties of lubricating greases. The method and procedure are identical to ASTM D2783. The method is technically equivalent to IP 239 and ISO 11008.

ISO 20623 specifies the test conditions for European laboratories for using the four-ball test rig for wear and load carrying tests. DIN 51350 covers the methods and procedures for determining the wear and load-carrying properties of oils and greases, all in one standard.

8.3.2 FALEX PIN AND VEE TESTS

ASTM D2670 "Standard Test Method for Measuring Wear Properties of Fluid Lubricants (Falex Pin and Vee Block Method)" is used to determine wear obtained with fluid lubricants under prescribed test conditions.

A test pin is placed between two stationary vee blocks and this assembly is immersed in the test oil. The pin is driven at 290 rpm and a prescribed load is applied to the vee blocks by a ratchet mechanism. Wear is determined and recorded as the number of teeth the ratchet system advanced to maintain constant load during the test. Certain fluid lubricants may require different test parameters depending upon their performance characteristics. The values stated in inch-pound units are to be regarded as the standard. A photograph of the machine and diagram of the pin and blocks are shown in Figure 8.2.

Again, according to the ASTM, "The user of the test method should determine to his or her own satisfaction whether results of the test procedure correlate with field performance or other bench test machines. If the test conditions are changed, wear values may change and relative ratings of fluids may be different".

ASTM D3233 "Standard Test Methods for Measurement of Extreme Pressure Properties of Fluid Lubricants (Falex Pin and Vee Block Methods)" covers two procedures for making a preliminary evaluation of the load-carrying properties of

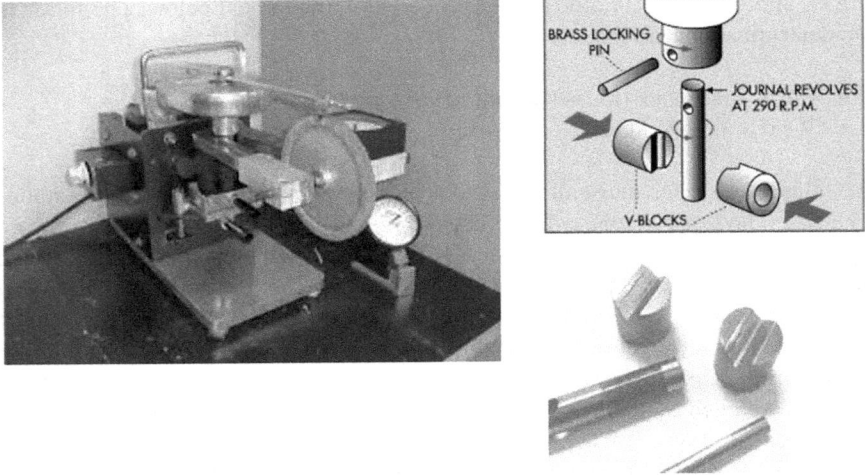

FIGURE 8.2   Falex machine, blocks and pin.

Source: Pathmaster Marketing Ltd.

industrial oils. Evaluations by both test methods differentiate between fluids having low, medium and high levels of EP properties. The user should establish any correlation between results by either method and service performance.

For both procedures, a test pin is placed between two stationary vee blocks and immersed in the sample oil. As with ASTM D2670, the pin is driven at 290 rpm and a load is applied to the vee blocks by a ratchet mechanism. In Test Method A, an increasing load is continuously applied. In Test Method B, loads are applied in 250 lbf increments with each load maintained for 1 minute. For both methods, there is a 5-minute run-in period at 300 lbf. The results are reported as the load at which the test pin or shear pin breaks or when the inability to maintain or increase the load occurs. If no failure occurs, the last load run is reported with a plus (+) sign after the value.

Relative ratings by both test methods on the reference fluids covered in the detailed procedures in ASTM D2670 are in good general agreement with four-ball weld-point relative ratings obtained on these same reference fluids, covered in ASTM D2783. Additional information can be found in Appendix X1 regarding coefficient of friction, load gauge conversions and load gauge calibration curve.

8.3.3   TIMKEN MACHINE TEST

ASTM D2782, IP 240 and DIN 51434 "Standard Test Method for Measurement of Extreme-Pressure Properties of Lubricating Fluids (Timken Method)" is used widely for the determination of EP properties, by means of the Timken Extreme Pressure Tester, for specification purposes. Users are cautioned to carefully consider the precision and bias statements in the test method when establishing specification limits.

The tester is operated with a steel test cup rotating against a steel test block. The rotating speed is 123.71 ± 0.77 m/min (405.88 ± 2.54 ft/min) which is equivalent to

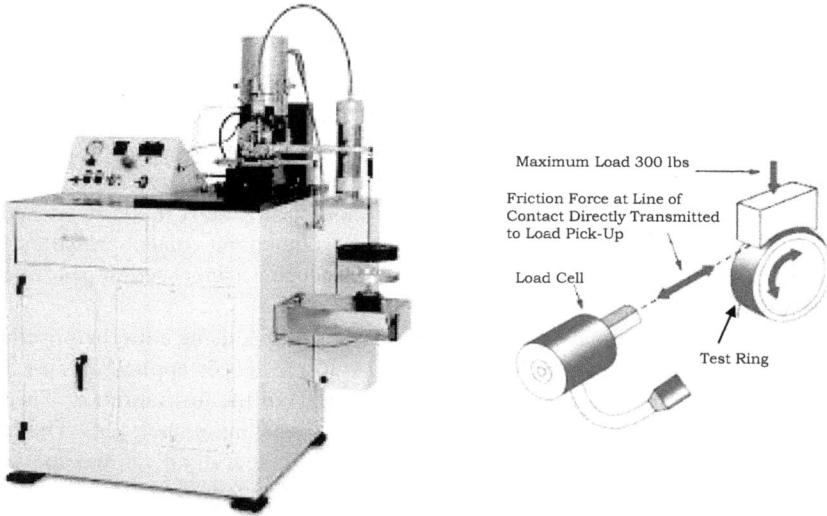

FIGURE 8.3   Timken machine.

Source: Pathmaster Marketing Ltd.

a spindle speed of 800 rpm. Fluid samples are pre-heated to 37.8°C (100°F) before starting the test. The test cup is loaded against the test block by means of a load lever. The test fluid is circulated around the test cup and block during the test. A photograph of the machine and a diagram of the cup and block are shown in Figure 8.3.

Two determinations are made:

- The minimum load (score value) that will rupture the lubricant film being tested between the rotating cup and the stationary block and cause scoring or seizure.
- The maximum load (OK value) at which the rotating cup will not rupture the lubricant film and cause scoring or seizure between the rotating cup and the stationary block.

The test method is suitable for testing fluids having a viscosity of less than about 5,000 cSt (mm²/s) at 40°C. For testing fluids having a higher viscosity, the fluid temperature should be increased to 65.6°C (150°F). The values stated in SI units are to be regarded as standard. Because the equipment used in this test method is available only in inch-pound units, SI units are omitted when referring to the equipment and the test specimens.

ASTM D2509 "Standard Test Method for Measurement of Load-Carrying Capacity of Lubricating Grease (Timken Method)" uses the same machine as ASTM D2782. The test temperature of the grease is 24°C ± 6°C (75°F ± 10°F). The test grease is not circulated around the test cup and block; a grease feed mechanism is used to apply the test grease at a uniform rate of 45 ± 9 g/min.

8.3.4   SRV Tests

The ASTM D6425 and DIN 51834 "Standard Test Method for Measuring Friction and Wear Properties of Extreme Pressure (EP) Lubricating Oils Using SRV Test Machine" test method covers an EP lubricating oil's coefficient of friction and its ability to protect against wear when subjected to high-frequency, linear oscillatory motion. The test method can be used at selected temperatures and loads specified for use in applications in which high-speed vibration or stop-start motions are present for extended periods of time under initial high Hertzian point contact pressures. This method has found application as a screening test for lubricants used in gear or cam/follower systems.

The test method is performed on an SRV test machine, using a test ball oscillated at constant frequency and stroke amplitude. A 300 N load is applied against a test disk that has been moistened with 0.3 millilitre (ml) of the test lubricant. The platform to which the test disk is attached is held at a constant temperature. The test is run for 2 hours. At the completion of the test, the wear scar diameter and coefficient of friction trace are reported along with the test parameters (temperature, break-in load, oscillating frequency, stroke, ball material and disk material). A photograph of the machine and the mechanism of oscillating motion are shown in Figure 8.4.

Another test that uses the SRV machine is ASTM D7421 "Extreme-Pressure Properties of Lubricating Oils Using a High-Frequency, Linear-Oscillation (SRV) Test Machine". The test method is based on ASTM D5706 (see later in this section) and DIN 51834 "High Frequency Oscillating Tribometer 'SRV'" and covers a procedure for determining the EP properties of hydraulic oils, gear oils and engine oils under high-frequency linear-oscillation motion using the SRV test machine. It can be

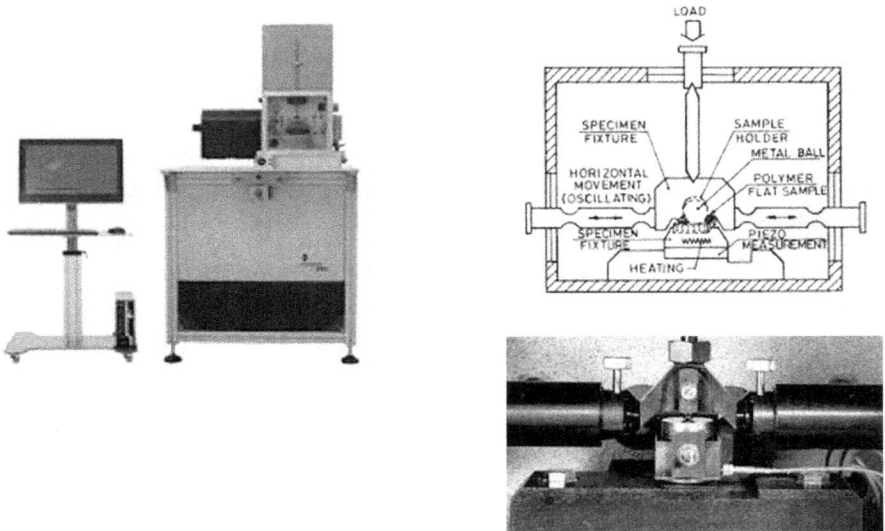

FIGURE 8.4   Optimol SRV machine.

Source: Pathmaster Marketing Ltd.

used to quickly determine them at selected temperatures specified for use in applications in which not only high-speed vibrational or start-stop motions are present with high Hertzian contact. The test method has found wide application in qualifying lubricating oils used in constant velocity joints of front-wheel-drive cars, in gears and in engine components.

The test is performed on an SRV machine using a steel test ball oscillating against a stationary steel test disk with lubricant between them. The test load is increased in 100-N increments until seizure occurs. The test load is increased every 2 minutes until the specimens weld together, indicating lubricant failure. The load immediately before the load at which seizure occurs is measured and reported as OK-load, which can be converted into Hertzian contact pressure. Because the test is relatively quick, it may be of value in a lubricant condition monitoring programme for those laboratories that can afford to purchase and run an SRV machine.

The SRV machine can also be used to evaluate the lubricant performance of greases. ASTM D5707 "Friction and Wear Properties of Lubricating Grease Using a High-Frequency, Linear-Oscillation (SRV) Test Machine" is used for determining wear properties and coefficient of friction of lubricating greases at selected temperatures and loads. These are specified for applications in which high-speed vibrational or start-stop motions are present for extended periods of time under initial high Hertzian point contact pressures. The test method has found application in qualifying lubricating greases used in constant velocity joints of front-wheel-drive cars and lubricating greases used in roller bearings.

The test is performed using a test ball oscillated under constant load against a test disk. The test load is 200 N, the test frequency is 50 Hz, a 1.00 mm stroke amplitude and test temperatures ranging from ambient to 280°C. The test duration is 2 hours. Other test loads from 10 to 1,400 N, frequencies from 5 to 500 Hz and stroke amplitudes from 0.1 to 3.30 mm can also be used, if specified. The wear scar on the test ball and coefficient of friction are measured. If a profilometer is available, a trace of the wear scar on the disk can be used to obtain additional wear information.

ASTM D7594 "Fretting Wear Resistance of Lubricating Greases Under High Hertzian Contact Pressures Using a High-Frequency Linear-Oscillation (SRV) Test Machine" can be used to determine anti-wear properties and coefficient of friction of greases to prevent "fretting" wear under linear oscillation with associated strokes and high Hertzian contact pressures at selected temperatures and loads. These are those specified for use in applications in which induced, high-speed vibrational motions are present for extended periods of time. It has found applications as a screening test for grease lubricants used in ball and roller bearings, roller or ball screw (spindle) drives or side shaft systems.

The test method is performed using a steel test ball oscillating under constant frequency, short stroke amplitude and constant load, against a stationary steel test disk with a grease lubricant between them to determine the coefficient of friction and wear scar diameter. Again the test duration is 2 hours, so may be suitable for a lubricant condition monitoring programme.

ASTM D5706 "Extreme-Pressure Properties of Lubricating Greases Using High-Frequency, Linear-Oscillation (SRV) Test Machine" is used to quickly determine the EP properties of lubricating greases at selected temperatures. These are those specified for use in applications in which high-speed vibrational or start-stop motions are present with high Hertzian point contact. The test method has found wide application in qualifying lubricating greases used in constant velocity joints of front-wheel-drive cars.

This test method is also performed using a steel test ball oscillating against a steel test disk with lubricant between them. The test load is increased in 100-N increments until seizure occurs. Again, the test load is increased every 2 minutes until the specimens weld together, indicating lubricant failure. The load immediately before the load at which seizure occurs is measured and reported. Test frequency, stroke length, temperature and ball and disk materials can be varied to simulate field conditions. The method is similar to ASTM D4172 four-ball wear test method.

8.3.5 PIN-ON-DISK TESTS

Pin-on-disk wear testing is a method of characterising the coefficient of friction, frictional force and rate of wear between two materials. Multiple configurations are available depending on the test goals and objectives. Common specifications include ASTM G99, ASTM G133 and ASTM F732.

Pin-on-disk wear testing can simulate multiple modes of wear, including: unidirectional, bidirectional, omnidirectional and quasi-rotational. Mass-loss evaluation and differential analysis of test fluids are typically performed post-test to characterise wear properties. In addition, a contact profilometer can be used to evaluate the changes in surface topography due to articulation. Metallurgical evaluation of the post-test wear scarring can also be performed. Testing can also introduce third body debris for accelerated wear evaluations.

A pin-on-disk tribometer consists of a stationary "pin" under an applied load in contact with a rotating disk. The pin can have any shape to simulate a specific contact, but spherical tips are often used to simplify the contact geometry. Coefficient of friction is determined by the ratio of the frictional force to the loading force on the pin. The pin-on-disk test has proved useful in providing a simple wear and friction test for low-friction coatings such as diamond-like carbon coatings on valve train components in internal combustion engines. A photograph of a pin-on-disk machine and the mechanism of rotating motion are shown in Figure 8.5.

ASTM G99 "Standard Test Method for Wear Testing with a Pin-on-Disk Apparatus" is used for many different types of wear of sliding materials, including lubricants in metal-to-metal contacts.

The amount of wear in any system will, in general, depend upon the number of system factors such as the applied load, machine characteristics, sliding speed, sliding distance, the environment and the material properties. The value of any wear test method lies in predicting the relative ranking of material combinations. Since the pin-on-disk test method does not attempt to duplicate all the conditions that may be experienced in service (for example, lubrication, load, pressure, contact geometry, removal of wear debris and/or presence of a corrosive environment), there is no certainty that the test will predict the wear rate of a given material under conditions differing from those in the test.

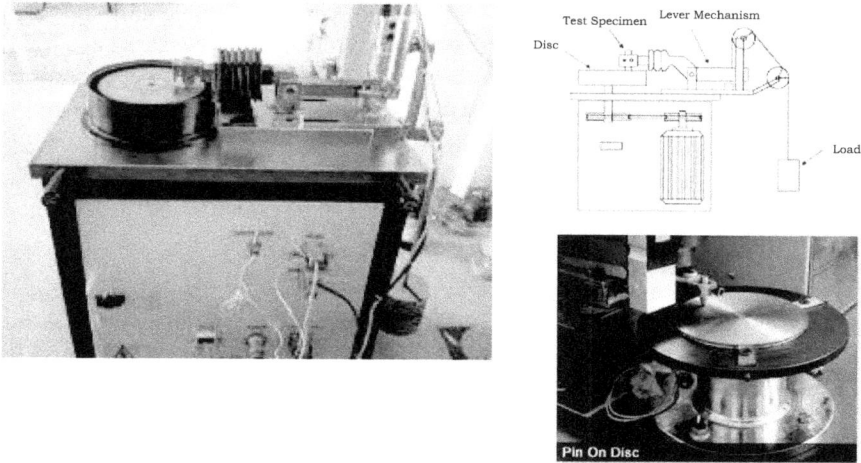

FIGURE 8.5   Pin-on-disk machine.

Source: Pathmaster Marketing Ltd.

Use of the ASTM G99 test method will fall into one of two categories:

* The test(s) will follow all particulars of the standard, and the results will have been compared with the ILS data.
* The test(s) will have followed the procedures/methodology of ASTM G99. but applied to other materials or using other parameters such as load, speed, materials and/or other factors.

In this latter case, the results cannot be compared with the ILS data, so it must be clearly stated what choices of test parameters and materials were used.

8.3.6   FE-8 BEARING RIG TEST

The DIN 51819 "Testing of lubricants – Mechanical-dynamic testing in the roller bearing test apparatus FE8- Part 3: Test method for lubricating oils – applied test bearing: axial cylindrical roller bearing" test is used for mechanical-dynamical tests of oils and greases in roller bearings. The determination of the wear protection of oils and greases takes place under practice-related conditions. In the wear test, the test duration and the loss of weight of the bearing components are used to classify the aptitude of the lubricants.

For oil tests, the test bearings are cylindrical roller thrust bearings. For grease tests, the test bearings are either tapered roller bearings or angular contact ball bearings. Two specific test bearings are assembled into each test head, the load as specified on the test method is applied to the test head that is then assembled onto the test rig. The friction torque is continuously recorded during the entire test as the summary friction torque of both bearings. Both bearing temperatures are recorded and allocated to the temperature controls of the heating and the cooling unit. Before and after the examination, the weights of the bearing elements are measured and used to evaluate the quality of the oil or grease.

Because of the amount of oil or grease used in FE-8 rig tests, the method is really suitable only for new oils or greases and is, therefore, not really suited for an oil or grease condition monitoring programme.

8.3.7 FZG Test

IP 334 "Load-Carrying Capacity Test for Oils (FZG Gear Machine)" is used to determine the scuffing properties of industrial lubricating oils as defined by the total damaged surface area of the gear teeth. The test is technically equivalent to Method B of CEC L-07-95. FZG is Forschungsstelle für Zahnräder und Getriebebau; Technische Hochschule Munchen (Research Institute for Gears and Gear Design; Technical University Munich).

Special gears are run in the test oil at a constant speed of 1,450 rpm for a fixed period of 21,700 revolutions (about 15 minutes) at successively increasing loads. The load on the gears is increased through 12 stages or until the failure criteria are reached. A photograph of an FZG gear machine and the arrangement of the loaded gears is shown in Figure 8.6. The test oil is filled into the gear case up to the centre line of the gears, which is usually about 1.25 litres.

Starting at load stage 4, the initial oil temperature is specified to be between 60°C and 80°C. However, the temperature of test oils will often rise during a load stage. When the temperature rises above 90°C ± 3°C, the test oil and gears must be cooled (using a cooling water system) to 90°C ± 3°C before running the next load stage. The gears are examined after each load stage. The test is complete when the total width of damage on all of the gear teeth equals the width of one tooth. The failure

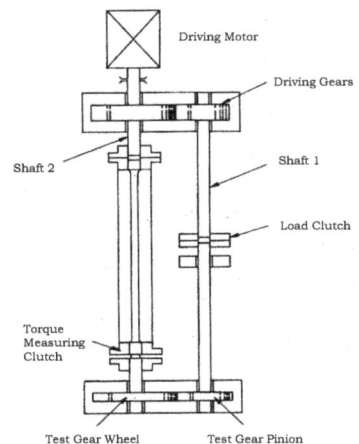

FIGURE 8.6   FZG machine.

Source: Pathmaster Marketing Ltd.

load stage is reported along with a description of the gear teeth damage. If the failure criteria are not met at the end of load stage 12, then the failure load stage is reported as greater than 12.

The test results are reported as the load stage in which damage occurs and the test conditions are reported as A/8.3/90 (Visual Method). ("A" refers to the design of the special gears.) This test is used for oils with mild EP/AW performance.

Again, the volume of oil required for the test means that it can only be used realistically in a condition monitoring programme for in-service gear, hydraulic and compressor oil samples from the largest systems.

8.3.8 Filterability Tests

AFNOR filtration tests are designed to evaluate mineral oils used for hydraulic applications. They measure the ability of an oil to pass through a fine filter without plugging under wet or dry conditions. The AFNOR NFE48-690 test method specifies the conditions for testing dry oils, and AFNOR NFE48-691 specifies the conditions under which oils containing a small amount of free water are tested. The wet test method is more severe. The tests are designed for samples of new oils, to determine their likely properties and performance in use.

A 320 ml sample of the test oil is poured into a 500 ml jar. If a wet filtration is to be run, 0.2% vol distilled water is added to the jar at this time. The jar is capped and is placed in an oven at 70°C for 2 hours. The jar is then removed from the oven, and the test sample is mixed for 5 minutes at 1,500 rpm. The jar is re-capped and returned to the oven for 70 hours. After 70 hours in the oven, the jar is removed and placed in a dark cabinet for 24 hours at room temperature. The test sample is then pressure filtered at 14.5 psi through a fine filter into a graduated cylinder. The times to filter several specified volumes of the test oil are recorded, and a ratio of these filtration times is calculated. This ratio is reported as the filterability index for the test sample.

The ISO 13357 filterability test has two parts; ISO 13357-1 and ISO 13357-2. ISO 13357 "Petroleum products – Determination of the filterability of lubricating oils – Part 1: Procedure for oils in the presence of water" is essentially the same as AFNOR T48-691. It specifies a procedure for the evaluation of the filterability of lubricating oils, particularly those designed for hydraulic applications, in the presence of water. The procedure only applies to mineral-based oils, since fluids manufactured from other materials (for example, fire-resistant hydraulic fluids) may not be compatible with the specified test membranes. The range of application has been evaluated with oils of viscosity up to ISO viscosity grade (VG) 100, as defined in ISO 3448. Within the range described, the filterability as defined is not dependent on the viscosity of the oil.

ISO 13357-2 "Petroleum products – Determination of the filterability of lubricating oils – Part 2: Procedure for dry oils" test method is essentially the same as AFNOR 48-690. Again, the method only applies to mineral-oil-based hydraulic fluids with viscosities up to ISO viscosity grade (VG) 100.

Similarly, the IP 448 test method has two parts: Part 1 for oils in the presence of water and Part 2 for dry oils.

The Denison Filterability Test, TP-02100, is also intended to evaluate the filterability (ability to be filtered) properties of petroleum base and synthetic hydraulic fluids. Many fluids used in industrial and mobile hydraulic systems do not filter easily, especially in the presence of small amounts of water contamination in the system. This may result in plugging of system filters and thereby drastically increase the contamination wear of pumps and other components in the system. This test permits the evaluation of this important quality of the hydraulic fluids with and without water.

According to method A-TP-02100, the test requires four 100 ml samples of test fluid, with and without water, to be filtered through a 47 mm diameter 1.2 microns (μm) filter paper, under a vacuum of 66 cm of mercury (0.88 bar). 2% v/v of distilled water is added to two samples of test oil, which are then shaken vigorously for 5 minutes. All tests are done at room temperature. The times required to filter 75 ml of each of the four test fluids are reported. The filters are inspected and any residues or other signs or contamination are also reported.

Filterability tests can be quite useful in an oil condition monitoring programme, as the volumes of test oil can be quite small and the tests do not take much time to do on in-service oils. Samples taken from hydraulic systems do not need to be prepared for test in the same ways as in the AFNOR, ISO, IP or Denison tests with wet oils.

However, in-service oil samples that contain moderate or high levels of particulates, whether dirt or wear metals, are likely to give poor filterability test results. If an in-service oil does not have moderate or high levels of particulates but still shows poor filterability, this could be a sign of oil and/or additive degradation. Other tests, discussed in Chapters 6 and 7, are likely to corroborate this finding.

8.3.9 SHEAR STABILITY TESTS

IP 294 "Shear Stability of Polymer-Containing Oils using a Diesel Injector Rig" covers procedures for evaluating the shear stability of polymer-containing oils in terms of the permanent viscosity loss resulting when a sample is mechanically stressed under the test conditions. The method is technically equivalent to DIN 51382 and has been adopted by CEC as CEC L-14-93.

The test fluid is mechanically stressed by being pumped through a circuit incorporating a diesel injector nozzle and a diesel injection fuel pump. The injector breaking pressure is set at 175 bar (2,540 psi). Samples are taken after the fluid has completed a predetermined number of passes through the circuit. Kinematic viscosities of the original sample and of samples taken during the test are determined at 40°C and 100°C. These figures are used to calculate on a percentage basis the extent of oil degradation due to shear forces in this test. Up to 30 passes are usually sufficient for gasoline and diesel engine oils, but hydraulic oils need 250 or even 500 passes.

ASTM D6278 "Shear Stability of Polymer Containing Fluids Using a European Diesel Injector Apparatus" uses the same equipment as IP 294, but different operating conditions and a different period of time required for calibration. Only 30 passes are used in the test and only kinematic viscosities at 100°C of the test oil before and after test are measured. The test method evaluates the percent viscosity loss for polymer-containing fluids resulting from polymer degradation in the high-shear nozzle device. Thermal or oxidative effects are minimised. According to the ASTM, the test is not intended to predict viscosity loss in field service in different field equipment under widely varying operating conditions, which may cause lubricant viscosity to change because of thermal and oxidative changes as well as by the mechanical shearing of the polymer.

ASTM D2603 (see later) has been used for similar evaluation of shear stability, but no detailed attempt has been made to correlate the results by the two methods. ASTM D5275 (see later) also shears oils in a diesel injector apparatus, but may give different results. This test method has different calibration and operational requirements than ASTM D3945, which was withdrawn in 1998.

As a consequence of the differences in the amount of shearing, depending on the stability of the polymer and the number of passes through the diesel injector, ASTM D7109 "Shear Stability of Polymer-Containing Fluids Using a European Diesel Injector Apparatus at 30 Cycles and 90 Cycles" was introduced. The test uses the apparatus as defined in IP 294 and CEC L-14-93 and passes the test oil through the diesel injector for 30 times and 90 times.

ASTM D5275 "Standard Test Method for Fuel Injector Shear Stability Test (FISST) for Polymer Containing Fluids" also evaluates the percentage viscosity loss for polymer-containing fluids resulting from polymer degradation in the high-shear nozzle device. Minimum interference from thermal or oxidative effects is anticipated. The test is not intended to predict viscosity loss in field service for different polymer classes or for different field equipment. The test covers the measurement of the percent viscosity loss at 100°C of polymer-containing fluids using fuel injector shear stability test (FISST) equipment. It was originally published as Procedure B of ASTM D3945, but was made a separate test after tests using a series of polymer-containing fluids showed that Procedures A and B of ASTM Methods D3945 often gave different results.

The ASTM has noted that, while ASTM D5275, ASTM D6278 and ASTM D7109 have found some utility in the evaluation of crankcase oils, the stress imparted to the sample has been found to be insufficient to shear polymers of the shear-resistant type found in aircraft hydraulic fluids.

ASTM D2603 "Sonic Shear Stability of Polymer-Containing Oils" permits the evaluation of shear stability with minimum interference from thermal and oxidative factors which may be present in some applications. According to the ASTM, within the limitations expressed in the scope of the test method, it has been successfully applied to hydraulic fluids, transmission fluids, tractor fluids and other fluids of similar applications. It has been found applicable to fluids containing both readily sheared and shear-resistant polymers. Correlation with performance in the case of automotive engine oils has not been established.

The test evaluates the shear stability of an oil containing a polymer in terms of the permanent loss in viscosity that results from irradiating a sample of the oil in a sonic oscillator. However, it is not intended to predict the performance of polymer-containing oils in service. Evidence has been presented that correlation between the shear degradation results obtained by means of sonic oscillation and those obtained in mechanical devices can be poor. This is especially true in the case of automotive engine oils. Other evidence indicates that the sonic technique may rate different families of polymers in a different order than mechanical devices.

ASTM D5621 "Standard Test Method for Sonic Shear Stability of Hydraulic Fluids" was developed using ASTM D2603 and has been found applicable to fluids containing both readily sheared and shear-resistant polymers. Correlation with performance in the case of hydraulic oil has been established. Also, evidence has been presented that a good correlation exists between the shear degradation that results from sonic oscillation and that obtained in a vane pump test.

Hydraulic oil is irradiated in a sonic oscillator for 40 minutes and the change in kinematic viscosity is determined using the ASTM D445. A standard reference fluid containing a readily sheared polymer is run frequently to ensure that the equipment imparts a controlled amount of sonic energy to the sample. The test equipment is based on a probe-type ultrasonic device with a fixed-frequency oscillator and ultrasonic horn run at a frequency of 24 kHz and power rating of 400W, with a cooling water bath or ice bath in order to maintain a desired temperature, such as 0°C.

IP 351 "Shear Stability of Polymer-Containing Oils Using the FZG Gear Rig" also covers a procedure for evaluating the shear stability of polymer-containing oils in terms of the permanent viscosity loss resulting when a sample is mechanically stressed under the test conditions described.

Special gear wheels are run in the lubricant under test in a dip lubrication system at a constant speed for a fixed time. The bulk oil temperature is controlled and the loading of the gear teeth is set according to the chosen condition. At the end of the test period, the oil is assessed for permanent viscosity loss and the difference in viscosity between the new and sheared oil is used to characterise the shear stability of the oil.

The FZG rig has for some years been used to evaluate the shear stability of gasoline and diesel engine oils. The test has been found to be particularly relevant to applications in which engine and transmission share a common lubricant. This method includes operating conditions appropriate to crankcase lubricants, transmission lubricants and hydraulic fluids. The test is likely to be especially suitable for products exhibiting high shear stability, when the diesel injector shear stability test (IP 294) may not be severe enough.

Three test conditions can be used to assess the shear stability of a test oil:

- Condition A: Suitable for engine oils. FZG load stage 3, oil temperature 90°C, 14 hour test.
- Condition B: Suitable for gear oils. FZG load stage 6, oil temperature 90°C, 20 hour test.
- Condition C: Suitable for hydraulic oils. FZG load stage 3, oil temperature 60°C, 14 hour test.

In all three conditions, the motor speed is 3,000 rpm and 800 g of test oil is used.

CEC L-45-99 "Viscosity Shear Stability of Transmission Lubricants (Taper Roller Bearing Rig)" uses a standard design four-ball EP tester with thermostatic control, but with a test head to accommodate a tapered roller bearing. The test procedure is used widely among OEMs and suppliers for engine oil, transmission oils and hydraulic oils. DIN 51350 and ISO 26422 tests are technically equivalent to CEC L-45-99.

A sample of fluid is used to lubricate a tapered roller bearing that is mounted in a specially made housing. The bearing is loaded axially with 5,000 N, rotated at 1,475 rev/min and maintained at a temperature of 60°C. The test is run for duration of 1,740,000 motor revolutions (approximately 20 hours).

The target parameter is the relative shear loss of the tested fluids, calculated by the viscosity at start and end of test. Kinematic viscosity measurement of the fluid both before and after test is done according to the DIN 51562 method (or equivalent), and the test result is expressed as the percentage reduction in viscosity. The test allows conclusions to be drawn on the permanent viscosity loss to be expected under operating conditions, for example, in gearboxes, which is caused by mechanical stressing of the oil. The method is technically equivalent to ISO 26422.

Comparison of the various methods of determining the shear stability of polymer containing oils over many years has found that the IP 294, DIN 51382, CEC L-14-93, ASTM D5275, ASTM D6278 and ASTM D7109 tests using the Bosch diesel injector are considered to be the least severe of the alternative methods. The sonic shear methods, ASTM D2603 and ASTM D5621, have been found to be more severe and are favoured by some US OEMs. The most severe test is the CEC L45-A-99, DIN 51350, ISO 26422 method, which is becoming the test of choice of many OEMs worldwide.

All these tests are used primarily to test new oils and in the development of improved formulations. The sonic tests may be useful in a lubricant condition monitoring programme, but as the FZG gear and tapered roller bearing tests take either 14 or 20 hours to complete, they may not be suitable for use in such programmes.

8.4   FUTURE MECHANICAL RIG TESTS FOR OILS AND GREASES

As noted in Chapters 6 and 7, new types of machines, equipment and systems that use lubricants are being developed and introduced all the time. New tests have been required to evaluate the properties and performance of new lubricants and fluids and the impacts they have on the new designs and arrangements of engines, motors, gearboxes, bearing and other components.

Most of these tests are still in development and are not yet standard methods. Some involve chemical and physical properties and performance, while others may involve mechanical rig tests. Many have been and will be used to develop new and improved lubricants and fluids. Many are likely to involve lengthy testing and evaluation times. As a consequence, they will not be suitable for use in a lubricant condition monitoring programme.

It would therefore seem likely that many of the mechanical rig tests that will be required in the future to monitor the condition of these new types of oils and greases in the new applications are those that have already been described in this chapter. Some new lubricant condition monitoring tests may be required in the future. When they have gained at least some utility, credibility and acceptance, they will be included in future editions of this book.

8.5 SUMMARY

A number of mechanical rig tests are used to determine various properties and performances of oils and greases as lubricants. Some of these might be useful in a lubricant condition monitoring programme, although several are not, most usually because they require amounts of in-service oils or greases that are too large to justify.

Although there are several additional mechanical test methods used to assess lubricants, including gearbox, hydraulic system and compressor tests, these have not been included in this chapter, as they are not relevant to lubricant condition monitoring.

9 Engine Tests for Lubricants

9.1 INTRODUCTION

In the author's opinion, the only true test of a lubricant's performance is whether it functions satisfactorily for several years in the machinery for which it was formulated. Lubricant specifications, laboratory tests, rig tests, engine tests and field trials are only useful as a guide or prediction of likely performance in service. In assessing the possible suitability of a specific lubricant for a specific application, it is critically important to remember to check that the predictive tests are the best ones for that application.

All of these laboratory, bench, rig and engine tests and field trials simply attempt to shorten the time taken to assess the properties or performance of lubricants, because the "real life" performance of lubricants can take many years to establish. If actual performance in service was the only method by which lubricants could be evaluated, the development of new or improved products could take a very long time indeed. It is, however, very important to remember that laboratory, bench, rig, engine and even field tests are useful mainly for establishing lubricants that are unlikely to perform satisfactorily in practical operations, not those that definitely will perform satisfactorily for many years.

Laboratory tests, usually for physical or chemical properties, are usually quick and comparatively inexpensive. Bench and rig tests, which are intended to assess performance properties, such as anti-wear, corrosion inhibition or deposit formation, are performed in specially designed equipment or machines that are smaller and/or less complex than real engines or gearboxes. Consequently, they tend to be shorter and cheaper than engine tests. Engine tests are performed in special buildings, to try to provide as much repeatability and reproducibility as possible, and are more realistic, but are longer and more expensive than laboratory or rig tests. Field tests try to reproduce "real life" operating conditions, but are lengthy and expensive.

The tests described in this chapter are almost never used to monitor the condition of a lubricant in use. They are included in this book in order to describe why they are not used in a lubricant condition monitoring programme.

9.2 ENGINE TESTS

In these tests, a representative engine is mounted on a test bed and provided with fuel, a cooling system and a means of absorbing the power produced, such as a dynamometer or "brake". Engines can be either full-sized, multi-cylinder commercial engines or small, single-cylinder engines developed specifically for laboratory testing. Each engine will be run to a specified procedure, continuously or intermittently,

DOI: 10.1201/9781003245254-9

and with programmed variations in speed and load. The conditions of operation (temperature, speed and/or load) are likely to be more severe than in "normal" service, so as to reduce the testing time. Usually, before the test starts and after it is completed, the engine will be dismantled completely, the parts will be rated and/or measured for cleanliness and wear and the oil will be tested for physical, chemical and other properties,

For tests used in setting performance specifications, a supervising organisation will define test procedures and reference oils against which the performance of the test oil will be compared. In some aspects of engine performance, fuel quality is an important parameter, so this will be carefully specified. For most engine tests, a stock of reference fuel must be kept and replenished occasionally. Engines and parts have to be carefully selected and controlled to ensure consistency. Supplies of engines and parts must be available even if the test engine is no longer used in production vehicles. Organisation of continuing and consistent parts supply is one of the most difficult tasks of supervising organisations.

Engines used for lubricant testing have tended to increase in size significantly over the last few years, and full-sized large truck engines are now being commonly run on test beds. Computer control is widely used for engine testing. The organisation, supply, control and rating of engine parts make engine testing both manpower-intensive and expensive.

Military organisations, engine manufacturers, lubricant suppliers and additive manufacturers are all active in engine test method development. Over the last 25 years, a large independent engine test industry has developed, running tests on a contractual basis for those companies or laboratories which do not have the resources or desire to conduct the tests themselves, or which have not installed a particular test in their laboratories. There are so many different tests that no one laboratory can run them all.

Some research laboratories have developed flexible single-cylinder engines that can be adapted to meet the configuration of commercial multi-cylinder engines. However, in general, development of engines specifically for test purposes has proved time-consuming and expensive, so doubts on their relevance to the real world have tended to arise. As a result, using commercial engines of increasing size and complexity has become common. Large engines require larger test beds and dynamometers, more fuel and more time for rating and rebuilding than smaller ones, so the costs of engine test evaluation are rising continually.

Many of the more commonly used engines used for lubricant testing, and the main performance properties they are used to evaluate, are listed in Tables 9.1 to 9.5. Different engines are used to evaluate gasoline and light-duty diesel engines compared with heavy-duty diesel engines.

9.3 FIELD TESTS

Field testing might be expected to be the most useful performance evaluation to which a lubricant could be subjected. In truth it is surprisingly difficult to obtain worthwhile results from field tests. To do so requires a great deal of time and effort, so a good field test is not cheap. With care, however, valuable results can be obtained. Some field

TABLE 9.1
Light-Duty Engine Tests Used for ACEA Specifications

Test Engine	Performance Evaluation
Peugeot/BMW EP6CDT	Gasoline direct injection engine cleanliness
Ford 4.6 litre, 8 cylinders	Sequence VH: Low temperature sludge
Mercedes-Benz M271 EVO	Black sludge
Mercedes-Benz M111	Fuel economy
Peugeot DV6C	Direct injection diesel oil dispersion at medium temperature
Mercedes-Benz OM646LA	Diesel engine wear
Volkswagen TDI3	Direct injection diesel piston cleanliness and ring sticking
Toyota 2NRFE 1.5 litre	Sequence IVB: Valve train wear
Ford 2.0 litre Econoboost	Sequence IX: Low-speed pre-ignition
Ford 2.0 litre, 4 cylinder GTDI	Sequence X: Chain wear
Mercedes-Benz OM646LA Bio	Effects of biodiesel

Sources: ACEA, Lubrizol, Pathmaster Marketing Ltd.

TABLE 9.2
Light-Duty Engine Tests Used for ILSAC and API Specifications

Test Engine	Performance Evaluation
Chrysler Pentastar 3.6 litre V6	Sequence IIIH: Oxidation and deposits
Toyota 2NRFE 1.5 litre	Sequence IVB: Cam/wear
Ford 4.6 litre, 8 cylinders	Sequence VH: Sludge and varnish
General Motors 3.6 litre V6	Sequence VIE: Fuel economy
General Motors 3.6 litre V6	Sequence VIF: Fuel economy
CLR 0.7 litre, single cylinder	Sequence VIII: Corrosion
Ford 2.0 litre Econoboost	Sequence IX: Low-speed pre-ignition
Ford 2.0 litre, 4 cylinder GTDI	Sequence X: Chain wear

Sources: ILSAC, API, Lubrizol, Pathmaster Marketing Ltd.

testing is always desirable when new technology has been developed, to try to ensure that the standard laboratory tests have adequately screened the new approach.

Field tests generally fall into one of three distinct types:

- Uncontrolled tests.
- Controlled tests.
- Caravan tests.

Uncontrolled testing is what many people think of as field testing. The oil in a fleet of vehicles, chosen indiscriminately from available sources, is changed to that under

TABLE 9.3
Heavy-Duty Engine Tests Used for ACEA Specifications

Test Engine	Performance Evaluation
Mercedes-Benz OM646LA	Diesel engine wear
Mack T-8E	Soot in oil
Mercedes-Benz OM501LA	Bore polishing and piston cleanliness
Cummins ISM	Soot induced wear
Mack T12	Wear of liners, rings and bearings
Mercedes-Benz OM646LA Bio	Biofuel impacted piston cleanliness and engine sludge

Source: ACEA, Pathmaster Marketing Ltd.

TABLE 9.4
Heavy-Duty Diesel Engine Tests Used for API Specifications

Test Engine	Performance Evaluation
Cummins 5.9l ISB	Tappet and cam wear, sludge, filter pressure drop
Cummins ISM	Valve train wear (soot related)
Mack E-TECH-V-MAC III	Viscosity increase (soot thickening)
Mack E-TECH-V-MAC III	Ring, liner and bearing wear. oil consumption
Caterpillar 1M-PC, 1K	High-speed super-charged piston engine deposits
Caterpillar C13	Piston deposits, piston ring sticking, oil consumption control
Cummins NTC400	High-speed deposits, wear, oil consumption
Navistar 7.3L	Air entrainment

Source: API, Pathmaster Marketing Ltd.

TABLE 9.5
Engine Tests Used in Japan

Test Engine	Performance Evaluation
Gasoline engines	
Toyota 3A	Valve train wear
Nissan KA24E	Valve train wear
Nissan VG-20E	Low-temperature detergency
Toyota 1-G FE	High-temperature oxidation
Diesel engines	
Nissan SD-22	Detergency
Mitsubishi 4D34T4	Valve train wear

Source: Pathmaster Marketing Ltd.

test and the fleet carries on with normal operations. Sometimes the collection of vehicles is split and two oils are compared, one of which may be a standard commercial oil or the oil normally used by the vehicle. Typical sources of vehicles are the personal cars of employees or the haulage trucks and tankers of an oil company or additive supplier. Testing may well continue for over 12 months, during which time any unusual signs of engine distress or other operational problems are monitored. A score is kept of a number of problems for the test oil versus the reference oil or against some indication of previous experience.

When failures occur, it is very difficult to decide if these are due to the test lubricant, the individual engine and its previous history or the driving regime to which it has been subjected. There is a feeling of confidence about the quality of the test oil if no failures are found, but then the severity of the test in relation to possibly more extreme service is often not questioned adequately.

The best that can be said for this type of testing is that, if the oil were to be seriously deficient in some important property (which is unlikely), then this should show up, provided the testing was sufficiently extensive. Uncontrolled tests should be regarded only as supportive, rather than definitive of oil quality.

In controlled field tests, the condition of the engines used is known exactly at the beginning of the test period and test conditions are controlled and monitored. Normally, these tests are started with brand new or reconditioned engines. If measurements of wear are to be a feature of the test, then each engine must be stripped down and measured before the test commences. If deposits only are to be assessed, then this stage is not necessary and the engines can be installed without this extra step. At the end of the test, the engines are removed from the vehicles, stripped down and completely rated for deposits, wear and general condition. In some cases they may have an intermediate inspection during the course of the test in which case they may be partially stripped in order to evaluate how the test is progressing. During the course of the test, the condition of the lubricant will be carefully monitored.

The precise objectives of these tests must be carefully defined. Is the primary concern to measure piston deposits, engine wear, oil consumption, general oil condition or to study some other criterion? Are average or mixed types of service conditions being considered or very severe operations? Having decided on the objectives of the test, a suitable fleet or fleets of vehicles must be found. The absolute minimum number of vehicles to be run on each oil in the test is three, but this can easily prove insufficient, and a realistic minimum is probably better set at five or six. The vehicles in the test need to be on similar types of duties and cover approximately the same mileages during the course of a year. Typical fleets for automobiles would be limousine or taxi companies, while for heavy-duty oil evaluation, a haulage (trucking) company would be a normal choice. Fleet owners can be quite co-operative if offered new engines for their vehicles and free supplies of oil!

Typically, the organiser of the field test provides the new engines for the vehicles and arranges supplies of test oil, reference oils and filters. The fleet owner undertakes to run the field test in accordance with agreed guidelines, to make trucks or cars available for intermediate and final assessments and to provide oil samples as necessary. He or she will also have to carry a slight risk that there may be engine problems, but the test organiser can protect against excessive down time by keeping a spare engine available. The test organiser will also have to make careful arrangements with the fleet operator to ensure that the correct oil is used at all times in each vehicle.

It is particularly important that some of the vehicles are run on a well-defined reference oil. There is often an inclination to save money by omitting the reference oil, but as field test results are usually "good", it is then impossible to tell whether the new oil is actually "better" or "worse" than any other. If the fleet operator is to provide oil samples in between oil changes, then details of sampling procedure must be agreed and suitable receptacles and equipment for sampling must be provided. Details of duties, lubricant consumption, mileages and any special situations must be recorded by the driver or the fleet operator. The test organiser should provide recording forms that require minimum effort to complete.

Controlled field tests are often run by an oil or additive company in co-operation with an equipment manufacturer. This would apply particularly if one of the purposes of the testing is to obtain OEM approval. In such cases the OEM will insist on defining the details (including severity) of the test, will often help to find a suitable fleet operator, will send representatives to examine the engine condition at test completion and will compare the results of test and reference oils. Even if an approval is not involved, the OEM can provide valuable advice on technical areas of special interest and will normally be highly interested in the test results.

The duration of a controlled test depends on the purpose. In Europe, for light-duty vehicles such as taxis, a typical period might be 80,000 km (50,000 miles). By this time any significant developments will usually have become apparent. For heavy-duty trucks, it is more usual to run for a longer period of up to 320,000 km (200,000 miles) which may often take two or three years. This would typically be the mileage up to a first major overhaul. In the United States, with longer typical journeys, test mileages can be considerably greater. Such long tests have inherent problems. Apart from the risk of accidents, there are risks that the lubricant technology may have been made obsolete for some reason, such as a change in a critical specification, or the fleet operator may have lost a contract to deliver goods to a particular location, resulting in a change to test severity. Careful choice of fleet is very important, but the test sponsor must be sensitive to changing circumstances and from time to time may need to abort a test early or modify the objectives. Maintaining goodwill with the fleet operator is always important.

It is quite common for one or two of the vehicles on each test oil to be lost from the programme, due to accidents, gasket leaks or non-oil-related breakdown. These engines are best eliminated from the test, which is why it is considered that three vehicles are not really sufficient for a meaningful trial.

Caravan tests are the most difficult to set up and the most expensive to run, but can give relatively precise results in a short period of time. The value and meaning of the results depend on the conditions and duration chosen, but in general such tests are very useful. Typical objectives of such tests would be to measure parameters such as fuel economy with different lubricants, or emission levels, but in principle a wide range of evaluations is possible.

In a caravan test, a fleet of identical vehicles traces out a standardised route with specified speeds and timings, and this route is repeated over and over again until the desired mileage has been accumulated. Usually, a mixture of slow urban driving in traffic with some fast highway sections will be specified, but this depends on the test objectives. As far as possible all vehicles are required to undergo the same driving conditions, so they often will follow each other in a "caravan" of vehicles. Sometimes vehicles are driven at high speed on test tracks to accumulate mileage quickly. Such high-speed testing alone is suitable for measuring high-temperature performance but is otherwise generally a poor predictor of real marketplace conditions.

Hired drivers and hired vehicles with dummy loads are normally used, with new or reconditioned engines being in place at the start of the test. In some cases, such as short-duration fuel economy testing, the reconditioning may only consist of sufficient flushing to remove all traces of the previous lubricant. The driving of the vehicles is very tedious and is usually continuous until the test mileage is achieved, so several drivers per vehicle may be needed. Drivers may be rotated between vehicles in order to remove any driver effects. Wind speed and acceleration effects may be important for cases such as fuel consumption measurements, and therefore it may be desirable to also rotate the positions of each vehicle in the caravan.

Referencing against results with a standard lubricant is an essential part of most caravan tests. The reference oil can be in some vehicles with test oils in others, or for short-duration tests the various oils can be run sequentially in the same engine. In this case a test could theoretically be run with a single vehicle provided it consistently followed the prescribed route, but it would be usual to average the results from several vehicles.

Due to its relatively short duration, a caravan test will not demonstrate effects which depend on ageing of oils or engine deposits, but provides comparative data between oils precisely and quickly.

9.4 SUMMARY

Engine tests and field trials are not used as part of a lubricant user's oil monitoring programme. For a lubricant condition monitoring programme, many of the tests used to assess the characteristics and performance of finished lubricants are not appropriate or suitable, because they take too long to run and/or require large amounts of test lubricants. This is particularly true for engine and large-scale bench tests, but is also true for some of the chemical and physical tests described in Chapters 6 and 7.

In one sense, a field trial might involve a lubricant condition monitoring programme conducted over several years. In fact, engine tests and field trials used as part of the development of a new or improved lubricant are part of an oil or grease and machine condition monitoring programme, for the benefit and assurance of the lubricant developer and manufacturer.

Engine tests, while very important for the formulation and development of new lubricants, are completely unsuitable for consideration in a lubricant condition monitoring programme.

10 Condition Monitoring of Engines

10.1 INTRODUCTION

An engine oil obviously affects the operational efficiency of the engine, and monitoring the properties of the oil is obviously important. Conversely, maintaining those engine components that affect the lubrication process is also important.

With both gasoline and diesel engines, it is necessary to have clean combustion and crankcase ventilation air, which means that air cleaners and the positive crankcase ventilation (PCV) system must be serviced regularly. A clogged air cleaner, while it may be effective in cleaning the air, can restrict the volume of air reaching the engine enough to reduce power output significantly. To maintain power, drivers have a natural tendency to open the throttle, which only adds to the difficulty, as the extra fuel is not burned. Diesel engines are quite sensitive to such a condition and react by generating soot and building up engine deposits. Soot in engine oil can become very damaging. In addition to keeping the filters serviced, it is important that the piping connecting the filter to the engine be unobstructed and leak-free. This is particularly important in installations having the air filter located a considerable distance from the engine.

The ignition systems of spark ignition engines and the fuel injection systems of both gasoline and diesel engines should be in good working order and properly adjusted to assure the cleanest, most complete combustion possible. Malfunction or incorrect adjustment of these systems can result in increased amounts of unburned fuel in the cylinders and dilution or more rapid build-up of contaminants in the oil, as well as increased emissions.

In most small engines, the oil is changed at a pre-defined number of miles or kilometres, whether it needs changing or not. This is known as an "oil drain interval" (ODI). Drain intervals can vary considerably depending on the engine and service, as will be discussed in the next section.

For example, a large diesel engine in central station use, with a relatively large crankcase oil supply, may operate for thousands of hours between oil changes. Such engines are usually in good adjustment, temperatures are moderate and the contamination rate is low in comparison to the volume of oil in the system.

The presence of an oil filter does not necessarily permit an extension of the ODI. Filters do not remove oil-soluble contaminants and water, which are important factors in deposit formation. Regular filter changes are, however, important in keeping the filter operable so that it can perform its function of removing insoluble contaminants from the oil.

DOI: 10.1201/9781003245254-10

10.2 GASOLINE AND LIGHT-DUTY DIESEL ENGINE OILS USED IN CARS, VANS AND TAXIS

The volumes of oils in the engines of most light-duty vehicles (cars, vans and taxis) are too small to take regular samples for analysis.

For these engines, the ODIs are either fixed or are determined using computer algorithms (calculations) that are based on various measurements of engine and vehicle operation.

A number of factors are likely to shorten ODIs for car and vans:

- Short-distance driving: Frequent trips of under 5 miles in cold conditions are likely to accumulate water and fuel in the crankcase when the oil temperature does not reach the thermostat setting.
- High-mileage engines: Engines which have been driven extensively may generate more blow-by gases, unburnt fuel and corrosive agents that enter the crankcase oil.
- Diesel engines: These can produce more soot and acidic blow-by products.
- Turbo-charged engines: High temperatures in the turbo-charger can tend to degrade oils and additives.
- Dust: Driving in dusty conditions (dirt/gravel roads) with an economy-grade oil filter could allow abrasive particles into the oil, turning it into more of a honing compound. This dirty oil can generate more wear metal particles, leading to more wear, possibly sludge and higher rates of oil oxidation.
- Flexible fuels: Alcohol–gasoline fuel blends are prone to accumulate water in the crankcase.
- High oil consumption: Although high oil consumption replenishes additives, it may also be associated with high blow-by of combustion gases into the crankcase.
- Hot running conditions: Driving in hot environments, particularly deserts, can lead to premature oil oxidation, volatility problems and rapid depletion of additives.
- Towing heavy loads: Has similar effects to hot running conditions.

Several other factors can lengthen ODIs:

- Synthetic and high performance engine oils: Premium-quality synthetic oils have excellent oxidation stability, thermal stability and shear stability.
- High-capture efficiency oil filters: These remove dirt and, if any, wear metals from the oil, thereby minimising or eliminating wear and catalytic oxidation.
- Long-distance driving: While this may seem counter-intuitive, relatively constant engine revolutions and optimum engine oil temperature and pressure mean less stress on the engine oil and fewer operating hours per distance travelled.
- New engines with tighter mechanical tolerances: These engines have much lower levels of blow-by after the first 500 to 5,000 miles and, currently, up to 50,000 miles or kilometres.

- Computer controlled algorithms: Measure the operation of the vehicle and environmental conditions to optimise the time at which an oil change is required, particularly for steady, careful driving.

Many car, van and even truck and bus manufacturers have now introduced computer-controlled sensors and algorithms to inform the vehicle's owner when to have the vehicle serviced.

The General Motors (GM) Oil-Life™ System, first introduced commercially in 1998, determines when to change the engine oil and filter, based on recordings of engine revolutions, operating temperature and other factors that affect the length of oil change intervals. The system does not actually monitor any single quality or physical property of the oil. The algorithm is based on GM's determination that nearly all driving conditions can be grouped into one of four categories: easy freeway driving; high-temperature, high-load service; city driving; or extreme short-term, cold-start driving. GM found that oil degradation in the first three categories was largely a function of the oil temperature. During extreme short-trip driving, the principle cause of oil degradation was water condensation and contaminants in the oil – the lower the oil temperature, the greater the contamination. The software automatically adjusts the oil change interval based on engine characteristics, driving habits and climate.

Mercedes-Benz (MB) version of the oil monitor is called ASSYST in Europe and the Flexible Service System (FSS) in the United States. The system uses an onboard computer to track multiple engine operating conditions. From research on oil quality through the span of an engine's life, MB discovered that the deterioration in oil is determined by such factors as driving habits (journey lengths), driving speed and failure to replenish low oil levels. The system therefore monitors the time between oil changes, vehicle speed, coolant temperature, load signal, engine rpm, engine oil temperature and engine oil level. It uses this information to determine the remaining time and mileage before the next oil change, and it displays the information in the vehicle's instrument cluster.

MB also discovered that oil degradation is correlated directly with its ability to conduct electric current. MB has now fitted engines with a digital oil quality dielectric sensor that is mounted above the oil pan together with an analogue oil level sensor. This sensor measures changes in capacitance, which effectively is a proxy for the amount and type of contaminants and oil degradation products present in the oil. An increase in dielectric constant (less resistance to electrical flow) indicates oil contamination and degradation. MB has been incorporating the sensor into its vehicles since 1998.

Bosch GmbH Multifunction Oil Condition Sensor determines oil level and oil condition. The oil level information allows the oil dipstick to be omitted from the vehicle. Monitoring the engine oil condition is primarily intended to optimise ODIs, but it also provides increased insight into the actual state of the engine, which enables the possible detection of approaching engine failures or change in lubricant quality. The oil condition sensor constantly measures the oil's viscosity, permittivity, conductivity and temperature. The measured viscosity and permittivity (or dielectric constant) are the primary values supporting the oil condition evaluation. Commonly, chemical oil deterioration is associated with an increase in viscosity, whereas mechanical wear (shear) and fuel dilution lead to a decrease in viscosity.

A novel microacoustic device determines the viscosity. This device uses the piezoelectric effect to electrically excite high-frequency mechanic (or acoustic) vibrations at a sensitive surface. When this sensitive surface comes into contact with the oil, the electrical device parameters, such as oscillation frequency and damping, are changed according to the oil's mechanical properties, especially viscosity. Thus, the viscosity can be electrically detected by measuring these parameters. In contrast to conventional viscometers, which are commonly used in laboratory applications, the microacoustic sensor does not contain any moving parts. In addition, its small size enables it to be easily incorporated into the multi-functional oil-level and condition sensor.

Delphi Corporation's INTELLEK® Oil Condition Sensor uses both a computer algorithm and a sensing element that directly measures various oil properties. The algorithm takes into account important factors affecting the rate of oil deterioration such as temperature, driving severity, oil level and oil type. It measures the temperature every 10 seconds to verify whether it reaches a specific normal operating temperature before the engine shuts off. It also records the number of times the engine turns on and off.

A proprietary capacitive sensing element is the core technology. It monitors the oil's conductivity, detects water and glycol contamination, oil temperature and determines the oil level. According to Delphi, the oil's conductivity is important because it characterises additive depletion and changes in viscosity and acid number. The INTELLEK Oil Condition Sensor tracks the many different parameters using onboard software to indicate when the oil is nearing the end of its service life.

The Continental Temic Microelectronic GmbH QLT Oil Condition Sensor was launched in 1996 to monitor engine oil quality, level and temperature. Two sensors simultaneously and continuously monitor diesel engine oils containing soot. The instrument also monitors nitric oxide and oxidation products in spark-ignited engines, as well as water and fuel contamination. Because these factors influence the oil's electrical properties and permittivity (ability of a material to resist the formation of an electric field within it), an effective oil condition sensor is achieved, according to the manufacturer. The QLT also has an integrated precision probe that allows it to measure critical temperatures and exact oil levels. It can track temperatures ranging from $-40°C$ to $160°C$. The oil level, up to 100 millilitre (ml), is calculated by a second capacitor.

The Voelker Sensors Inc. Oil Insyte sensor uses a patented technology based on the electrical properties of an oil-insoluble polymeric bead matrix. The Oil Insyte employs an in-line method for continuous oil condition monitoring with an LCD readout providing detailed information about oxidation, additive depletion, soot contamination and oil temperature. The technology does not require external calibration standards and reports oil condition independent of viscosity. According to the manufacturer, the sensor measures key indicators of oil degradation and allows the conventional analyses approach of oil monitoring (sampling and analysis) to be combined into a single more efficient analysis. No assumptions are required as to the condition of the engine or the initial baseline quality of the oil.

The soot detection feature of the sensor determines the amount of undispersed agglomerated soot (compared with dispersed finely divided soot) present in the oil. Depending on the oil's additive package, the same amount of undispersed soot can be present at 1% to 2% (for the base oil without dispersants) as a fully formulated motor oil with more than 7% soot.

The Lubrigard Oil Condition Monitoring Sensor unit is designed to be fitted by original equipment manufacturers (OEMs) to new cars and trucks to warn the operator of abnormal lubricant conditions. According to the manufacturer, it indicates when an oil or filter change is necessary or when the oil should be inspected or tested. The sensor was designed to optimise ODIs and to detect problems such as coolant leaks, metallic wear debris and oil degradation by direct measurement. It is particularly useful for measuring high concentrations of soot in diesel engine oils.

The sensor is based on the dielectric loss factor, also known as Tan Delta. According to Lubrigard, this method is more sensitive to changes in contamination than other dielectric measurements. At the same time, it is tolerant of normal differences in operating temperatures and lubricant formulations. To compensate for temperature variations, a temperature sensor communicates with the unit's microcontroller. The system monitors soot, water, coolant, oxidation and/or wear particles, and it will work in both gasoline and diesel engine oils.

Symyx Technologies Inc. developed a "Solid-State Oil Condition Sensor" that uses a solid-state micromechanical resonator and a special signal-processing algorithm to measure important physical properties of lubricants. This sensor measures three independent physical properties in real time: viscosity, density and dielectric constant. This is significant because the direct measurement of a lubricant's physical properties can provide important information about changing lubricant and engine health.

Eaton Corporation has developed a "Fluid Condition Monitor" (FCM) that can monitor multiple fluid properties. The Eaton FCM is an in-situ real-time sensor based on impedance spectroscopy, a technology that measures multiple electrical properties of a fluid. It uses very small alternating current (AC) signals, which do not permanently disturb the fluid or the electrodes used in the measurement. Eaton's FCM technology is differentiated by two critical attributes: it measures surface properties of the fluid in addition to bulk properties, and it has more degrees of freedom to enable the independent tracking of multiple lubricant parameters.

Measuring bulk properties reveals information about the conductivity (concentration and charge of ions) and dielectric constant (size, shape and polarisability of the base fluid and its additives). Measuring the surface properties provides a quantitative measure of the physical and chemical properties of a fluid at the fluid-to-metal interface. This is a powerful technique when it is correlated to the real and measured physiochemical property changes occurring in ageing or stressed motor oils.

These, and other systems, sensors and algorithms, are now used in almost all vehicles in Europe, the United States and Japan to determine when the vehicle needs to be serviced, particularly its engine oil and filter.

10.3   HEAVY-DUTY DIESEL ENGINE OILS
USED IN TRUCKS AND BUSES

10.3.1   Oil Drain Intervals

Sixty years ago in the United States or Western Europe, a truck engine oil would probably have been changed as frequently as every 500 miles. Now, higher-quality engine oils, cleaner fuels, improved filter technology and more dependable engines allow ODIs as high as 100,000 miles, kilometres or more.

However, typical ODIs remain at around 25,000 miles, and little attention is paid to changing this standard in view of the diverse environments and other factors that trucks encounter. For example, two identically produced vehicles may experience a very different oil life; one may reach close to 50,000 miles, while another might require an oil change at 15,000 miles. This variance in engine oil life is the result of many factors from three main areas:

- Engine design, age and conditions: Engine design characteristics and numerous running conditions can affect oil life factors from exposures to contaminants and other conditions.
- Driving patterns and conditions: Where and how the truck is driven.
- Oil properties: Quality and formulation performance of the engine oil.

The most important factor linked to engine oil life is engine fuel efficiency. It is improved by combustion efficiency, which can determine the type and amount of materials that are "blown" by the piston rings and into the engine sump. Piston "blow-by" is usually the primary source for ingression of contaminants into the oil. This can include dirt, water, soot, fuel, nitrogen oxides (NOx) and partially burned hydrocarbons (HC). Other engine design factors, such as seal efficiency, temperature control and emissions control methods, influence the type and concentration of contaminants in the oil.

Some measurable engine characteristics, such as total operating hours and mileage, affect engine oil life. Maintaining a healthy oil-flow system is achieved using good oil filtration and tight seals. Poor filtration can produce a chain reaction of damaging effects on the engine, shorter ODIs and higher operating costs. Filtration is the counter to contamination. It is therefore important that the dirt-holding capacity of the oil filter should be appropriate for the anticipated or needed ODI.

Seals will also have an effect on ODIs. Effective seals will keep oil in the engine system while leaking seals will not. Engines with leaking seals will require more frequent top-up, thereby apparently extending ODIs. However, leaking seals damage the environment and more frequent top-up adds to fleet operator's costs.

Another engine characteristic that has an effect on ODIs is the oil capacity or sump size. In essence, with an increased volume of engine oil circulating within the engine, there will be a decrease in contamination concentration. A larger sump size also means a larger amount of oil additives to deal with the operating environment. More oil also means less thermal distress on the oil. As a result, engine manufacturers that offer the largest sump size generally allow for the industry's longest recommended ODIs.

While all engines are designed to provide healthy conditions for the oil to flow, even the best designed models have some level of anticipated contamination over time, either generated from internal or external sources. Consequently, it is the responsibility of the truck owner to ensure that optimal maintenance and healthy conditions are established.

10.3.2 Wear Metals

The principal wear metals of interest to owners and operators of fleets of trucks and buses are iron, aluminium, copper and zinc. Particles of these metals in in-service engine oils can come from parts that are starting to fail or from various kinds of contamination. They can act as abrasives and accelerate wear in other parts of the engine.

The method of choice for wear metal testing is Inductively Coupled Plasma-Atomic Emission Spectrometry (ICP-AES). This type of analysis typically covers about 24 chemical elements, including wear metals, dirt and metals used as lubricant additives. The standard test method is ASTM D5185. Many lubricant user and independent laboratories modify this method for faster sample throughput by making their dilutions on a volume-to-volume basis rather than by weight. Volume-based dilutions are not suitable for determining absolute values for metal content, but they can be used to measure trends over the lifetime of oil from one specific reservoir.

ASTM D5185 requires the use of internal standards: reference metals added in known quantities to each lab sample to serve as calibration and quantitation markers. Internal standards must be elements that are not otherwise present in the samples so that the signal comes only from the reference standard. Cobalt is still the most commonly used internal standard, but other internal standard metals recommended for this method can present problems. Cadmium, one such standard, now shows up as a wear metal in some samples. Rare earth elements, including yttrium and scandium, can affect results when they interact with some viscosity improvers.

ICP-AES instruments in lubricant analysis laboratories must be set up to run non-aqueous samples. Also, it is very important that samples are stirred just before they are run to make sure that the wear particles are included in the sample stream instead of settling to the bottom of the sample container. Newer ICP-AES instruments can achieve the needed precision using smaller samples, which speeds up sample throughput.

10.3.3 Fuel Contamination

Gas chromatography (GC) has largely replaced Fourier transform infrared spectroscopy (FTIR) for fuel contamination testing in lubricating oils. Older types of diesel fuels contained amounts of sulphur that were much higher than the amounts in newer low-sulphur diesel fuels, and the FTIR methods used this sulphur as a marker. Low-sulphur fuels cause less pollution, but they rendered FTIR ineffective as a test method for fuel dilution in engine oils.

Some laboratories measure viscosity as an initial screening for fuel contamination, but a high soot content can keep viscosity high, counteracting the viscosity-reducing effects of fuel dilution. Flash point tests can be used as another indicator of fuel contamination.

The ability of GC analysis to distinguish between gasoline, diesel and biodiesel makes it a more reliable and valuable test. It has a linear response which allows the amount of each component present in the sample to be quantified. Very importantly, GC requires no sample preparation, saving time between a sample arriving in the laboratory and the results being reported.

The earlier standard GC method for fuel dilution testing takes about 35 minutes per sample, because the sample flows through the instrument column at the same speed from start to finish. ASTM D7593, a newer standard first published in 2014, uses a back-flush configuration to reduce sample run time to about 2 minutes. In this method, fuel contaminants, which are usually more volatile, reach the detector early in the run, and the rest of the sample is back flushed off the column.

10.3.4 COOLANT CONTAMINATION

Coolant contaminants increase the acidity of a lubricant, and they promote corrosion. GC has begun to replace ICP-AES for detecting coolant contamination, because sodium, potassium, boron and molybdenum are now used in oil additives as well as in coolant formulations. Consequently, ICP-AES analysis is now less valuable for determining coolant contamination. GC distinguishes lubricants from coolants by identifying the various organic components of each.

FTIR only detects high levels of coolant contamination, and colorimetric methods for coolant contamination testing are expensive, labour-intensive and prone to yielding false positives.

A new standard method, ASTM D7922, for contamination by ethylene glycol, a common coolant base, was first published in 2014. In this method, phenylboronic acid is added to the sample to derivatise the ethylene glycol and increase its volatility. The volatile derivative separates from the oil matrix and collects in the headspace at the top of the sealed sample container, where it can be removed for injection into the GC instrument. This method also works with propylene glycol anti-freeze coolants.

10.3.5 OTHER CONTAMINANTS AND PRODUCTS OF DEGRADATION

The ICP-AES test that analyses for wear metals also finds evidence for dirt (including abrasive minerals such as silica), inorganic components from coolants (sodium, potassium and boron) and metallic lubricant additives.

FTIR and GC pick up organic contaminants that ICP-AES is not designed to detect. FTIR can also be used to check for water and soot. Some laboratories that run large numbers of acid number (AN) and base number (BN) tests use FTIR as an initial screening method to reduce the number of samples that they need to analyse using the titration tests. Instead of using one marker on the FTIR spectrum, these laboratories use chemometrics methods that use the whole spectrum for the analysis.

10.3.6 LUBRICANT FORMULATIONS AND ADDITIVES

Oil analysis also helps to monitor components that are supposed to be present in the lubricant. It can be used to track additive level trends and alert the end user when it is time to top-up a particular additive. Oil analysis also is used to check for signs that the lubricant is starting to degrade and is due for replacement.

ICP-AES analysis can track trends in metals commonly used in oil additives, for example, calcium, magnesium, sodium, sulphur and phosphorous. The presence of other metals in delivery of a new oil may indicates incorrect blending, incorrect packaging or batch cross-contamination.

FTIR can be used to check in-service oils for the depletion of organic additives, as well as to confirm that the correct oil is being used in a specific truck or bus. It can also be used to determine whether an incorrect oil has been filled into a fleet operator's bulk storage tank.

For truck and bus fleets, reference (also called differential) FTIR, which evaluates the spectral difference between a new oil and an in-service oil sample of the same type and grade, can be used to check for oxidative or thermal degradation. Another approach is a non-reference method, which measures trends for in-service oil samples compared with a baseline sample of the same oil.

10.3.7 ALARM LIMITS FOR HEAVY-DUTY DIESEL ENGINE OILS

There are no fixed alarm limits for metal and non-metals in heavy-duty diesel engine oils. Important variables that influence alarm limits include:

* Truck or bus type and size.
* Engine size and horsepower.
* Oil sump size.
* Oil filter size, type operational method.
* Truck or bus operating environment.
* Oil viscosity grade and specification.

A set of example alarm limits for one type of oil in one type of heavy-duty diesel engine made by one manufacturer in one type of operation in one specific environment is shown in Table 10.1. Table 10.1 should only be used as a general guide. Because many OEMs issue their own flagging limits, some of which are not always publicly available, the limits shown should not be considered as representing a set of specific OEM limits. Also, the flagging limits will depend on the use of the heavy-duty diesel engine, whether on-highway or off-highway.

An example laboratory analysis report, which shows trends over an almost two-year period, is shown in Table 10.2. It is worth noting that the engine oil was changed four times during the period of the reporting. Two of the sets of reported results contained alarm flags.

10.4 OILS USED IN HYBRID AND ELECTRIC VEHICLES

Electric vehicles use one or more batteries to power one or four electric motors. (Designs with four electric motors have one in each wheel of the vehicle.) The batteries must be recharged after every journey, or occasionally may need to be recharged during a longer journey. Hybrid electric vehicles use either a gasoline or a diesel

TABLE 10.1
Example Alarm Limits for a Heavy-Duty Diesel Engine Oil

Property	Units	Alarm Limits		
		Normal	Abnormal	Critical
Aluminium (Al)	ppm	<20	20–30	>30
Chromium (Cr)	ppm	<10	10–25	>25
Copper (Cu)	ppm	<30	30–75	>75
Iron (Fe)	ppm	<100	100–200	>200
Lead (Pb)	ppm	<30	30–75	>75
Nickel(Ni)	ppm	<10	10–20	>20
Silicon (Si)	ppm	<20	20–50	>50
Silver (Ag)	ppm	<3	3–15	>15
Sodium (Na)	ppm	<50	50–200	>200
Tin (Sn)	ppm	<20	20–30	>30
Titanium (Ti)	ppm	<10	10–20	>20
Fuel dilution	%wt	<2	2–6	>6
Soot	%wt	<2	2-6	>6

Source: Pathmaster Marketing.

engine to power a generator, to produce electricity. Gasoline engines tend to be used in cars, vans and taxis, while diesel engines tend to be used in trucks and off-highway machines.

Three alternative configurations can be used for gasoline-electric and diesel-electric hybrid vehicles: series, parallel and series–parallel. With series types, the gasoline engine powers a generator, which feeds electricity to either a battery or an electric motor, which then powers the wheels through a conventional hypoid gearbox. The battery can also feed electricity to the electric motor. The speed of the engine is disconnected from the speed of the wheels.

In the parallel configuration, either the gasoline engine or the electric motor can drive the wheels, using a double gearbox arrangement. The electric motor is fed from a battery, which is charged separately while the vehicle is stationary. With the series–parallel types, the gasoline engine can either drive the wheels directly or can power the generator indirectly, with the generator feeding electricity to either the electric motor or the battery. Again, the electric motor can also drive the wheels, using a double gearbox arrangement.

In all three configurations, the gasoline engine can be lubricated using conventional gasoline engine oils. However, the use of newer, low-viscosity oils can deliver additional benefits. Hybrid drivetrains take advantage of normal driving to increase efficiency. At lower driving speeds, hybrids can operate exclusively on the battery pack and electric motor. Regenerative braking functions to recharge the battery, by capturing the kinetic energy while slowing down and stopping. Start-stop systems shut down the engine while stationary, cutting down the energy required for idling.

TABLE 10.2

Example Laboratory Analysis Report of Heavy-Duty Diesel Engine Oil

Date sampled	19/10/15	11/4/16	14/11/16	22/5/17	18/12/17	11/6/18
Date received	21/10/15	13/4/16	16/11/16	25/5/17	20/12/17	13/6/18
Date reported	23/10/15	14/4/16	18/11/16	26/5/17	22/12/17	14/6/18
Lab. number	nd	nd	nd	nd	nd	nd
Oil brand	nd	nd	nd	nd	nd	nd
Oil type	nd	nd	nd	nd	nd	nd
Oil grade	15W-40	15W-40	15W-40	15W-40	15W-40	15W-40
Oil added (litres)	–	15	–	12	–	–
Oil changed	Yes	No	Yes	No	Yes	Yes
Metals (ppm)						
Aluminium (Al)	2	4	1	2	3	2
Chromium (Cr)	1	5	1	<1	1	<1
Copper (Cu)	9	11	5	3	13	4
Iron (Fe)	8	35	11	7	12	11
Lead (Pb)	<1	1	<1	<1	1	<1
Nickel (Ni)	<1	<1	<1	<1	<1	<1
Silver (Ag)	<1	<1	<1	<1	<1	<1
Tin (Sn)	<1	2	<1	<1	1	1
Titanium (Ti)	<1	<1	2	3	3	<1
Contaminants (ppm)						
Potassium (K)	<5	**126**	11	17	25	13
Silicon (Si)	6	14	7	5	7	6
Sodium (Na)	6	**344**	29	23	31	18
Additives (ppm)						
Barium (Ba)	<1	<1	<1	<1	<1	<1
Calcium (Ca)	1,999	2,166	2,006	2,070	2,162	2,036
Boron (B)	14	21	<5	29	19	21
Magnesium (Mg)	159	165	134	179	197	146
Molybdenum (Mo)	10	66	22	10	12	4
Phosphorous (P)	1,026	963	1,085	1,117	1,035	1,064
Zinc (Zn)	1,233	1,136	1,312	1,271	1,186	1,269
Contaminants						
Water (%wt)	<0.05	0.08	<0.05	<0.05	<0.05	<0.05
Coolant	No	**Yes**	No	No	No	No
KV at 100°C (cSt)	14.8	15.3	12.9	14.5	15.1	13.2
Fuel (%)	<1	<1	1	<1	<1	<1
Soot (%)	0.8	1.0	0.7	0.5	0.5	1.1

Source: Pathmaster Marketing.

Notes

nd = not disclosed.

Electric-only and gasoline-only power options allow the series–parallel design to provide the most fuel-efficient operation. The system can act as a series drive at lower speeds, but then turn to gasoline-only at higher speeds. However, in the parallel and series–parallel configurations, the gasoline engine is required to "kick in" instantly when the computer control detects the higher speed required. Because response times need to be rapid, lower-viscosity engine oils produce less resistance to the engine.

Lower-viscosity engine oils are also beneficial for the series configuration. Because the engine is not connected directly to the drivetrain, it can be run at a constant speed, to charge the battery pack. Constant engine speed does not put the same stresses on the engine oil as with conventional gasoline or diesel engines operating at variable speeds.

A number of lubricant manufacturers in Japan and Europe are now offering 0W-16 viscosity engine oils for gasoline-electric hybrid cars. Toyota and Honda are now recommending the use of 0W-16 oils for selected hybrid vehicles, although they will allow 0W-20 viscosity grades if the lower-viscosity oils are not available. It is likely that 0W-16, 0W-12 and even 0W-8 viscosity grades will become more widely available for hybrid vehicles.

Toyota's Prius hybrid has a series drivetrain, which uses a power split unit that consists of a planetary gear set, the generator, the electric motor and an oil pump. These components have a higher system voltage, a greater presence of copper and electronics, higher temperatures and speeds and varying friction requirements. Lubricants are being developed that are better able to protect and ensure the smooth functioning of the electrified drivetrain components. The main considerations are the electrical conductivity of fluids and related safety concerns, corrosion protection, high-speed bearing protection, thermal transfer properties, material compatibility and oxidation and sludge control.

The electrical conductivity of fluids is an issue because, if it is too high, there is a risk of current leaking, but if it is too low, static charge can build up, resulting in electrical arcing in oil. This leads to the degradation of the fluid, compromising its protective properties. Additionally, conductivity increases as an oil oxidises.

Some gasoline-electric hybrids use a dual clutch transmission, a multiple-speed automatic gearbox that uses an e-motor in addition to the twin clutch for seamless increases or decreases in speed. Because of the electrical components and wiring involved, a gearbox fluid would have to offer copper corrosion protection while at the same time being compatible with the insulation and other materials of construction. Several of the early models of these vehicles used an automatic transmission fluid (ATF) in the combined transmission/motor system. Although many ATFs have adequate electrical resistivity, they were not optimised for use in electric and hybrid vehicles. A significant amount of research work is now in progress to develop fluids that are both lubricants and coolants in these applications. Work is also in progress to reduce the temperatures generated in the electrical windings of motors.

All electric heavy-duty and long-haul trucks are not currently considered to be a commercially viable option, because the batteries take up space that would have been used to transport good. Significantly improved batteries may help to solve this problem in the future.

Meanwhile, diesel-electric hybrid trucks are being developed, trialled and introduced to the market by many of the leading truck and bus manufacturers. Some of these hybrid trucks use the series–parallel arrangement, so that the batteries power the truck in and around town, while the diesel engine powers the vehicle between towns and cities. This arrangement also allows the truck to make deliveries in town, silently and at night, so as not to disturb the neighbourhoods.

All electric and diesel-electric hybrid buses have also been introduced to the market. One of the advantages of an all-electric bus is that the batteries can be recharged in the bus depot, overnight. This is also the case for battery-powered in-town delivery vans.

One of the identified problems with electric and hybrid vehicles is that electric motors run hot, so need to be cooled. Obviously, the cooling fluids need to be compatible with the electrical circuits of the motor. Some system designers have found that low-viscosity engine oils and higher-viscosity driveline fluids are not necessarily suitable for this purpose. Research work is currently in progress to identify technically and commercially acceptable solutions to this problem.

At the time of writing, it was becoming apparent that many OEMs have unique designs for electric or hybrid vehicle powertrain configurations, so developing functional fluids for these vehicles is presenting challenges for lubricant companies and additive manufacturers. Lubricants for use in hybrid and electrical vehicles have greater technical requirements than those for use in gasoline or diesel engines. For example, electric motors can be either "dry" or "wet". With dry electric motors, the cooling fluid is outside the electrical windings, while with wet electrical motors, the cooling fluid surrounds the electrical windings. In the latter case, the cooling fluid may be required to also lubricate the motor's gearbox.

While the growing adoption of electric and hybrid vehicles is a fundamental shift away from gasoline and diesel vehicles, condition monitoring of the lubricants, cooling fluids and equipment used in them is unlikely to change much in the foreseeable future. For cars, vans, taxis and motorcycles, sensors such as those described in Section 10.2 are likely to be used in much the same way. Flagging limits and oil/fluid change intervals may be slightly different. In some vehicles, ODIs may even be slightly longer than now. In most electric and hybrid vehicles being manufactured at the time of writing, the high-voltage battery and the power electronics and charger module systems are cooled using a system whose temperature is monitored and which motorists are asked to check the levels of coolant visually at regular intervals. The battery and charger module temperatures are also monitored as part of the vehicle monitoring programmes. However, in many cars and vans, these coolants can only be changed or topped-up by a qualified technician, and motorists are advised to seek the assistance of a qualified workshop. The algorithms required for electrical system coolant change intervals are still being devised and may take some years to be fully developed and implemented. For electric and hybrid trucks and buses, condition monitoring programmes appear to be likely to be similar to those used currently for gasoline- or diesel-powered vehicles.

It is also becoming increasingly evident that additional sensors may be required in electric and hybrid vehicles to monitor the electrical properties of the oils, and possibly, the greases used in these vehicles. However, which of the many electrical properties described in Chapter 7 that will need to be monitored in the future and whether this will require one or several sensors have yet to be determined.

10.5 TWO-STROKE AND FOUR-STROKE OILS USED IN MOTORCYCLE ENGINES

Motorcycles with two-stroke engines are smaller than cars or vans, so have the same issues for oil and machine condition monitoring except for one very important factor. The lubricating oil is added to the fuel used in the engine and is therefore burnt, so there is no in-service oil to sample or monitor.

Motorcycles with four-stroke engines have small sumps, so the same types of oil change interval estimations or algorithms are used for these engine oils and engines as for cars and vans.

10.6 OILS USED IN MARINE ENGINES

10.6.1 TWO-STROKE AND FOUR STROKE DIESEL ENGINE OILS

Marine diesel engines range in size from large to huge. They are used for two main purposes; for propelling the ship and for generating electricity, which assists in powering the ship's propulsion plant. The efficiency of any machinery on board ship is directly related to its performance. In order to get the best out of marine engines, it is very important to monitor their performances.

Two-stroke marine diesel engines are generally used in the biggest ships, such as oil tankers, dry bulk cargo vessels and container ships. As with two-stroke motorcycle engines, the engine oil is injected directly into the cylinders (so is burnt with the fuel) and is circulated round the engine's bearings. Consequently, monitoring the engine oil is not particularly helpful, except to monitor the viscosity, oxidation performance, particle counts and wear metals contents of the circulating oil to check for possible problems with the bearings.

Monitoring the engine's operation and condition involves other parameters. Engine control parameters, such as fuel injection timing, exhaust valve timing, variable turbocharger vane opening angles, lambda control and others, can be monitored, and any variation is corrected to achieve the most efficient combustion.

Engine parameters are the best way of finding any fault or variation in engine performance. Variations in temperature, pressure and power produced by each cylinder should be monitored frequently and adjustment should be done as necessary to achieve efficient combustion.

Digital pressure monitoring using a digital pressure indicator (DPI) is an electronic method of monitoring the power and performance of an engine. DPI data can be used to plot and then interpret the variation in cylinder performance and corrective action can be taken.

The latest marine diesel engines are now continuously monitored using intelligent combustion monitoring (ICM), which measures the in-cylinder pressure in all cylinders in real time. This computer-controlled system offers a broad range of data processing tools for evaluating performance and for helping to determine engine malfunctions, such as extensive blow-by, exhaust valve operation, fuel injection and lubricant injection.

Vibration analysis is a powerful tool for monitoring an engine's condition. As a machine begins to fail and the interior working surfaces degrade, varying levels of vibration emanate from the machine, particularly bearings, as wear progresses. Spike energy is the high frequency levels of the vibration being produced by the "ringing" of the internal surfaces rubbing together. To get the maximum return from vibration analysis, a baseline should be established when the machine is first put into service.

As wear continues, the level of vibration or "noise" will trend upward. By understanding the natural frequency of the metals that are degrading, it is possible to make calculated estimations at the amount of wear as well as the relative life remaining in the component. This can be related to trending wear debris with oil analysis, which has the added benefit of looking at wear debris shapes (using ferrography) to monitor wear and diagnose the root cause. It is harder to do this when simply using vibration. When using vibration to monitor wear generation or a machine's failure modes, it is very important to capture the vibration results from the same place on the equipment each time it is to be monitored.

Another mechanical method for measuring the performance of engine cylinders is by applying indicator drum and plotting graph on cards. Two types of cards are used for this purpose: power card and draw card. With the help of these two diagrams, the compression pressure, peak pressure and engine power can be determined.

Log book monitoring is the most basic, but commonly ignored, method for monitoring engine performance. The log book record for engine room machinery is kept onboard for years on ship. The log book of current month and of previous months should be compared for recorded parameters, which will give the exact variation of engine parameters. If the variation is high, engine controls and operating parameters need to be adjusted or overhauled.

Engine emissions can also be used as a monitoring tool. Marine engines release exhaust smoke as a result of combustion. The colour and nature of the exhaust should be monitored continuously and engineers must know which exhaust trunk discharge is dedicated for which engine. A change in exhaust smoke is a prominent indication of a problem in the combustion chamber.

Many of these monitoring methods can also be used on four-stroke marine diesel engines. These tend to be smaller than the large two-stroke marine diesel engines, so are used on smaller cargo vessels, passenger and cruise ships, ferries and fishing vessels. They are also used on land for generating electricity.

Four-stroke marine diesel engines are often larger versions of the engines used in trucks, buses and off-highway equipment (see later in this chapter). As such, the same oil monitoring methodologies, discussed earlier, can be used on these engines. In many cases, the oil samples are taken just before a ship reaches the next port, are analysed very rapidly by a contracted laboratory and the results, together with any flags are communicated to the ship's engineers before the vessel leaves port.

Monitoring peak pressure using a mechanical peak pressure gauge is normally applied with four-stroke generator diesel engines, where a gauge is used for each cylinder and the pressure generated during combustion is noted. With the same gauge, the compression pressure of the cylinder is also measured when the unit is not firing.

The variation in the peak pressures generated is then taken into account for identifying units, adjusting fuel racks and overhauling combustion chamber parts in order to achieve efficient combustion.

A number of companies now offer portable test kits and mini test laboratories for use onboard ships.

Kittiwake, now owned by Parker Hannifin Corp., and trading as Parker Kittiwake, markets a range of on-site oil test kits based around an electronic test cell for water and/or BN. Features include fully digital display and five-year battery life. Where appropriate, tests incorporate a new range of EasySHIP reagents that greatly reduce HAZMAT and transportation considerations. DIGI tests provide quantitative results for trending, they generally require replacement consumables available in comprehensive packs. ECON tests are entry-level qualitative go/no go type test where the test also forms the replacement reagent pack. Various configurations are available in either an aluminium or ABS carry case complete with consumables for 25/50 tests. The range covers water, BN, viscosity, insolubles, salt and AN.

The company also markets an FTIR3 IR Field Analyser, with proven correlation to ASTM D7418. The analyser provides automated sampling and analysis, machine tagging, alarm functions, test history and results for a comprehensive set of in-service oil parameters including water, oxidation, sulphation, nitration, phosphate, EP additive depletion and glycol. The analyser can be used with the Parker Kittiwake Heated Viscometer and ANALEX fdMplus for additional viscosity and wear debris data.

Parker Kittiwake's Oil Test Centre (OTC) is a rugged, self-contained oil test laboratory for key oil quality parameters covering viscosity, water, BN, AN and insolubles. Specifically for onboard ships, the company markets a Fuel and Lube Test Cabinet, developed from a 15-year history in the deep sea marine market monitoring both in-service lubricants and bunker fuel deliveries. It is a single cabinet containing tests selected with an accuracy appropriate to the application, including Electronic Lab tests for viscosity, density and fuel compatibility, DIGI tests for water and BN and ECON tests for salt, insolubles and pour point.

On-Site Analysis Inc. recently expanded into the maritime industry with the introduction of the OBA ShipCheck analyser. Weighing just 95 pounds, the analyser conducts diagnostic analysis on a number of fluids, including engine lubricants, grease, gear lubricants and scrape down oils. OBA ShipCheck can automatically identify the presence of 20 metals and measure physical property levels for contaminants such as glycol, fuel, water, nitration, oxidation and BN.

The OBA ShipCheck connects directly to a Web-based data analysis system, allowing corporate fleet maintenance management to track the condition of an entire fleet on a real-time basis and receive e-mail alerts when emerging problems are identified.

10.6.2 Liquefied Natural Gas Engine Oils

Annex VI of the International Maritime Organisation's Marine Pollution Convention came into force on January 1, 2020. The convention, known as IMO 2020, restricts atmospheric emissions of sulphur oxides (SOx) from ships, mainly due to the combustion of heavy fuel oil. Many major ports and almost every national port authority have indicated that they will punish ship owners severely for ships that do not comply with the new convention.

The global upper limit on the sulphur content of ships' fuel oil will be reduced from 3.50% to 0.50%. The reduced limit is mandatory for all ships operating outside designated Emission Control Areas (Baltic Sea, North Sea, North American area and US Caribbean Sea area) where the limit is already 0.10%. The new limit will mean a 77% reduction in overall SOx emissions from ships, equivalent to an annual reduction of approximately 8.5 million metric tonnes of SOx. Particulate matter will also be reduced.

Four options are available for ship owners to comply. Ships can be fitted with exhaust gas scrubbers to remove the SOx, they can burn low-sulphur fuel oil or bio-diesel, or they can burn liquefied natural gas (LNG). According to Drewry Shipping Consultants, in 2019 the cost of including a scrubber in the exhaust system of a new crude oil tanker was between US$2.5 and US$3.0 million, and the cost of retro-fitting a scrubber to an existing tanker was between US$4.0 and US$4.5 million. Industry analysts believe that most ships will burn low-sulphur fuel oil, although some new ships might be fitted with exhaust gas scrubbers. However, some shipowners may either convert to using natural gas or order new ships whose engines are designed to burn natural gas. For obvious reasons, almost all current LNG carriers power their engines using natural gas.

Gas burns more cleanly than most liquid fuels, but also burns hot and dry, which results in higher stress on the lubricating oil in gas engines. Because of the higher combustion temperature, oxides of nitrogen are formed more readily compared with the combustion of fuel oils or biodiesel. This is called nitration, and it can affect the performance of the engine oil.

Traditionally, oils for engines burning high-sulphur fuel oil have had a BN of around 70. In recent years, slow steaming to reduce fuel consumption, which leads to more condensation of sulphur acids in the cylinders, has required oils with BNs of 100 or even 140. Industry consultants believe that between 85% and 90% of large ships will switch to low-sulphur fuel oils. However, because there is only limited experience of using these fuels in two-stroke marine engines, OEMs are recommending that operators of two-stroke marine engines start with engine oils having a BN of 40 and monitoring the results. According to additive manufacturers, too much alkalinity in the cylinders, caused by an oil feed rate that is too high, can lead to excess deposits. Conversely, too low a BN can allow sludge to turn to varnish in cylinder hot-spots.

A small, though increasing, number of ships are either being converted to operate on natural gas or are being built with engines powered by natural gas. The condition monitoring of the engine oils and engines is, therefore, similar to those for natural gas engines, discussed in the next section.

10.7 NATURAL GAS AND BIOGAS ENGINE OILS

Various designs of stationary natural gas or biogas engines are used, including vertical in-line and V-type four-stroke engines, two-stroke V-type engines integral with a horizontally opposed reciprocating compressor and dual crankshaft, vertically opposed, two-stroke engines. Their primary functions are either to generate electricity or to power a gas compressor.

These engines burn a variety of gases including, but not necessarily limited to:

- Sour natural gas, containing sulphur compounds.
- Sweet natural gas, containing no sulphur and very little carbon dioxide.
- Wet natural gas, containing relatively high quantities of component gases such as butane.
- Landfill, digester or sewage gas, composed primarily of methane and carbon dioxide and which frequently contains halogens such as fluorine and chlorine, but can also contain ammonia, silicon compounds and/or sulphur compounds. This is also known as biogas.

The chemical compositions of these gases vary considerably, in contrast to gasoline and diesel fuels, which are more limited in composition based on their octane rating, cetane number or sulphur content. The gas composition strongly influences the stationary gas engine oil (SGEO) required for a particular engine.

Liquid fuels also contribute to the protection of the valves inside the cylinder head of four-stroke engines, while gaseous fuels do not protect the valves. This means that the lubricating oil and its ash are required to protect the intake and exhaust valves in a gas engine. While some ash is needed, too much ash becomes a problem, as it can build up inside the combustion chamber, leading to cylinder liner wear, valve problems, pre-ignition and detonation.

The latest types of gas engines have greater horsepower output (and brake mean effective pressure, BMEP) with higher combustion temperatures, which results in higher local lubricating oil temperatures inside the engine. Modern engines are designed for better fuel economy, stricter emissions and smaller size to power ratios. Smaller engines have smaller oil volumes, adding to increased stress on the lubricating oil, which means faster oil degradation and shorter oil service life. Higher compression ratios make modern engines more sensitive to ash build-up.

In most countries where gas engines operate, exhaust emissions have become a serious concern. To control or eliminate these emissions, some of the current engine designs require catalytic converters, which limit the additive types and the formulated percentage levels that can be used in the engine oils.

These lubricants vary with engine design and operating conditions. They range from simple uninhibited mineral oils, to medium-to-high ash, alkaline and oxidation inhibited detergent oils, to totally ashless, yet highly detergent, types.

The primary difference between natural gas and other internal combustion engine oils is the necessity to withstand the various levels of oil degradation caused by the gas fuel combustion process, which results in the accumulation of oxides of nitrogen. Nitration must be monitored regularly if both lubricant and engine life are to be maintained. Sulphated ash content is another consideration unique to natural gas engine oils.

Lubricating engines that are burning biogas is a particularly difficult problem. Several alternative types of engine oil need to be used, depending on the composition of the biogas, which must be analysed before selecting which SGEO to use. OEM recommendations are a useful initial guide. Factors to be considered in lubricating oil selection include the sulphated ash content of the oil, depending on the gas composition and quality, the load on the engine and the engine make, model and engine history. Other factors include the desired oil service life (with regard to the maintenance schedule), the acid neutralisation capacity needed, the engine sensitivity to combustion chamber component deposit formation or the required corrosion protection based on the mode of operation. Operations can range from continuous at high loads (with no cooling period) to start/stop (for peak load shaving electricity generation, with regular intermittent cooling).

Gas engine oils for biogas applications are usually divided into four groups, based on additive chemistry:

- Low ash engine oils for pre-treated gases that have low chlorine and ammonia contents, where the silicon content matches OEM specifications and the hydrogen sulphide content is less than 200ppm. These oils are suitable for biogas service but are formulated for sweet natural gas service. They are mainly SAE 40 viscosity grades, with ash contents less than 0.6 %wt, good oxidative stability and nitration resistance, but with limited acid neutralisation capacity.
- Low ash engine oils formulated for burning contaminated biogas. They are also SAE 40 viscosity grades with ash contents less than 0.6 %wt, but with an additive package that provides control over acid formation using advanced detergents and dispersants, corrosion inhibitors and metal passivators. They are used in applications with moderate-to-high chlorine and ammonia contents, silicon contents according to OEM specifications, hydrogen sulphide greater than 200ppm.
- Mid-ash engine oils formulated for contaminated biogas applications and extended ODIs. They are SAE 40 viscosity grades with ash contents between 0.6 and 1.0 %wt and higher base numbers to provide a higher acid neutralisation capacity. They are used when the biogas has low-to-moderate chlorine content, moderate-to-high ammonia content, silicon content according to OEM specifications, more than 200 ppm hydrogen sulphide.
- Mid-ash engine oils formulated for burning highly contaminated biogas applications. They are also SAE 40 viscosity grades with ash contents between 0.6 and 1.0 %wt and advanced detergent/dispersant combinations, corrosion inhibitors and metal passivators to protect engine components from corrosive wear from high chlorine and sulphur contents.

Several common condition monitoring techniques are used for natural gas engines. An analysis of the compression pressure/crank angle or pressure–time (P-T) curve is one common technique. Natural gas engines have some cycle-to-cycle combustion variations. By measuring the P-T curves, it is possible to determine such conditions as high fuel consumption, uneven internal pressures, high temperatures and unbalance causing detonation. All of these can affect the life of upper cylinder components and the effectiveness of the lubrication and emission control systems.

An analysis of the pressure–volume (P-V) curve can be used to balance cylinders, detect valve train problems and determine frictional losses, by comparing engine horsepower to compressor horsepower. Also, analysing reciprocating vibration patterns can provide an understanding of certain mechanical conditions, such as burned valves or gas leaks.

Oil analysis tests which should be considered part of a regularly scheduled predictive maintenance and condition monitoring programme for natural gas engines include:

- Viscosity.
- BN.
- AN.
- Sulphated ash content.
- Glycol contamination.
- Water contamination.
- Insolubles content.
- Wear metals contents.
- FTIR analysis.

BN is not often used as a test for natural gas engine oils unless the application operates under dual fuel conditions (where the engine uses either diesel or natural gas as fuel under various operating conditions). If the operation requires that diesel fuel is used for up to 50% of running time, BN testing should be included as an oil analysis requirement. Because most natural gas engine oils are formulated as low-to-medium ash oils, the BNs will generally be in a range of 3 to 7. These levels may not be sufficient to protect engines using dual fuels.

BN is also an important oil analysis test when the fuel in use contains high levels of sulphur and/or organic halogens. When high sulphur sour gas or landfill gas is in use, the typical natural gas oils available may not sufficiently protect the engine from acid compounds. In these cases, the engine operator may need to shorten oil drains or select an oil with a higher BN, which will provide a higher level of alkalinity. Potential lubrication problems caused by the use of the kinds of fuel described should be discussed with both the engine manufacturer and the lubricant supplier.

Two-stroke natural gas engine operation tends to form various deposits such as varnish, sludge and an ash residue which remains after the oil is burned during operation. The varnish and sludge are controlled by the detergent/dispersant additive combination. However, the detergent additives tend to leave a grey, fluffy ash residue after the oil has been burned. This ash residue comprises metal sulphates from calcium, phosphorus, zinc, magnesium, boron and barium detergent compounds.

Therefore, lubricant formulators must ensure that the additive concentrations are high enough to help prevent valve recession, but not so high as to cause unwanted and harmful deposits or cause catalysts to become ineffective. Valve recession is the premature wearing of the valve seat into the cylinder head. The sulphated ash residue helps to prevent premature valve recession by "cushioning" the valve seat area.

Excessively high concentrations of certain additives, such as those that contain zinc, sulphur or phosphorus, can also be harmful to catalyst equipped natural gas

TABLE 10.3

Example Flagging Limits for Natural Gas Engine Oils

	Flagging Limit	
Test	Waukesha	Caterpillar
Viscosity	−20%/+30% change from new oil	3 cSt or more at 100°C above new oil
BN	30%–50% of new oil	50% of new oil
AN	2.5–3.0 above new oil	<4.0
Oxidation	25 absorption per cm (ABS/CM)	FTIR 100% allowable or 25 ABS/CM
Nitration	25 absorption per cm (ABS/CM)	FTIR 125% allowable or 25 ABS/CM
Insolubles	>1.0%	–
Water content	>0.2%	>0.5%
Glycol content	Any detectable amount	Any detectable amount
Wear metals	Based on trend analysis	Based on trend analysis
Chlorine	900 ppm	–

Source: Noria Corporation, with permission.

engines. These additives may deactivate the exhaust catalyst by forming glassy-amorphous deposits which prevent the exhaust gas from reaching the active surfaces of the catalyst, which in turn makes control of harmful emissions difficult.

Natural gas engine manufacturers list the levels of sulphated ash and the additive concentrations that are acceptable for use in their engines. For specific recommendations concerning sulphated ash content and additive levels, the engine operator should contact both the engine manufacturer and the lubricant supplier.

FTIR is most commonly used to monitor nitration and oxidation in gas engine oils. As nitration progresses, more nitrogen compounds are formed, which can lead to an increase in the acidity of the oil, leading to oxidation, corrosion and the wear of internal engine surfaces. FTIR also provides reliable information on soot, fuel dilution, glycol concentration and water. The typical FTIR absorbance to monitor is at approximately 1,630 cm^{-1}. This wavenumber has a few interferences from additives such as viscosity index improvers and dispersants, but for the most part, it is an adequate representation of the amount of nitration compounds found in the sample.

When it is necessary to switch lubricating oil, users should rely on the used oil analysis trending of the in-service product to tell how an oil behaves under the specific conditions. Example flagging limits used by Waukesha and Caterpillar are shown in Table 10.3. Other OEMs will have different flagging limits.

10.8   INDUSTRIAL AND OFF-HIGHWAY ENGINE OILS

Gasoline and diesel engines are also used in a wide range of industrial and off-highway equipment, including back-up electricity generators, powering gas compressors, mining haul-packs, construction equipment, agricultural equipment, forestry equipment and railroad engines. As such, they are almost always larger versions of the gasoline and diesel engines used in cars, vans, trucks and buses.

Many of these engines are sufficiently large, with large oil sumps, to allow regular oil condition monitoring. Obviously, monitoring engine operation should also be part of a condition monitoring programme. Engine temperature, coolant level, noise, vibration, fuel consumption and operating hours should all be monitored.

The ASTM has published a guide to oil monitoring in auxiliary power plant equipment. ASTM D6224 "Standard Practice for In-Service Monitoring of Lubricating Oil for Auxiliary Power Plant Equipment" covers gear and circulating oils, hydraulic oils, turbine-type oils (but not steam and gas turbine oils, which are covered in ASTM D4378, discussed in a later chapter), air compressor oils, phosphate ester electrohydraulic oils and diesel engine oils.

The guidance for diesel engine oils contained in ASTM D6224 can be applied to larger diesel engines used in other applications. Oil condition monitoring tests for diesel engines are:

- Viscosity at 100°C.
- Water content.
- Flash point.
- Insolubles.
- Foaming characteristics.
- Base number.
- Glycol content.
- Fuel dilution.
- Wear particle concentration.
- Wear debris analysis.
- Elemental analysis.

A more detailed discussion of ASTM D6224 will be presented in Chapter 12 on hydraulic systems.

Two oil condition monitoring tests of specific relevance to heavy-duty diesel engines are glycol content and fuel dilution. Any glycol detected in the engine oil is indicative that the anti-freeze coolant has leaked into the oil. In addition to changing the oil, the cause of the leak should be investigated and remedied as a matter of urgency. Similarly for fuel dilution, the cause of the leakage of fuel into the oil should be investigated and remedied. It is also advisable to change the oil.

10.9   RAILROAD ENGINE OILS

Railroad, also called railway in many countries, locomotives use either direct electricity or a diesel engine as a primary source of power. For electric models, direct current from an electric line powers traction motors that drive the locomotive. In diesel locomotives, a diesel-fuelled engine usually drives a generator, which supplies electricity to power the traction motors.

The main global manufacturers of railroad locomotives include GE Transportation, which was spun off from GE (General Electric) and bought by Wabtec in 2019, and Progress Rail, a division of Caterpillar, which markets its locomotives under the EMD (Electro-Motive-Diesel) brand. Both companies are based in the United States, but supply locomotives worldwide. Siemens Mobility and Alstom, which acquired Bombardier Transportation early in 2021, are based in Europe. Both companies also sell worldwide. GE and EMD make diesel locomotives, while Siemens and Alstom make electric locomotives. CRRC Corporation, based in China, and Indian Railways are state-owned companies that make both diesel and electric locomotives.

Railroad locomotives typically generate between 1,200 and 6,000 horsepower. Most of the GE and EMD locomotive diesel engines are rated at 4,500 horsepower. The engine oil is required to lubricate the piston rings and liners, connecting rod and main bearings, the valve train and the turbocharger bearings. In addition to lubricating, the engine oil must cool and maintain engine cleanliness by removing combustion by-products and deposits, while also neutralising acidic combustion products such as sulphuric and nitric acids. Formulations generally include detergent, dispersant, anti-wear, antioxidant, viscosity index improver and pour point depressant additives. Railroad diesel engine oils are increasingly required to meet tougher emission standards, so low-sulphur diesel fuels are being used increasingly.

Railroad engine oils have traditionally been zinc-free because EMD engines had silver-plated wrist-pin bearings that could be damaged by zinc-containing additives such as zinc dialkyldithiophosphate (ZDDP). GE engines did not use silver-plated bearings, but railroad companies did not want to stock two different engine oils, so zinc-free engine oils became the standard. Although EMD stopped using silver in their bearings in the early 2000s, the zinc-free oil standard continued to protect the legacy fleet and remains a part of the industry today.

The performance of railroad diesel engine oils is classified by generation numbers 1 to 7, with specifications defined and approved by the Locomotive Maintenance Officers Association. Though a North American organisation, LMOA generations are used around the world to specify railroad diesel engine oils. The most recent LMOA Generation 7 engine oil specifications have been established to comply with US Environmental Protection Agency (EPA) emissions regulations for Tier 4 locomotive engines.

SAE 20W-40 viscosity oils are typically used in locomotive diesel engines. Depending on the sulphur content in the fuel, engine oils have a BN of between 9 and 13. Generation 7 oils typically have a BN of 11 as determined by ASTM D2896. These are now standard in North America. Generations 5, 6 and 7 oils are used in other countries. Generation 7 oils provide improved BN retention and combustion acid control and have better ability to withstand the additional stresses of exhaust gas recirculation systems and lower oil consumption in GE Evolution Series Tier 4 engines. They can also be used with low sulphur (less than 500 ppm) and ultra-low sulphur (less than 5 ppm) fuels.

Railroad engines in newer locomotives could also benefit from lower viscosity and/or synthetic engine oils, similarly to the heavy-duty truck and off-highway vehicle markets. However, this would mean railroad operators needing to stock at least two grades of engine oils, as older generations of railroad diesel engines have very high oil consumption compared with a typical car or truck. The cost of stocking and using two grades of engine oils and the risk of topping-up with the wrong grade might be outweighed by the benefits of longer oil drain intervals, better fuel economy and improved engine durability.

Railroad operators usually require engine oils to be changed every three or six months. However, oil change intervals can be longer or shorter, depending on the type of railroad, engine horsepower, type of operation and operating environment. GE's guidelines recommend testing in-service oils every seven to ten days under condition monitoring programmes that include tests for kinematic viscosity, BN, pentane insolubles, water content, soot content, anti-oxidant content and wear metals. Although GE does suggest flagging limits for engine oils, these vary with different Generation engines.

10.10   SUMMARY

Sampling and testing engine oils is not practical for small engines in small vehicles. Instead, oil change intervals are either fixed or are determined using computer algorithms that monitor a number of variables associated with either or both the engine oil and the engine. These computer calculations are designed to detect and predict potential mechanical damage to or failure of the engine. Complete sets of algorithms for electric and hybrid cars, vans and motorcycles are still being developed.

For larger gasoline, diesel or gas engines, there are numerous methods and tests for monitoring the condition of both the engine oil and the engine. For obvious reasons, the tests and methods are not the same for all engines operating in different applications and environments. Additionally, the flagging limits are also OEM, application and operating environment-dependent.

11 Condition Monitoring of Gears and Transmissions

11.1 INTRODUCTION

To adjust the speed of an engine or motor to the required speed(s) of a machine, vehicle or item of equipment requires a transmission system of some type. The effective lubrication of these transmissions is just as important as the lubrication of the engines driving them and, usually, requires special lubricants.

Gears are used to transmit motion and power from one rotating shaft to another or from a rotating shaft to a reciprocating element. Gearboxes contain both gears and bearings, because the rotating shafts must rotate smoothly, whatever load is put on the gears. As gear teeth mesh, they roll and slide together. This combination of sliding and rolling occurs with all meshing gear teeth, regardless of the type of gears. The two factors that vary are the amount of sliding in proportion to the amount of rolling and the direction of slide relative to the lines of contact between the tooth surfaces.

Monitoring the oils and greases used to lubricate the different types of gears and bearings in transmissions can be critically important and may require different methodologies and tests to those used for engines. For example, if the gearbox in a helicopter fails, the aircraft cannot fly, and it will fall to earth like a stone.

Chapter 11 describes and discusses the different types of gears and their lubrication, monitoring gears used in automotive applications and in industrial applications and the different types of bearings and their lubrication. The lubrication and monitoring of wind turbine gears and bearings are also explored in depth.

11.2 TYPES OF GEARS AND THEIR LUBRICATION

Numerous types of gears are used in a multitude of applications. The most frequently used types are:

- Spur.
- Bevel.
- Spiral bevel.
- Helical and double helical.
- Skew.
- Worm.
- Hypoid.
- Rack and pinion.

In each of the huge range of applications, gears are required to transmit power under large variations in loads, speeds (both input and output), sizes and input and output

DOI: 10.1201/9781003245254-11

shaft orientations. Gears operate in all types of conditions and in either mobile or fixed operations.

Illustrations of each of the types of gears are shown in Figures 11.1 to 11.3. Each type has specific advantages and disadvantages.

With spur gears, the axes of the driving gear and pinion are parallel, the gear teeth are cut straight and the gears make a theoretical line contact between pairs of meshing teeth. They usually have one or two teeth in mesh at all times. The direction of sliding is at right angles to the lines of contact. The arrangements can be either external (as shown in Figure 11.1) or internal, where the driving gear is inside the driven gear. Spur gears can be considered as a special type of bevel gear, in which the angle between the shafts is zero. The load-carrying capacity of spur gears is comparatively low, and they tend to be relatively noisy.

The axes of the driving gear and the pinion of a bevel gear intersect. Again, the gear teeth are cut straight and mesh in line contact, although both the driving gear and the pinion are cone-shaped. The contact lines can be extended to meet at the same point, in which case the gear teeth are developed on conical pitch surfaces. Again, the arrangement can be either external or internal and their load-carrying capacity is comparatively low. They are used in applications in which the rotating shafts are at right angles.

With spiral bevel gears, the gear teeth are curved. The flanks of the gear teeth mesh gradually and have a higher contact ratio than spur or bevel gears. Because of the curved shape of the gears, the theoretical lines of contact slant across the tooth

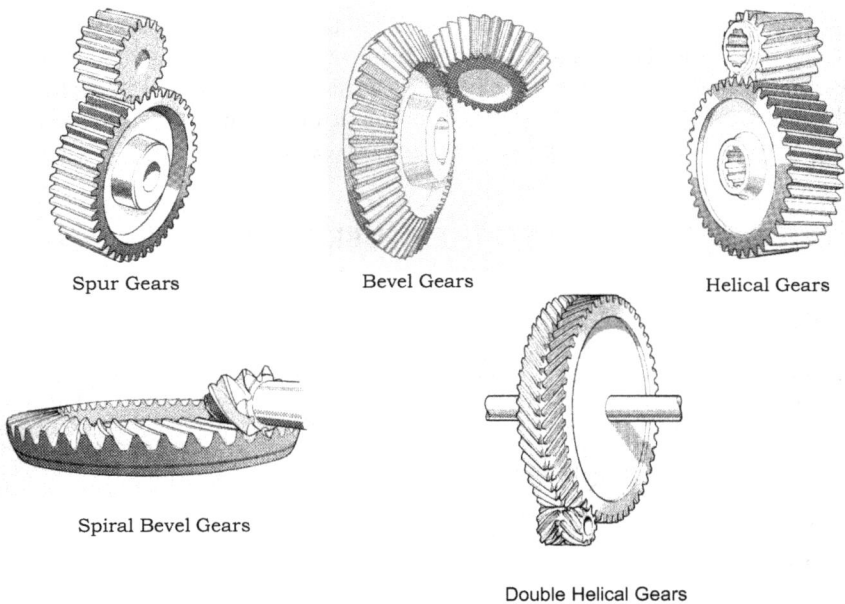

Spur Gears

Bevel Gears

Helical Gears

Spiral Bevel Gears

Double Helical Gears

FIGURE 11.1 Spur, bevel, helical and spiral bevel gears.

Source: Pathmaster Marketing Ltd.

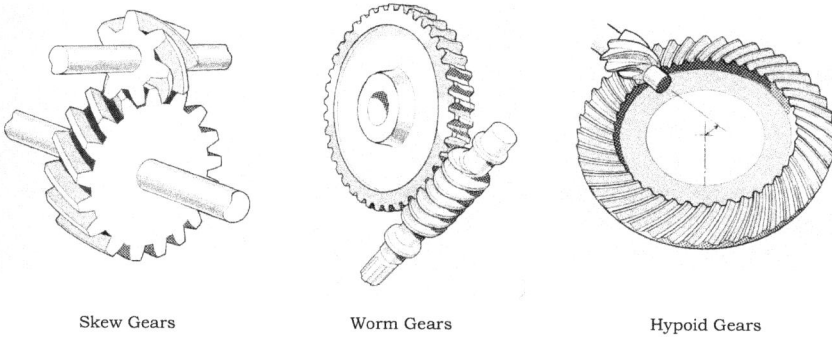

Skew Gears   Worm Gears   Hypoid Gears

FIGURE 11.2   Skew, worm and hypoid gears.

Source: Pathmaster Marketing Ltd.

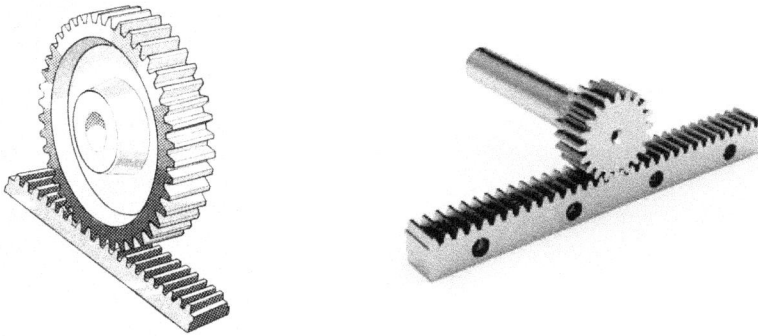

Rack and pinion gears with pinions of different sizes

FIGURE 11.3   Rack and pinion gears.

Source: Pathmaster Marketing Ltd.

surfaces. The direction of sliding, therefore, is not at right angles to the lines of contact and some sliding occurs across the lines of contact. This means that shock loads and noise generation are lower than with spur or bevel gears. It also means that the axial loads are higher, so shaft bearing design is more complex.

This is the same with helical gears, although in this case the axes of the driving gear and pinion are parallel. The gear teeth are curved or helix and the flanks of the gear teeth mesh gradually and have a higher contact ratio than spur or bevel gears. As with spiral bevel gears, shock loads and noise generation are lower and axial loads are higher, so shaft bearing design is more complex. Double helical gears, shown in Figure 11.1, eliminate this problem by having a pair of joined helical gears with oppositely curved teeth. As a result, the axial thrust loads are eliminated, so shaft bearing design is easier. However, gear design and manufacturing is much more complicated.

With skew gears, the shaft axes are neither parallel nor intersecting. The gear teeth are helical, so there is only point contact between mating teeth at any instant. This results in excessive stress concentration and very high slipping. These gears are only useful for low speeds and loads in gearboxes with unusual input and output shaft arrangements.

Worm gears have large and small helical gears having non-intersecting axes. The smaller diameter (worm) gear has a continuous helix (or helices). Tooth loads are relatively low, but rubbing speeds are very high, so frictional losses and heat generation are high. The same sliding and rolling action occurs as with spur gears, as the teeth pass through the mesh. Usually, the sliding and rolling action is relatively slow, because of the low rotational speed of the worm wheel. In addition, rotation of the worm introduces a high rate of sliding. The combination of two sliding actions produces a resultant slide which in some areas is directly along the line of contact. Transmission ratios are high (5:1 to 70:1) and noise generation is low, but power transmission efficiency is lower than with other types of gears.

Hypoid gears are displaced spiral bevel gears, in which the input and output axes do not intersect. The gear teeth have a modified shape, with pitch lines that are hyperbolic in form. In addition to the usual rolling action, hypoid gears have a combination of radial and sideways sliding that is intermediate between spiral bevel gears and worm gears. The greater the shaft offset, the more the sliding conditions are to those found in worm gears. The load-carrying abilities are much better than bevel gears, so hypoid gears are able to tolerate high specific loads. Their operation is much smoother than bevel gears, although they are generally used only in automotive applications.

A rack and pinion is a type of linear actuator that comprises a circular pinion engaging a linear gear, the rack. It functions to either translate rotational motion into linear motion or vice versa. Driving the pinion into rotation causes the rack to be driven linearly. Driving the rack linearly will cause the pinion to be driven into a rotation. Obviously, in all applications the rack has a finite length, so the amount of movement is limited. Rack and pinion gears are available in three variations.

Straight teeth gears have the tooth axis parallel to the axis of rotation. Their operation is similar to those of spur gears. Helical tooth gears include teeth that are twisted along a helical path in the axial direction. Helical teeth gears provide continuous engagement along the tooth length and are often quieter and more efficient than straight tooth gears. Roller pinion drives use bearing supported rollers that mesh with the teeth of that rack in order to provide minimal to no backlash.

The maximum force that can be transmitted in a rack and pinion mechanism is determined by the tooth pitch and the size of the pinion. A rack and pinion is commonly found in the steering mechanism of steered vehicles, such as cars, vans, trucks and busses. Rack and pinion provides less mechanical advantage than other mechanisms, but less backlash and greater feedback or steering "feel". The mechanism may be power-assisted, usually by hydraulic or electrical means.

Different factors affect the lubrication of each of these types of gears. For spur, bevel, spiral bevel and helical gears, normal rotational speeds and loads generate a hydrodynamic lubricant wedge between the gear teeth. Tooth contact time is too short to allow the wedge to be squeezed out of the contact zone, except that higher speeds squeeze the wedge out of the contact. Tooth wear will occur if the lubricant does not have boundary or extreme-pressure (EP) additives. At low speeds, sliding velocities are insufficient to result in incipient welding of asperities.

Worm gears tend to slide along their line of contact. The exact shape of the contact line and the direction of slipping depend on the form and degree of meshing of the worm and driven gear. Maintaining a hydrodynamic oil wedge is virtually impossible, so boundary lubrication conditions will exist at all times. Specific tooth loadings are much lower than in hypoid gears and correct meshing of the gears is very important.

With hypoid gears, sliding speeds and loads are such that they operate beyond the limiting hydrodynamic conditions. EP additives have to be used to minimise scoring, abrasion and adhesion, but unsuitable EP additives may cause corrosive wear.

Another factor in the lubrication of gears is the numerous types of materials used:

- Mild steel.
- Carburised steel (case hardened).
- Through hardened steel.
- Tufftrided steel.
- Cast iron.
- Bronze.
- Phosphor bronze.
- Brass.
- Acetal.
- Nylon.
- Polyester.
- Polycarbonate.

Both ambient temperature and lubricant temperature in the gear contact are important. Peak contact temperatures are difficult to measure, so maximum temperature and maximum increase in temperature are usually specified. For spur, bevel and helical gears, the maximum operating temperature should be 65°C and the temperature increase from ambient should be no more than 30°C. For worm gears, these temperatures are 90°C and 55°C respectively and for hypoid gears they are 95°C and 55°C, respectively.

Higher tooth speeds allow use of lower-viscosity oils. Lower-viscosity oil is also used at higher speeds to minimise drag when oil-bath lubrication is used or to allow ease of oil circulation in pressure-fed systems. Lower-viscosity oils also permit more rapid spreading over teeth, before meshing. However, oil viscosity needs to be a compromise for gearboxes containing gears operating at different speeds.

Average tooth loading is much less important than peak tooth pressure. Vibration or shock loading can lead to very high peak pressures, which may rupture oil films on gear teeth. EP additives are essential if peak pressures are high. Application of continuous loads above the breakdown value of the boundary oil film may have catastrophic effects.

Excessive case hardening can cause brittle fracture of teeth. High carbon steels should not be case hardened because excessive surface carbon can lead to brittleness. Incorrect grinding or honing can cause micropitting.

Gear lubricants can be applied by dipping, splashing, spray jet or mist. Gear oils applied by spray jet or atomiser are more likely to oxidise. Oil sprays or mists help to dissipate frictional heat from the gear teeth, at the expense of higher rates of oxidation or thermal degradation. Gearbox lubrication systems may involve pumps, filters, coolers and pipes, all of which affect the type of gear lubricant required.

The operating environment is also an important factor in the lubrication of gears. There could be the presence of excessive amounts of water or dust. Water contamination can lead to corrosion of gears, gearbox and bearings and can reduce the load-carrying properties of the lubricant. Dust can lead to excessive wear of gear teeth, causing loss of power transmission and higher operating temperatures.

This means that the properties required of gear lubricants are numerous:

- Viscosity: High enough to maintain either a hydrodynamic or boundary film, but not too high to generate heat from internal friction.
- Viscosity index: To keep the viscosity within limits at normal operating temperatures.
- Adhesion: Resistance to removal from gear teeth by wiping or rotation.
- EP properties: For high specific or shock loads.
- Chemical stability: Resistance to oxidation and thermal degradation.
- Anti-corrosion: Protection of all surfaces.
- Demulsibility: Rapid separation of water contamination.
- Anti-foaming: To maintain the lubricant film.

These required properties impact the methods used to monitor gear oils, gear greases and gearboxes.

11.3   MONITORING OF AUTOMOTIVE GEARS

Automotive gearboxes and drive-train components, similar to automotive engines, are relatively small with limited volumes of gear oils. Routine sampling is impractical.

OEMs tend to regard automotive gearboxes as virtually indestructible, so monitoring of the gears or lubricants is unnecessary. Automotive gearboxes are usually "filled-for-life" with very high-performance gear oils (the additive levels in automotive gear oils are higher than those for industrial gear oils) and if an automotive gearbox fails in-service, it is often replaced with a new one.

Larger manual gearboxes used in trucks and buses are sometimes repaired following problems or failures, usually due to a failure to transmit power from the engine to the drivetrain. Occasionally, severely noisy gearboxes or those experiencing unusual vibrations may be repaired. However, for operational efficiency and productivity, it is often more cost-effective to replace the gearbox, unless it can be repaired in a matter of a few hours.

11.4   AUTOMATIC TRANSMISSIONS

An automatic transmission is a type of motor vehicle transmission that can automatically change gear ratios as the vehicle moves, freeing the driver from having to shift gears manually. As with other transmission systems on vehicles, it allows an internal combustion engine, best suited to run at a relatively high rotational speed, to provide a range of speed and torque outputs necessary for vehicular travel. The number of forward gear ratios is often expressed in the same way as for manual transmissions, for example, 4, 6, 7, 8 or 9 speed.

The most popular form found in automotive vehicles (cars, vans, trucks and buses) is the hydraulic automatic transmission. Similar but larger devices are also used for heavy-duty commercial and industrial vehicles and equipment. This system uses a fluid coupling in place of a friction clutch and accomplishes gear changes by hydraulically locking and unlocking a system of planetary gears. These systems have a defined set of gear ranges, often with a parking pawl that locks the output shaft of the transmission to keep the vehicle from rolling either forward or backward. Some machines with limited speed ranges or fixed engine speeds, such as some forklifts and lawn mowers, only use a torque converter to provide a variable gearing of the engine to the wheels.

In addition to traditional hydraulic automatic transmissions, there are other types of automated transmissions, such as a continuously variable transmission (CVT) and semi-automatic transmissions. These also free the driver from having to shift gears manually, by using the transmission's computer to change gear, if, for example, the driver was redlining the engine. Despite superficial similarity to other transmissions, traditional automatic transmissions differ significantly in internal operation and driver's "feel" when compared with semi-automatics and CVTs. In contrast to conventional automatic transmissions, a CVT uses a belt or other torque transmission scheme to allow an "infinite" number of gear ratios instead of a fixed number of gear ratios. A semi-automatic retains a clutch like a manual transmission, but controls the clutch through electrohydraulic means. The ability to shift gears manually, often via paddle shifters, can also be found on certain automated transmissions.

To overcome the heating and churning action within the transmission casing, an oxidation-inhibited and non-foaming oil is obviously essential. To maintain cleanliness within the transmission, a dispersant is added and a viscosity index improver prevents excessive thinning of the fluid when it becomes hot. (These additives are often combined in the form of a dispersant polyester type of viscosity modifier.) The oil must also be compatible with the elastomeric seals of the transmission and must not corrode any of the metallic components. It should also inhibit the transmission against internal corrosion from atmospheric moisture when it is left standing.

Of vital importance to the smooth operation of the transmission is the effect of the oil on the friction surfaces of the clutches and bands. The fluid must not attack the friction surfaces and is also required to provide the correct frictional characteristics as the clutches and bands come into engagement. The required frictional characteristics are specified by the transmission manufacturers and must not change significantly over the life of the fluid. They are evaluated in bench clutch pack friction tests and in "shift feel" testing on the road.

For many years, there was a major difference between the frictional requirements of Ford and those of General Motors and most other manufacturers. To produce a compact transmission with a possibility for eventual "fill-for-life" capability, Ford, until the mid-1970s, used relatively small frictional surfaces and required these to have a high holding capacity with a fast and positive lock-up on the clutches and bands to prevent overheating from slippage. This required a static coefficient of friction which was higher than that of the dynamic coefficient. Other manufacturers, with the notable exception of Borg-Warner and Toyota (up to 1984), preferred to have smoother and more prolonged changes produced by a fluid with a static coefficient of friction lower than that of the dynamic coefficient.

This property is achieved by the use of additives known as friction modifiers in the fluid. Following complaints of "clunk" on low-speed changes in new cars, Ford modified their policy in 1978 and introduced a friction-modified fluid for factory-fill purposes. In 1987, Ford introduced MERCON automatic transmission fluids (ATFs). There are now only minor differences between the frictional requirements of General Motors' DEXRON specifications and those of the Ford MERCON fluids.

For the larger commercial vehicles, there are automatic and semi-automatic transmissions for which the fluid requirements are set by makers, such as Allison and Caterpillar. The frictional requirements of these ATFs are similar to those for passenger car applications.

Tests for new ATFs include physical properties, bench tests in transmission and test rigs and a performance evaluation in field trials. Specifications include evaluations for:

- Viscosities at 40°C and 100°C.
- Brookfield viscosities at various temperatures to −40°C.
- Flash point.
- Copper corrosion.
- Rust protection.
- Foam tests.
- Elastomer seal compatibility tests.
- Wear test (vane pump).
- Friction tests on clutch and band materials.
- Transmission oxidation test.
- Transmission cycling test.
- Shift performance on the road.

The physical tests for these oils are the same as those used for engine oils or gear oils. Various tests in full-sized transmissions are used to evaluate ATFs. Most manufacturers have proprietary tests, run on both laboratory test beds and on the road.

With the latest 6, 8 and 9 speed automatic transmissions of all sizes, their mechanical tolerances are so precise that it is exceedingly difficult for a standard service workshop to maintain or repair them. In almost all cases, if an automatic transmission develops problems or fails, it is more cost-effective (although quite expensive) to simply replace the unit and return the faulty unit to the OEM for repair or refurbishment.

11.5 MONITORING INDUSTRIAL GEARS

Almost all manufacturing plants, worldwide, use a gearbox in one or more of their machines.

The lubrication of industrial gears presents significant challenges, because of high loads, heat, aeration and the presence of chemically active particles, such as brass and steel. Contaminants can be present that weaken the lubricating film strength and interfere with critical lubricant film clearances.

Misalignment, unbalance and looseness of gears are a risk to varying extents, depending upon the configuration of the machine train, what the gearbox is coupled to and the coupling mechanism. As with most machines, vibration analysis and oil analysis are required for proactive control over the root causes of mechanical wear and failure. The proactive aspects of integrated condition monitoring apply to most of the machine types.

With gearboxes, the most common failure modes are gear tooth wear and gear tooth fracture. Detecting wear-related gear faults with vibration analysis can be challenging because there are so many competing vibration signals, particularly at slow speeds, where the amplitude of the vibration signal may not be strong enough to overcome the noise factor. However, oil analysis provides excellent resolution in detecting contact fatigue, abrasion and adhesive wear, making it very suitable for the early detection of these failure modes. Although ferrography can be expensive and time-consuming, it is particularly good at detecting gear tooth wear. Particle counting as a pre-cursor to ferrography can be very useful.

If a gear tooth breaks away from the shaft, it is unlikely that the failure will produce a detectable concentration of wear particles, especially if the failure is caused by a sudden impact or defective material. However, monitoring the gear mesh frequency with vibration analysis will almost always detect a broken gear tooth.

Occasionally, a broken gear tooth will cause more substantial damage to the meshing gears and may cause gear misalignment, leading to failure of the bearings. (Almost all industrial gearboxes have two or more bearings, usually rolling bearings.) When the failure of a gearbox involves failure of the bearings, diagnosis of which came first (bearing failure will lead to gearbox failure) is particularly difficult and requires expert tribological analysis. However, when used together, vibration monitoring and oil analysis make a powerful combination for managing the condition of industrial gearboxes.

Many laboratories monitoring industrial gearboxes have believed for many years that a high ISO cleanliness level, as measured through particle counting, was considered normal. In many cases, particle count testing was not even performed on gearbox applications. It is estimated that it can cost nearly ten times more to remove contamination than it takes to keep contamination out in the first place.

An efficient breather is one of the most common methods of contamination control for gearboxes. Unfortunately, it is common for a manufacturer to deliver a gearbox with a simple vent port and plug. These plugs are not sufficient to keep out contamination at the size levels that can cause harm to the internal components. Several aftermarket breathers are widely available. Spin-on filters are a good choice for a breather, provided the appropriate micron level of filter is used. It is generally recommended that a 3 μm filter be used in a breather application.

Desiccant breathers allow for air exchange in a gearbox. While the gearbox breathes, the air passes through a 3 μm particulate filter and a desiccant moisture absorbing filter. These breathers help to keep particles out, remove moisture from the environment and remove any moisture that may build up internally due to condensation.

Expansion chambers allow for a complete enclosure from the environment while still allowing the component to "breathe". Expansion chambers have an internal bladder that expands and contracts with the requirements of the gearbox.

Hybrid style breathers are a combination of desiccant breathers and expansion chambers. The advantage to these breathers over expansion chambers is that the presence of the desiccant media will allow for water vapour removal that may have become present through condensation. These types of breathers are ideal for applications where frequent wash downs occur, high humidity areas and outside applications.

Controlling contamination in gearboxes also involves effective seals. A common approach to seals on high-speed and low-speed shafts of gearboxes is the use of lip seals. Occasionally these lip seals will use grease in order to help keep out contamination. While lip seals are capable of keeping out contamination, they are certainly inferior to labyrinth seals.

Labyrinth style seals are most often associated with pump applications. The move to these types of seals for gearboxes is becoming more of a common decision as industry learns of the overall destructive nature of contamination and the superior performance of labyrinth seals.

Another source of contaminant ingress into a gearbox is through checking oil levels. Unfortunately, the two most common methods for checking oil levels are either using the supplied dipstick or via a level port that must be removed for level confirmation. Both of these methods have the potential to introduce unwanted contamination to the system.

Modifications that may be considered for level checking include the addition of a bull's eye style sight glass into those areas where a level port exists or adding a stand pipe style level gauge to the drain or auxiliary side port of the gearbox. However, adding a stand pipe level gauge does not fully address the possibility of contamination, as contaminant ingress is possible through the vent hole of the level gauge itself. Applications that use an external level gauge should also have the gauge vented back to the case or to the breather assembly via a tee-style fitting.

In addition to keeping contaminants out, removing them once they get in or as they are formed inside the gearbox must also be performed. The first option is portable filtration. Ideally, quick connect fittings will be used in order to connect the filter cart without opening the gearbox to the environment. A common choice of quick connect fittings is the ISO B industrial interchange. The best approach to this is to use a female coupling on either the discharge or return and use a male coupling on the opposite end. This will help to reduce the chances of connecting the filter cart lines in reverse. It has been found, however, that the female ISO B industrial interchange fittings are twice the cost of the male. This has resulted in many end users simply using male fittings on both the suction and returns for the gearboxes and applying female fittings to both lines on the filtration unit.

The second option is a permanent offline kidney loop filter system on the gearbox. This will allow for continuous condition and contaminant removal from the lubricating oil. During the installation of this type of filter system, it is important to include the installation of an appropriate sample port for overall condition monitoring via oil analysis.

Research studies have shown that rolling element bearing life can be increased up to seven times by changing from a 40 μm filter to a 3 μm filter. The results also showed that a gearbox must be clean after assembly or a fine filter will not be effective. This occurs because permanent damage to gears and bearings can occur in only 30 minutes during running in, and there is not enough time for a filter to remove any built-in contamination.

The studies found wide variations in gearbox performances, with some gearboxes experiencing severe contamination and failure, while others operate with contaminant-free oil and run successfully for years. It is believed the varying performances are largely due to differences in the design of the gearbox and lubrication system and differences in assembly cleanliness. Consequently, careful attention is required throughout the manufacturing, assembly and initial running in process, to assure reliable gearboxes.

The interior of gear housings should be painted with white epoxy sealer to provide a hard smooth surface that is easy to clean, seals any porosity in the housing and seals in debris such as casting sand. All components for assembly should be properly stored in a dry area prior to assembly. All gears and bearings should be covered and bearings should be stored on their sides.

All components should be cleaned prior to assembly. Initial cleaning should be done in an area separate from the clean room, followed by final cleaning in the clean room just prior to assembly. All components should be carefully inspected to ensure they are clean and rust-free before assembly.

Gearboxes should be assembled in a clean room separate from any manufacturing processes such as machining, grinding, welding or deburring. Windows and doors should be adequately sealed to prevent contamination ingression and the ventilation system should be filtered so it provides clean, draft-free air. The floor should be painted and sealed so it is easily cleaned and the overhead structure should be painted and dust-free. Powered vehicles should not be allowed in the clean room, because they invariably introduce contaminants. All work should be done by skilled technicians properly trained in gearbox assembly. All tools should be clean, appropriate for gearbox assembly, in good condition and properly calibrated.

The ASTM has published a guide to oil monitoring in auxiliary power plant equipment. ASTM D6224 "Standard Practice for In-Service Monitoring of Lubricating Oil for Auxiliary Power Plant Equipment" covers gear and circulating oils, hydraulic oils, turbine type oils (but not steam and gas turbine oils, which are covered in ASTM D4378, discussed in Chapter 13), air compressor oils, phosphate ester electrohydraulic oils and diesel engine oils. A detailed discussion of ASTM D6224 will be presented in the next chapter, on hydraulic systems.

Condition monitoring for industrial gearboxes and gearbox lubricants is relatively straightforward. Measurements and tests include:

- Operating temperature and temperature differential.
- Vibration and/or noise.
- Kinematic viscosity.
- Low-temperature Brookfield viscosity.
- Acid number.

- Remaining antioxidant life.
- Water content.
- Sediment and insolubles.
- Wear, contaminant and additive elements.
- Cleanliness level.

Unfortunately for gearbox owners, the flagging limits for these tests are likely to be different for each type of gearbox and the numerous factors that affect gearbox operation.

Operating temperature is a major factor that affects gearbox lubrication. Both ambient temperature and lubricant temperature in the gear contact are important. Peak contact temperatures are difficult to measure, so maximum operating temperature and maximum increase in temperature through the gear contact are usually specified. For spur, bevel and helical gears, the maximum operating temperature should be 65°C and the maximum increase in temperature should be 30°C. That means the bulk gear oil temperature should be no more than 35°C. For worm gears, the two temperatures should be 90°C and 55°C, and for hypoid gears, they should be 95°C and 55°C.

Another important factor that affects gearbox lubrication is the surface speed of the gear teeth. Higher tooth speeds allow use of lower-viscosity oils. Lower-viscosity oil is also used at higher tooth speeds to minimise viscous drag when oil-bath lubrication is used or to allow easier oil circulation in pressure-fed systems. Lower-viscosity oils also permit more rapid spreading of oil over the gear teeth, before meshing. However, gear oil viscosity needs to be a compromise for gearboxes containing gears operating at different speeds.

Average tooth loading is much less important than peak tooth pressure. Vibration or shock loading can lead to very high peak pressures, which may rupture oil films on gear teeth. As a result, EP additives are essential in gear oils or greases if peak tooth pressures are high. Obviously, application of continuous loads above the breakdown value of the boundary oil film may have catastrophic effects.

Gear tooth surface finish is another important factor. Excessive case hardening can cause brittle fracture of teeth, so high-carbon steels should not be case hardened because excessive surface carbon can lead to brittleness. In addition, incorrect grinding or honing of gears can cause micropitting.

The method of gear lubrication can affect the performance of a gear oil or a gear grease. Application of the lubricant can be either dipping, splashing, spray jet or mist. Gear oils applied by spray jet or atomiser are more likely to oxidise. Oil sprays or mists help to dissipate frictional heat from the gear teeth, at the expense of higher rates of oxidation or thermal degradation. Gearbox lubrication systems may involve pumps, filters, coolers and pipes, all of which affect the type of gear lubricant required.

The operating environment is the final (but no less important) factor affecting gearbox lubrication. The main issues are the presence of excessive amounts of water or dust. Water contamination can lead to corrosion of gears, gearbox and bearings and can reduce the load-carrying properties of the lubricant. Dust can lead to excessive wear of gear teeth, causing loss of power transmission efficiency and higher operating temperatures. Both can affect the oxidation and thermal stabilities of gear oils and gear greases.

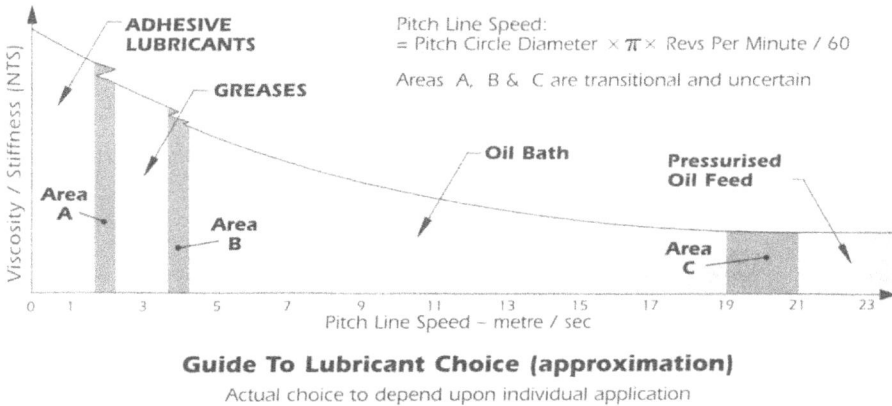

Guide To Lubricant Choice (approximation)
Actual choice to depend upon individual application

FIGURE 11.4   Selection of gear lubricant.

Source: Pathmaster Marketing Ltd.

As a result, the properties required of gear oils are a suitable viscosity and vis-cosity index, excellent resistance to oxidation and thermal degradation, suitability for high and low temperatures (for inside and outside use), good load carrying (anti-wear and/or EP) performance, excellent rust and corrosion inhibiting abilities, low foaming tendency, good demulsibility and compatibility with all gearbox materials, including ferrous and non-ferrous metals, seals and paints.

The properties required of gear greases are slightly different. They include a suit-able consistency to allow penetration between the meshing teeth, good adhesion, low oil separation, low starting torque, noise dampening properties and protection against ingress of contaminants into the gear contacts. They must also be compatible with all gearbox materials, including ferrous and non-ferrous metals, seals and paints.

A guide to when to use a gear oil and when to use a gear grease and the methods to apply them is given in Figure 11.4.

11.6   BEARINGS

There are two main types of bearings: plain bearings and rolling bearings. Each type is sub-divided:

- Plain Bearings:
  - Sliding: Relative motion is linear.
  - Journal: Shaft rotates inside a journal in the form of a cylinder.
  - Thrust: Used to take axial loads on rotating shafts.
- Rolling Bearings:
  - Ball Bearings: Deep groove, angular contact, self-aligning, thrust and two-row angular contact (split ring and filling slot).
  - Roller Bearings: Cylindrical, tapered, spherical, barrel, thrust (cylindri-cal, taper and spherical) and split (spherical and cylindrical).

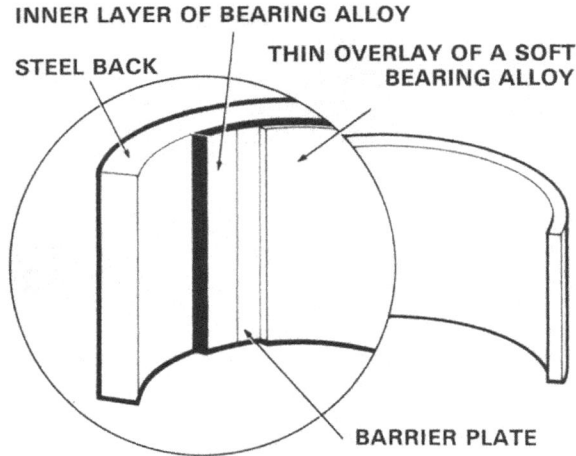

FIGURE 11.5   Journal bearing.

Source: Pathmaster Marketing Ltd.

Plain bearings either locate two parts of a mechanism or transmit forces between two parts of the mechanism. They allow relative motion between the parts. Journal bearings support a rotating shaft. Plain bearings can be oil- or gas-lubricated or self-lubricated. Plain bearings can be horizontal or vertical and hydrostatic or hydrodynamic. A diagram of a typical journal bearing is shown in Figure 11.5.

Low-friction materials are used to minimise friction during start-up, when metal-to-metal contact occurs. These materials include Babbit white metal (tin-based alloys containing copper and antimony), lead-based white metals, copper–lead alloys, copper–tin leaded bronze, phosphor bronze, aluminium bronze, high-tensile brasses or steel-backed three-layer bearings.

All types of rolling bearings have sealed and shielded variants. Roller bearing elements are made of hardened and polished "bearing" steel. Ball bearings are of lower cost, have lower friction and higher speed capabilities, but have lower load capacities. Both types of rolling bearings can be pre-greased and sealed. Diagrams of a ball bearing and a roller bearing are shown in Figures 11.6 and 11.7, respectively.

Rolling bearing cages are usually steel, brass or engineering plastics. Aluminium or phenolic resin is used for high speeds, precision or high temperatures. Pressed steel cages are the most common. Brass cages have high strength and durability. Glass fibre reinforced polyamide 66 cages are of low weight, flexible and low cost, but can only be used at bearing temperatures below 110°C

Split roller bearings are assembled around the shaft. They do not require any shaft fittings to be removed in order to slide the bearing on from the free end. Although they have a have high initial cost, they have significantly lower maintenance costs.

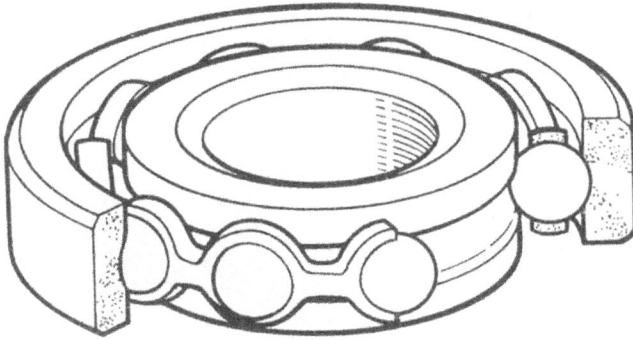

FIGURE 11.6   Ball bearing.

Source: Pathmaster Marketing Ltd.

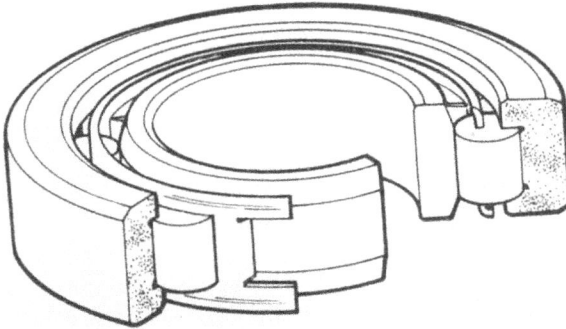

FIGURE 11.7   Cylindrical roller bearing.

Source: Pathmaster Marketing Ltd.

Rolling bearings are designed to operate with a dynamically generated elasto-hydrodynamic lubricant film between the rolling surfaces. Selection of the correct viscosity of oil depends on the size and type of bearing, its operating temperature and speed and specific application conditions. About 90% of all rolling bearings are grease-lubricated. In grease-lubricated bearings, the viscosity is taken from the base oil of the grease.

Rolling bearings are precision machine elements found in a wide variety of applications. They are typically very reliable even under the toughest conditions. Under normal operating conditions, bearings have a substantial service life, which is expressed as either a period of time or as the total number of rotations before the rolling elements or inner and outer rings fatigue or fail. According to research, less than 1% of rolling bearings do not reach their expected life.

The accurate diagnosis of a bearing failure is imperative to prevent repeat failures and additional costs. When a bearing does fail prematurely, it is usually due to causes that could have been avoided. For this reason, the possibility of reaching conclusions about the cause of a defect by means of studying its appearance is very useful.

Most bearing failures, such as flaking, pitting, spalling, unusual wear patterns, rust, corrosion, creeping and skewing, are usually attributed to a relatively small group of causes that are often interrelated and correctable. These causes include lubrication, mounting, operational stress, bearing selection and environmental influence.

The purpose of lubricating a bearing is to coat the rolling and sliding contact surfaces with a thin oil film to avoid direct metal-to-metal contact. When done effectively, this reduces friction and abrasion, transports heat generated by friction, prolongs service life, prevents rust and corrosion and keeps foreign objects and contamination away from rolling elements. Grease typically is used for lubricating bearings because it is easy to handle and simplifies the sealing system, while oil lubrication is more suitable for high-speed or high-temperature operations. Lubrication failures usually occur due to:

- Using the wrong type of lubricant.
- Too little grease or oil.
- Too much grease or oil.
- Mixing of grease or oil.
- Contamination of the grease or oil by contaminants or water.
- Incorrect grease service life.

In addition to the normal bearing service life, it is also important to take into account the normal grease service life. Grease service life is the time over which proper bearing function is sustained by a particular quantity and category of grease. This is especially crucial in pump, compressor, motor and super-precision applications.

In the process of mounting and installing a bearing, it is critical to use proper tools and ovens or induction heaters. A sleeve to impact the entire inner ring face being press fit should be used. The shaft and housing tolerances should be checked. If the fit is too tight, too much pre-load will be created. If the fit is too loose, too little pre-load will be created, which may allow the shaft to rotate or creep in the bearing. The proper diameters, roundness and chamfer radius should also be checked.

Misalignment or shaft deflection must be avoided. This is particularly significant in mounting bearings that have separable components such as cylindrical roller bearings, where successful load bearing and optimal life are established or diminished at installation.

Maintaining the radial internal clearance (RIC) of the bearing that was established in the original design is also very important. The standard scale in order of ascending clearance is C2, C0, C3, C4 and C5. The proper clearance for the application is important in that it allows for the challenges of lubrication, shaft fit and heat.

A proper film of lubricant must be established between the rolling elements of the bearing. Reducing internal clearance and impeding lubricant flow can lead to premature failure. With regard to shaft fit, it is inevitable that there can be a reduction in the RIC when the bearing is press fit. Also, in the normal operation of bearings, heat is produced, which creates thermal expansion of the inner and outer rings. This can reduce the internal clearance, which will reduce the optimal bearing life.

It is very rare to find a bearing that has been improperly designed into an application. However, factors within the larger application may change. If loads become too high, overloading and early fatigue may occur. If loads are too low, skidding and improper loading of the rolling elements occur. Early failure will be the result in both situations. Similar issues arise with improper internal clearance. The first indication of these situations will be unusual noises and/or increased temperatures. Bearing temperatures typically rise with start-up and stabilise at a temperature slightly lower than at start-up, normally 10°C to 40°C higher than room temperature. A desirable bearing temperature is below 100°C.

Typical abnormal sounds from bearings indicate certain issues in the bearing application. While this is a subjective test, it is helpful to know that a screech or howling sound usually indicates too large an internal clearance or poor lubrication on a cylindrical roller bearing. A crunching felt when the shaft is rotated by hand normally suggests contamination of the raceways. Operational stresses in the application can also impact bearing life. It is essential to isolate vibrations in associated equipment, as they can cause uneven running and unusual noises.

Even with the best design, lubrication and installation, bearing failures will occur if the operating environment is not taken into consideration. While there are many potential issues, the primary ones include:

- Dust and dirt, which can aggressively contaminate a bearing. Special attention should be given to using proper sealing techniques.
- Aggressive media or water, for which correct sealing is again vital. Using special seals that do not score the shaft is recommended.
- External heat. The ambient operating temperature mandates many choices in radial internal clearance, high-temperature lubricants, intermittent or continuous running and other factors that affect bearing life.
- Current passage or electrolytic corrosion. If current is allowed to flow through the rolling elements, sparks can create pitting or fluting on the bearing surfaces. This can be corrected by creating a bypass circuit for the current or by using insulation on or within the bearing. This should be an inherent design consideration in applications such as wind turbines and all power-generating equipment.

The temperature of a bearing can be obtained easily using a non-contact infrared imager to quickly obtain a multi-point temperature profile. This inspection can be performed with little to no disruption to the plant's operations and can be used as a screening tool as part of a daily or weekly inspection. As with most predictive technologies, thermography's real value comes not from an instantaneous measurement but the trending of data and the further analysis of that data. The constant feedback from these inspections can be used to analyse any changes.

While there are no standard of references for the temperature profile of a specific machine component, there are some expectations. The variables that feed these expectations include but are not limited to speed, load, viscosity, contamination, size and lubricant base oil type. There are so many variables that can contribute to the generation of heat that it would be nearly impossible to consider them all to generate such a profile.

The first step in the overall prevention of bearing failure lies in considering bearing technologies that are most suitable to the application with regard to specifications, recommendations, maintenance strategies, fatigue life and wear resistance of the bearing. Premature bearing failure within a proper application is typically attributed to one or more of the above causes.

11.7   WIND TURBINES

Most wind turbines tend to be in remote locations and for which physical access to the gears, bearings and control systems is difficult and expensive. Routine sampling of lubricants is not at all practical. The complex internal components of a typical wind turbine are illustrated in Figure 11.8. Lubricants are required for the main shaft bearings, the gears and bearings in the speed-increasing gearbox, the generator bearings, the hydraulic system motor bearings and the yaw gears. In addition, the shaft bearings and gearbox gears and bearings are subject to cyclic loads as each of the turbine blades passes the tower that supports the turbine.

Condition monitoring programmes for wind turbines generally include time-scheduled oil analysis. Unfortunately, servicing of wind turbines is often limited to specific times each year. When the first large wind turbines were installed over 20 years ago, oil samples were only taken every 6 months or so. Trend analysis for important oil quality parameters was difficult and erratic.

Consequently, oil analysis and condition monitoring for both on-shore and off-shore wind turbines are being done increasingly using on-line oil and machine

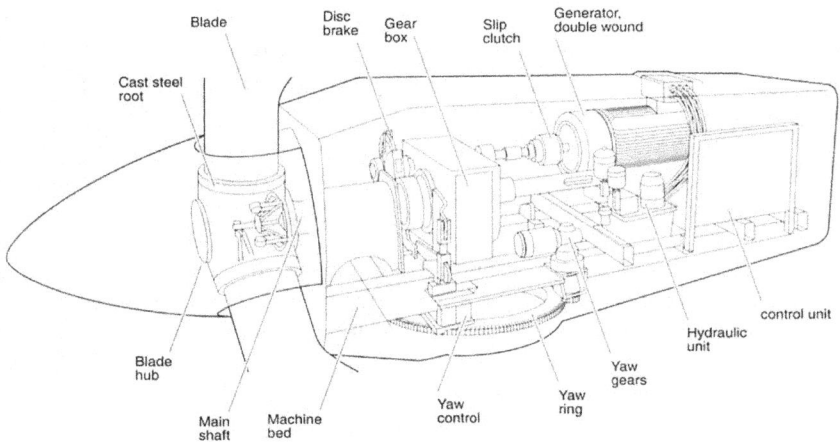

FIGURE 11.8   Wind turbine internals.

Source: Pathmaster Marketing Ltd.

monitoring sensors, similar to those being used in automotive engines and marine systems. The oil sampling intervals are thereby reduced from every six months or so to either daily or continuously. This allows a more precise trending of important oil and machine parameters.

Wind turbines in desert environments are exposed to airborne dust during the hot season and moisture during the rainy season. Offshore turbines are constantly exposed to moisture.

Micropitting is widespread in wind turbine gearboxes and is detrimental because it reduces gear accuracy, may cause gears to be noisy and may escalate into other failure modes such as macropitting, scuffing or bending fatigue. Micropitting refers to the formation of very small, micro-scale craters (or pits) on the surface of an active gear tooth flank or rolling bearing. Micropitting is a fatigue failure of the surface of a material and is also known as grey staining, micro spalling or frosting. The difference between pitting corrosion and micropitting is the size of the pits after surface fatigue. Pits formed by micropitting are approximately 10 to 20 μm in depth, and micropitted metal often has a frosted or grey appearance. Gear operation is inherently prone to micropitting due to the repetitive Hertzian contact between the tooth involute surfaces.

The precise causes of micropitting have not been established, despite many years of research work. It is currently thought to be due to a combination of several factors, including inadequate lubrication, overly aggressive EP additives, poor surface finish, poor tooth load distribution, incorrect metal hardening and surface defects. White etching cracking in wind turbine bearings and gears is thought to be similar to micropitting, in that it appears to have similar causes and can also lead to macropitting and fatigue damage. Both micropitting and white etching cracking are currently thought to be the result of hydrogen embrittlement, which may be the result of a build-up of static electricity in gears and bearings in wind turbines. A great deal of research work is currently in progress in universities, additive companies, lubricant suppliers and OEMs to understand the causes and mechanisms of action of hydrogen embrittlement, micropitting and white etching cracking.

Because Hertzian fatigue is one of the two failure modes that gearing is designed against, any micropitting which developed towards the end of the gearbox design life would be considered normal contact fatigue of the steel. However, micropitting and white etching cracking that lead to macropitting or fatigue damage contaminate wind turbine lubricants with large amounts of iron particles. The use of accurate and smooth surfaces, surface-hardened gears and splines and high-viscosity lubricants containing suitable anti-wear and EP additives can minimise internally generated wear debris. External and internal spline teeth should be nitrided and force-lubricated to prevent fretting corrosion. Annulus gears should be carburised or nitrided rather than through-hardened because through-hardened gears are relatively soft and prone to generating wear debris.

For wind turbine gearboxes and bearings, material selection and heat treatment choices that determine material hardness are especially important design considerations. Filters cannot remove particles once they become embedded in softer gearbox components. Hard particles embedded in through-hardened annulus gears can cause polishing (fine-scale abrasive wear) on mating planetary gears. This degrades gear accuracy and adds wear debris. Nonmetallic bearing cages should not be used because they are susceptible to particle embedment that can result in severe abrasive wear on rollers.

Wind turbines should not be parked for extended periods. Otherwise, fretting corrosion may occur on gear teeth, splines and bearings. In addition, wet clutches and brakes should have separate lubrication systems to avoid contaminating the gearbox with their wear debris.

Most wind turbine gearboxes have labyrinth seals that provide long life and adequately seal in oil, but may allow contaminant ingression. Therefore, V-rings should be used as external seals. They are effective, but should have metal shields to protect them from damage.

Maintenance in wind turbines is difficult and risky. The gearbox, lubrication system, work platforms and nacelle housing should be designed to ensure maintenance tasks are readily accomplished in a safe manner. Breathers, filters, drain ports, fill ports, sample ports, dip sticks, magnetic plugs and inspection ports should be designed and located for easy maintenance and minimisation of contamination.

All maintenance that involves opening the gearbox or lubrication system should be performed with good housekeeping procedures. Oil should be added from a filter cart connected to the gearbox with quick-connect couplings to minimise contaminants in the new oil and to minimise contaminant ingression during the transfer.

Many experiments have shown that water in oil promotes micropitting. The mechanism is unproven, but may be related to hydrogen embrittlement, as noted earlier. All lubricants are susceptible to water contamination, but ester-based lubricants and mineral oils with anti-scuff additives are especially prone to absorbing water and generally give lower fatigue life with high water contents. To remove water, the lubricant should be changed or processed, or an offline filter with water absorption capability should be added.

Lubrication systems should be properly designed and carefully maintained to ensure gears receive an adequate amount of cool, clean and dry lubricant. Modern filters are compact and provide fine filtration and long life without creating large pressure drops. Offline filters provide fine filtration during operation and during turbine shutdown. Once the oil is clean, it should stay clean provided the gearbox and lubrication system were properly designed and seals, breathers and maintenance are adequate.

As with industrial gearboxes, wind turbine gear and transmission systems should be assembled in a very clean environment. The cleanliness of the new wind turbine gearbox oil is also important. A useful guide is that the ISO rating of the new oil should be 16/14/11 maximum. It has been found that, after initial testing of the gearbox, this will have risen to around 17/15/12. In service, the gear oil should have a maximum ISO cleanliness of 18/16/13. This is significantly cleaner than conventional industrial gearboxes.

An oil monitoring programme can help prevent gear and bearing failures by showing when maintenance is required. Lubricant monitoring should include spectrographic and ferrographic analysis of contamination, particle counts and analysis of acidity, viscosity and water content. Used filter elements should be examined for wear debris and contaminants.

The most important oil properties that need to be monitored for wind turbines are:

- Viscosity.
- Acid number.
- Oxidation.
- Water content.
- Additive contents, particularly phosphorous.
- Elemental contents, particularly wear metals.

All these can now be done using electronic sensors from a number of suppliers, including the Bosch "Multifunction Oil Condition Sensor", the dielectric sensors used by Mercedes-Benz and Delphi Corporation and the "Fluid Condition Monitor" (FCM) developed by Eaton Corporation, described in Chapter 10. Even infrared spectroscopy can now be done online.

In a conventional oil analysis laboratory, Fourier transform infrared (FTIR) spectroscopy can be used to monitor oxidation, water content and additive elements. An online sensor developed by the Technical University of Munich has been found to be able to provide test results for acid number, oxidation, water content, phosphorous, silicon, zinc and copper that agree closely with those obtained by conventional FTIR. The sensor consists of a collimated infrared light source, a cuvette, a linear variable filter (LVF) and a linear pyroelectric detector array with application-specific integrated circuit (ASIC). The LVF is a spectral device that uses a Fabry–Pérot structure to provide virtually constant resolution over the required wavelength region.

Compared with conventional FTIR spectrometers, the LVF spectrometer has a reduced resolution and a limited spectral range. It operates in a range of about 1,800 to 900 cm^{-1} at a wavelength dependent resolution of about 36 at 1,800 cm^{-1}. (Standard FTIR spectrometers normally operate in a wavelength range of 4,000 to 500 cm^{-1} at a spectral resolution of 4 cm^{-1} or better.) Even so, the new sensor has been shown to provide acceptable correlation with laboratory test results, using data from several hundred oil samples. Extracting the oil quality parameters from the measured infrared spectra requires sophisticated data processing, using a multivariate regression model.

Data from all these sensors, all of which can be fitted to the bypass filtration line of a wind turbine gearbox, can be fed into Internet-based data analysis systems, allowing electrical system maintenance management to track the condition of an entire array of wind turbines in real time and receive e-mail alerts when emerging problems are identified.

11.8 CHAINS

Chains are another method of transmitting power. They have a number of advantages and disadvantages compared with other methods:

- Their action is positive over sprockets, as there is no slippage.
- They can carry very heavy loads with negligible stretch.

- They are able to be used in adverse environments, such as high temperatures or when subject to moisture, rainwater or dust.
- Their power transmission is comparatively efficient.
- Flexure occurs between bearing surfaces that are designed for resistance to wear.
- There is controlled flexibility, but in only one plane.

Chains can be grouped into three classes, two of which have several sub-types.

Roller chains, also known as bush roller chains, consist of a series of short cylindrical rollers held together by side links. The chain is driven by a toothed wheel called a sprocket, which is rotated by a motor. A diagram of a typical roller chain design is shown in Figure 11.9. The most familiar roller chains are those used on bicycles and motorbikes. All roller chains are constructed so that the rollers are evenly spaced throughout the chain. Links can be either roller links or pin links, alternately spaced throughout the length of the chain. The roller link is an assembly of steel shaped like the figure eight, with a hole at each end to receive the bushings. The centre distance between the holes is kept within a close tolerance to maintain a uniform pitch in an assembled chain. Each bushing is a hollow cylinder, with the outside and inside surfaces hardened to resist wear. The rollers are also hollow cylinders, hardened and finished to a precise diameter. Each is of the proper length to permit it to turn freely between the link plates. Roller chains are designed and constructed with the aim of reducing friction. The rollers engage the sprocket teeth of the driving and driven wheels, thereby transferring the sliding action to the internal

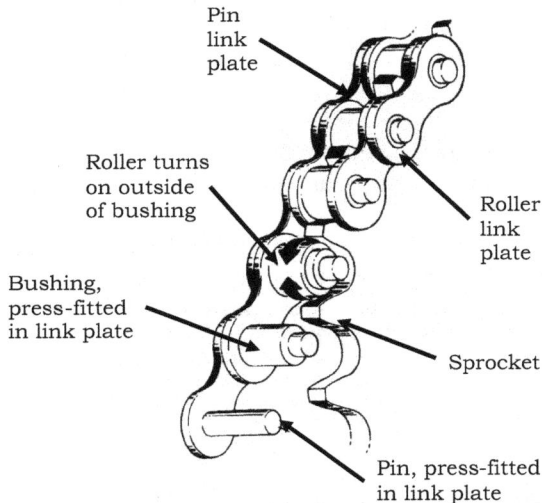

FIGURE 11.9   Roller chain.

Source: Pathmaster Marketing Ltd.

members of the chain. They are a fairly straightforward and simple method of transmitting mechanical energy.

However, unless roller chains are lubricated correctly, they are likely to wear out over time. When they do, the pitch increases, making them looser and more compliant, which can add instability in the form of "play" into the system. The primary factors that limit chain life are corrosion, wear and fatigue. Corrosion can also accelerate the effects of wear and fatigue. To combat this, make sure that the chain is protected. Some manufacturers offer options to improve fatigue resistance including specially manufactured components such as hardened pins and rivets as well as optimised chain dimensions for given applications. Most usually, when a chain becomes between 1.5% and 2% longer than its original length, it begins to ride up the sprocket teeth, and this is the point at which sprocket and chain wear accelerates. Some manufacturers offer corrosion-resistant chains, made from one of several types of materials including stainless steel, nickel-plated steel, titanium or combinations of stainless and special polymers.

Silent chains consist of a series of toothed linked plates and guide plates assembled on joint components in such a way that articulation occurs between adjoining pitches. They are designed to provide smooth and quiet operation at high speeds accompanied by long service life. The link plates contain notches, flats and projections in the apertures to position the joint in the link plates. Guide plates are provided in silent chains to prevent lateral movement. There are four types of silent chains:

- Round pin with bushing: The joint components consist of a plain cylindrical pin with a bushing added to increase the bearing area of the chain joint.
- Round pin and segmental bushing: The joint components are plain cylindrical pins with segmented bushings. Each bushing covers less than half the circumference, with lugs between the bushing segments to prevent rotation in the links.
- Rocker pins: Differently shaped and keyed components are in contact in such a way that a rocking motion occurs between the pins when the chains flex at the joint.
- Roller pins: The joint components are two pins with cylindrical surfaces that roll on each other.

The third type of chain are engineering steel chains which are able to bridge large spaces between parallel shafts. They are usually quite heavily loaded, move at relatively low speeds, are very rugged, have the toughness and elasticity to absorb heavy shock loads and are made of fabricated and machined steel. They are designed to perform a variety of other functions to meet known or anticipated operating conditions. There are many types of engineering steel drives available because of their wide application.

- Straight sidebar steel chains with and without rollers: These have steel bushings and pins and are assembled using interference fits between those parts and the sidebars. The inside link, made up of two sidebars and two bushings, is called the bushing link. The outside link is called the pin link.

- Offset sidebar steel chains with and without rollers: Each link is identical. The link components consist of sidebars, a bushing and a pin with a cotter.
- Bar link chains: These are sometimes called block and bar or steel block chains. They usually consists of two outer sidebars, one centre bar and two pins making up a two-link section. A bar link chain usually does not have bushings, the centre bar flexing directly on the pin. The sprocket contact is with the ends of the centre bars. Chains of this type are frequently used in tension linkage applications, in slow-moving conveyors and in very large versions to raise and lower river and channel lock and dam gates.
- Welded steel chains: These are similar to fabricated steel chains except that links are made as integral weldments rather than being held together by means of tight fits and locking surfaces. They are often used for slow-moving conveyor drives.
- Open barrel Steel Pintlr Chains: These are cost-effective lightweight conveyor chains. The basic chain consists of one piece offset steel links and pins. The pins are fixed against rotation by mechanical locks or interference fits.

Lubricants for roller chains should have a viscosity low enough be able to penetrate into the critical internal surfaces, have a sufficiently high viscosity (or appropriate additives) to maintain the lubricant film under the applied loads, be free of contaminants, adhere to the chain and be able to maintain lubrication under the prevailing conditions, whether high or low temperatures, rain, snow, dirt or dust.

Application of the lubricant to a chain can be either manual, drip, oil bath, slinger disc, oil stream or oil spray. The oil viscosity required will depend on the operating temperature. Most chain lubricants will contain anti-wear and/or EP additives, corrosion inhibitors and tackiness additives.

Monitoring the chain lubricant in service is impossible, as there is too little lubricant to collect for testing. The only way of monitoring the performance of the lubricant is to observe and measure the wear, fatigue and/or corrosion of the chain. The first signs of corrosion or wear should be a flagging limit to replace the lubricant.

11.9 COUPLINGS

Couplings are mechanical devices used to transmit power or torque from one shaft to another shaft. Some couplings are arranged in a straight line, while others can be used when the input and output shafts are at an angle. In addition to transmitting power, couplings are used to connect two components that are manufactured separately, to introduce extra flexibility when space is restricted, to introduce protection against overloads or to reduce the transmission of shock loads from one shaft to another.

A sleeve coupling, also known as a muff coupling, shown in Figure 11.10, is simply a thick hollow cylinder, the internal diameter of which is manufactured so that the shafts fit perfectly into the sleeve. The driving and driven shafts need to align in a straight line, but do not have to be the same diameter, in which case the sleeve's internal diameters are different at each end. Two or more threaded holes are provided in the sleeve as well as in both of the shafts, so the shafts don't move longitudinally

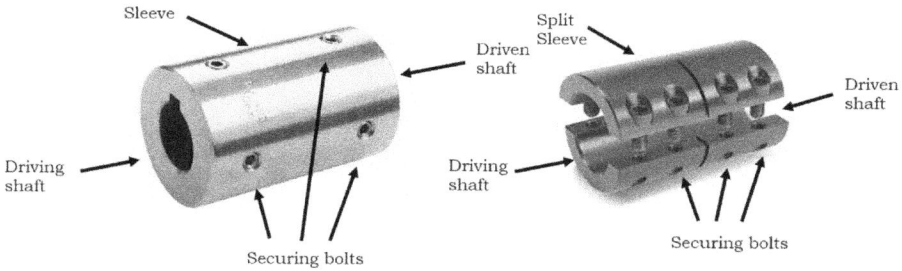

FIGURE 11.10   Sleeve (Muff) couplings.

Source: Pathmaster Marketing Ltd.

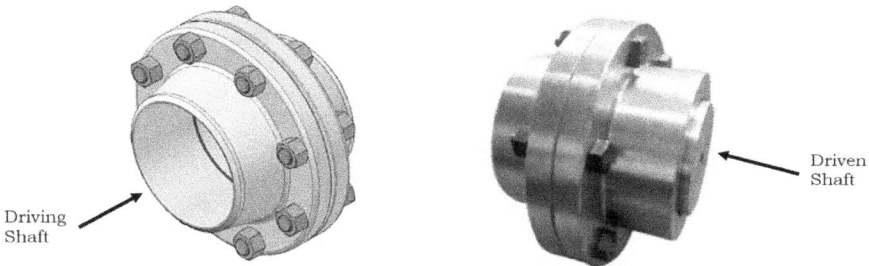

FIGURE 11.11   Flange couplings.

Source: Pathmaster Marketing Ltd.

when keyway bolts are inserted into them. They also ensure that the shafts and sleeve don't slip. Sleeve couplings are easy to manufacture, because there are few parts. They are used where the shafts don't require any alignment and torque loads are light-to-medium duty.

Split sleeve couplings are very similar, except that the sleeve is split into two parts, as also shown in Figure 11.10. The semi-cylindrical parts fit over the shaft, which means that both input and output shafts should ideally be the same diameter. Threaded holes are provided on the sleeve so that both the shafts can be joined with steel bolts or studs. These couplings offer the advantage that they can be assembled and disassembled without changing the position of the shafts. They are used for medium- to heavy-duty loads with moderate speed.

A flange coupling, shown in Figure 11.11, is also easy to manufacture and fit. There are flanges on either side of the two sleeves, with an equal number and location of threaded holes. The flanges are joined together with bolts and nuts. A tapered key section is also provided on the hubs and shafts so that the hubs don't loosen up or move backward and stay attached to the shafts. Again, the shafts need to be in perfect alignment.

A gear coupling, two examples of which are shown in Figure 11.12, is a modified version of the flange coupling. The flange and hub are separate parts assembled

FIGURE 11.12   Gear couplings.

Source: Pathmaster Marketing Ltd.

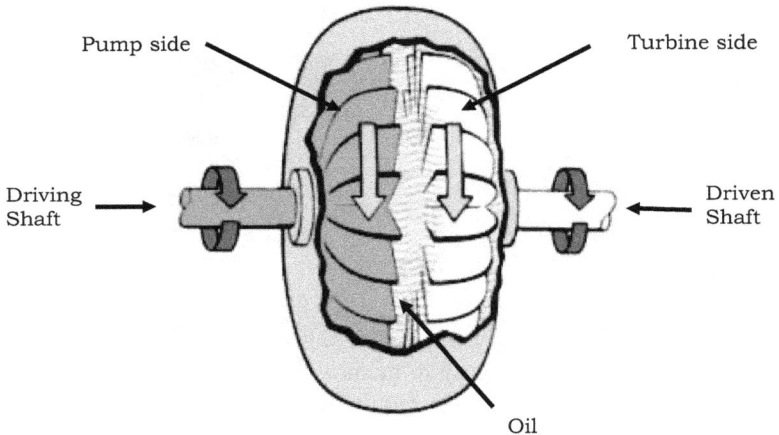

FIGURE 11.13   Fluid coupling.

Source: Pathmaster Marketing Ltd.

together, instead of one part as in a flange coupling. The hubs are externally splined, although these are sufficiently thick and deep so as to resemble gear teeth. The flanges have internal teeth, with a gear ratio of 1:1. This type of coupling can be used in applications where there could be a small degree of shaft misalignment, such as heavy-duty industrial uses with a higher torque transmission requirement.

A fluid coupling, shown in Figure 11.13, consists of two parts, a pump and a turbine. Both of these have blades similar to those of a steam, gas or water turbine. The pump side is rotated by the driving shaft, while the turbine side is on the driven shaft. Oil is used as a fluid medium to transfer the kinetic energy of the driving shaft to the driven shaft. The oil enters the pump through its centre, and when the driving shaft rotates, it is pushed outwards by centrifugal force. The coupling's casing is designed to divert the motion of the oil into the turbine blades, so that the driven shaft starts to rotate. Fluid couplings are able to tolerate a limited degree of shaft misalignment. When an extra part called a "reactor" is introduced between the pump and the turbine, a fluid coupling is called a "torque converter", such as those used in automatic transmissions. Fluid couplings are used widely in marine and industrial applications where controlled start-up of power transmission is essential. Fluid couplings can also act as hydrodynamic brakes, dissipating rotational energy as heat through frictional forces (both viscous and fluid/container). When a fluid coupling is used for braking, it is also known as a "retarder".

Constant velocity joints (also known as homokinetic or CV joints) allow a drive shaft to transmit power through a variable angle, at constant rotational speed, without an appreciable increase in friction or play. They are mainly used in front wheel drive vehicles. Modern rear wheel drive cars with independent rear suspension typically use CV joints at the ends of the rear axle half-shafts and increasingly use them on the drive shaft.

Several types of CV joint have been developed. A Rzeppa joint (invented by Alfred H Rzeppa in 1926) consists of a spherical inner shell with six grooves in it and a similar enveloping outer shell. An exploded diagram of a Rzeppa CV joint is shown in Figure 11.14. Each groove guides one ball. The input shaft fits in the centre of a large, steel, star-shaped "gear" that nests inside a circular cage. The cage is spherical but with ends open, and it typically has six openings around the perimeter. This cage and gear fit into a grooved cup that has a splined and threaded shaft attached to it. Six large steel balls sit inside the cup grooves and fit into the cage openings, nestled in the grooves of the star gear. The output shaft on the cup then runs through the wheel bearing and is secured by the axle nut. This joint can accommodate the large changes of angle when the front wheels of a front-wheel drive car are turned by the steering system.

A Birfield joint is based on the Rzeppa joint, but with the six balls confined using elliptical tracks rather than a cage. They have improved efficiency and are widely used in modern cars for the outboard driveshaft joints. A Weiss joint consists of two

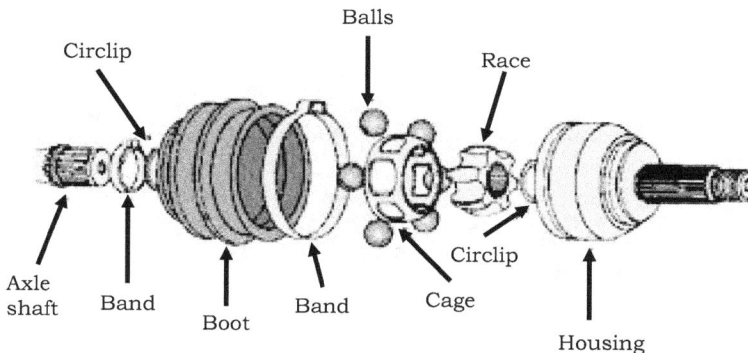

FIGURE 11.14   Constant velocity joint.

Source: Pathmaster Marketing Ltd.

identical ball yokes which are positively located (usually) by four balls. The two joints are centred by means of a ball with a hole in the middle. Two balls in circular tracks transmit the torque while the other two preload the joint and ensure there is no backlash when the direction of loading changes. Its construction differs from that of the Rzeppa joint in that the balls are a tight fit between two halves of the coupling and that no cage is used. The centre ball rotates on a pin inserted in the outer race and serves as a locking medium for the four other balls. When both shafts are in line, that is, at an angle of 180°, the balls lie in a plane that is 90° to the shafts. If the driving shaft remains in the original position, any movement of the driven shaft will cause the balls to move one half of the angular distance. For example, when the driven shaft moves through an angle of 20°, the angle between the two shafts is reduced to 160°. The balls will move 10° in the same direction, and the angle between the driving shaft and the plane in which the balls lie will be reduced to 80°.

Tripod joints are used at the inboard end of car drive shafts. They have a three-pointed yoke attached to the shaft, which has barrel-shaped roller bearings on the ends. These fit into a cup with three matching grooves, attached to the differential. Since there is only significant movement in one axis, this simple arrangement works well.

A universal joint (or Hooke's joint) is not a constant velocity joint, but is a type of coupling that can be used anywhere. It can transmit power even at high parallel or angular misalignments. It consists of a pair of hinges close together perpendicular to each other connected by cross shafts. This arrangement means that the speeds of the driving shaft and the driven shaft will not be same at every angle. Universal joints are used in machines where there are space restrictions or high flexibility is needed.

Fixed couplings (sleeve, muff and flange types) do not need lubrication, although they may need to be protected from corrosion in outdoor applications. However, even small misalignment is likely to result in some form of damage to the bearings of either the motor (with the driving shaft) or the equipment (with the driven shaft), so these should be monitored at regular intervals. Vibration or noise monitoring of bearings is probably the best way to check for misalignment. (Laser position monitoring during assembly should detect any misalignment at this stage of the equipment's life.)

Either lubricating oils or greases can be used for flexible couplings (gear, CV or universal joint types). Unless specifically advised by the coupling designer, couplings for the majority of industrial components are grease-lubricated. Coupling components are protected primarily by an oil film which bleeds from the grease and seeps into the loading zone. Flexible couplings require protection from the low-amplitude relative motion that develops between their components. Other concerns include centrifugal stress on the lubricant (particularly grease), which causes premature separation of the oil from the thickener, poor oil distribution within the housing and oil leakage from the housing. The low-amplitude relative motion, articulation speed and tendency towards a sliding rather than rolling action inhibit the development of hydrodynamic (full-film) lubrication. Greases having high-viscosity base oils, EP additives and metal-wetting agents are recommended to overcome the boundary (mixed film) conditions that often exist in flexible couplings. High oil viscosity also slows the leakage rates. Centrifugal forces in flexible couplings can be very high, being greater at higher distances from the rotational axis. Greases that resist oil separation are preferred.

Fluid couplings transfer kinetic energy from the input shaft to a fluid and then to the output shaft when transmitting power. Small amounts of misalignment are accommodated solely by clearances between the moving parts. It is also possible to compensate effectively for shock loading and high torque starting loads, as there is no solid connection between input and output shafts. However, because of viscous drag, the speed of the driven shaft will never reach the speed of the driving shaft. The difference in speed between the input and output shafts is known as slippage. The minimum input speed required to overcome the viscous drag is known as the stall speed. Equipment with large static loads, such as a steam or gas turbine, incorporate a fluid coupling to minimise the initial stress on the driving shaft. When the stall speed is exceeded, the output shaft will begin to accelerate, but will do so at a constrained rate due to its moment of inertia (resistance to angular acceleration).

The dissipation of energy that makes fluid couplings so tolerant of shock loading creates the potential for rapid and extreme increases in fluid temperature. The energy dissipated during stall and slip is converted to heat through the viscous shearing of the fluid (internal fluid friction). Consequently, resistance to oxidation and thermal degradation are important qualities of oils used for fluid couplings. A high viscosity index (VI) is also useful to prevent severe decreases in operating viscosity at temperature spikes and excessively high operating viscosity at low temperature conditions. Depending on operating conditions and power transmission requirements, oils having kinematic viscosities of between 2.5 and 72 cSt at 40°C are used for fluid couplings.

These fluids must also resist foaming due to the severe agitation caused by the impeller's movement and its impact upon the runner vanes. Rust protective properties help preserve the coupling's metal components. Hydrocarbon-based fluids are superior in this regard to other fluids, but their performance can be further improved through rust inhibiting additives. Seal compatibility is also important for long-life usefulness.

Low-viscosity multi-grade engine oils, automatic transmission fluids or high VI hydraulic fluids are generally preferred for fluid couplings. Increasing density of the fluid increases the amount of torque that can be transmitted at a given input speed. In applications in which a change in viscosity with temperature is undesirable, thereby affecting power transmission efficiency, oils with a high VI should be used.

Constant-velocity joints are protected by a rubber boot, a "CV gaiter", usually filled with molybdenum disulphide grease. Cracks and splits in the boot will allow contaminants in and grease to leak out, causing the joint to wear quickly due to inadequate lubrication, abrasion or scoring from dirt and/or corrosion from water or moisture. Wear of the boot is often in the form of small cracks, which appear closer to the wheel because the wheel produces most of the vibration and up and down motions. Cracks and tears in the areas closer to the axle are usually caused by external factors, such as packed snow, stones or uneven rocky off-road paths. Ageing and chemical damage can also cause boot failure.

Condition monitoring of lubricants used in couplings depends greatly on the size and application of the coupling. Small couplings, such as CV joints used in automotive applications, are very difficult to monitor, as the volume of lubricant is too small. Several years ago, motorists were advised to re-grease CV joints every six months. Now, modern CV joints are filled-for-life and if they fail for any reason, they are simply replaced.

Larger couplings used in industrial applications, particularly gear and fluid couplings, may benefit from lubricant condition monitoring. The amount of lubricant in the coupling and any signs of lubricant leakage should be checked at regular intervals, typically monthly. Gear couplings probably require the most maintenance. Typical re-lubrication intervals are six months to one year, depending upon application severity and experience.

If gear couplings or universal joints are cleaned of old grease during an inspection, care must be taken to ensure the complete removal of any cleaning products prior to re-greasing. Care should also be taken to avoid introducing contaminants, such as dirt or moisture. Greases can be monitored using the methods described in Chapters 5 and 15. The oil in fluid couplings can be monitored in the same way as for hydraulic fluids, described in the next chapter.

Vibration analysis or thermography can be used while the coupling is in use, to identify couplings that are not in alignment, as even the sturdiest foundations shift over time. It is advisable to check for proper alignment whenever intrusive maintenance or repairs are performed on the coupled components.

11.10 SUMMARY

Condition monitoring of gears and transmissions and their lubricants varies widely and is highly dependent on the type, size, application and operating conditions involved. Automotive gear and transmission oils and greases tend not to be routinely monitored and only larger automotive gearboxes and transmissions are monitored for vibration, noise and/or temperature. Larger industrial gearboxes and gear oils and greases are increasingly monitored, to minimise the risk of failure, particularly of gears and transmissions in wind turbines.

Condition monitoring of lubricants used in chains and couplings also depends on their type, application and operating conditions. Lubricants used in some applications cannot be monitored routinely, because their amounts are too small. Instead, the operation and mechanical condition of the chain or coupling and the equipment they connect must be monitored. Some couplings are not lubricated. Again, in larger industrial applications, the lubricants are monitored, but so are the couplings, motors, pumps and other equipment.

12 Condition Monitoring of Hydraulic Systems

12.1 INTRODUCTION

Hydraulic systems, like many other industrial systems, are increasingly being operated at higher speeds, at higher pressures and with higher power outputs and shorter cycle times, to give higher productivity. Control valves have tighter clearances to improve performance, and to reduce cost, weight and space, reservoirs are shrinking in size. Original equipment manufacturers (OEMs) are designing systems with reduced noise levels and lower carbon footprints. This is the case across many industries, ranging from injection moulding and steel mills to off-road and construction equipment.

Since the hydraulic fluid is an integral part of the equipment's operational components, making all of these evolutionary changes in hydraulic system design places further demands on hydraulic oils.

End users share some responsibility in the increased stress on the hydraulic oils. Their attention to leak reduction inadvertently results in fewer top-up additions of new oil. Because of the economics of production, many hydraulic machines are operated by users at a higher production rate than originally designed.

The combination of system design changes and more severe operating conditions results in more oxidative and mechanical stresses with less recovery time for the oil. Oxidation is now the most common form of fluid failure followed by the depletion of anti-wear protection. In addition, there are new degradation mechanisms being observed, such as spark discharge and micro-dieseling.

12.2 HYDRAULIC OIL AND HYDRAULIC SYSTEM ISSUES

Hydraulic systems transmit and apply large forces, over relatively short distances, providing high degrees of flexibility and control. They are suitable for either fixed or mobile applications, such as machine tools, robots, cranes, forestry equipment, excavators, aircraft, ships, steel mills and underground mines. They are part of a major technical discipline known as fluid power technology and rely in the virtual incompressibility of fluids.

An hydraulic system comprises a number of components. A pump converts mechanical energy into hydraulic energy, pipes (both fixed and flexible) transmit pressurised hydraulic fluid, and one or more units, such as actuators or motors, convert the hydraulic energy into mechanical work. The system has a control circuit, with valves that regulate flow, pressure, direction of movement and applied forces and a fluid reservoir that allows for the separation of any water or debris before the

DOI: 10.1201/9781003245254-12

fluid is returned to the system through a filter. A schematic diagram of a typical hydraulic system is shown in Figure 12.1.

Several types of pumps can be used in hydraulic systems. A list of these is shown in Table 12.1. The most commonly used pumps are rotary vane, gear and axial piston pumps, illustrative drawings of which are shown in Figures 12.2 to 12.4. Each pump type has advantages and disadvantages, which impact the ways in which they should be monitored. This is discussed later in this chapter.

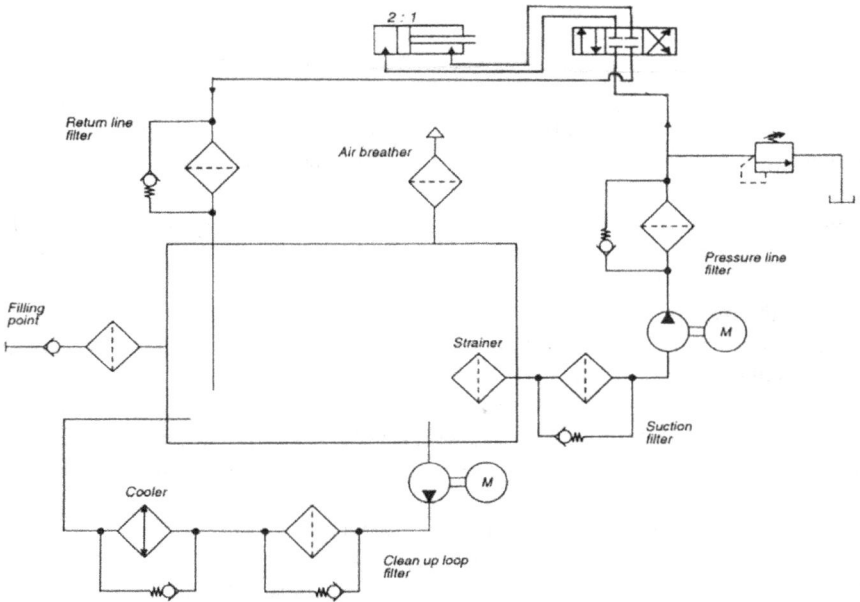

FIGURE 12.1   Typical hydraulic circuit diagram.

Source: Pathmaster Marketing Ltd.

TABLE 12.1

Types of Hydraulic System Pumps

Method of Imparting Energy to the Fluid	Pump Type
Volumetric displacement	Rotary vane
	Gear
	Screw
	Axial piston
	Diaphragm
Addition of kinetic energy	Centrifugal
	Regenerative
Electromagnetic force	Electromagnetic

Source: Pathmaster Marketing Ltd.

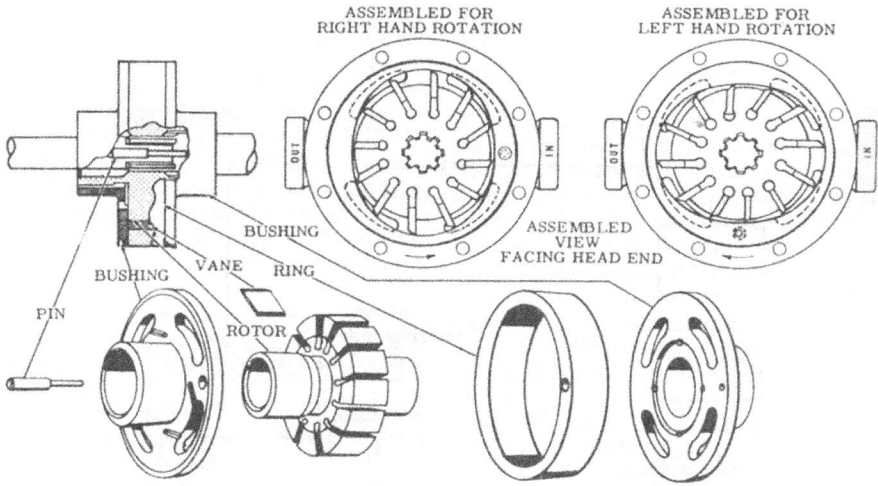

FIGURE 12.2   Rotary vane pump.

Source: Pathmaster Marketing Ltd.

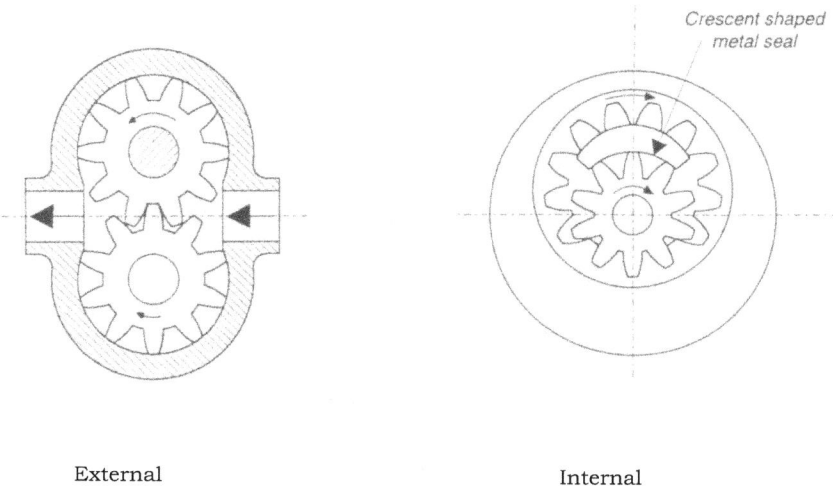

External   Internal

FIGURE 12.3   Gear pumps.

Source: Pathmaster Marketing Ltd.

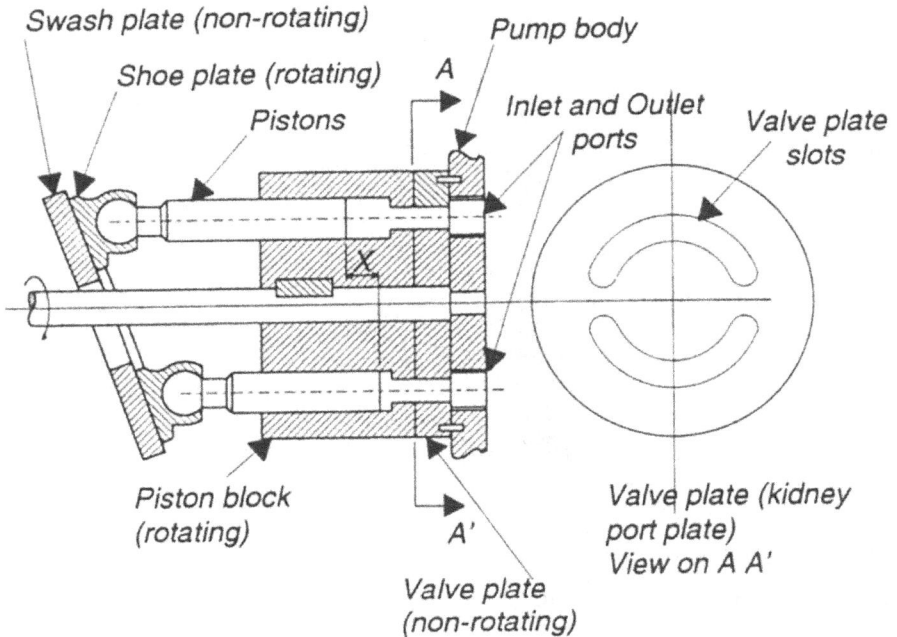

FIGURE 12.4 Axial piston pump.

Source: Pathmaster Marketing Ltd.

Hydraulic fluids are required to fulfil a very wide range of functions, listed in Table 12.2. A number of the required properties pose special problems for both fluids and systems.

Oxidation was originally defined as a reaction involving combination with oxygen. However, its definition has now been expanded to include any reaction in which electrons are transferred from a molecule. Oxidation is the predominant reaction a lubricant undergoes in service and accounts for significant lubricant performance problems. Oxidation is the major source of viscosity increase, varnish formation, sludge and sediment formation, additive depletion, base oil breakdown, filter plugging, loss of foam resistance, loss of demulsibility, acid number (AN) increase, rust and corrosion.

Hydraulic systems are particularly prone to performance problems due to oxidised oils. This is the result of using more sensitive and precise servo-valve controls that yield the improved operational performance. These control valves however have very low tolerance levels for oil contamination and in particular, varnish.

Oxidation is the most prevalent mode of hydraulic oil degradation, especially in modern hydraulic machines running at elevated temperatures, speeds and pressures. Oxidation is accelerated with heat and metal catalysts such as iron and copper.

Thermal degradation is a reaction occurring at temperatures in excess of 300°C. There are several distinct modes of thermal degradation:

TABLE 12.2

Functions and Properties of Hydraulic Fluids

Function	Required Property
Power transfer and control medium	Low compressibility (high bulk modulus), low foaming tendency, low volatility
Lubricant	Viscosity to maintain fluid film, anti-wear performance, low temperature fluidity, oxidation and thermal stability, hydrolytic stability, demulsibility, anti-corrosion performance, material compatibility, filterability, cleanliness
Heat transfer medium	Good thermal capacity and conductivity
Sealing medium	Adequate viscosity and shear stability
Special functions	Fire resistance, friction modification, radiation resistance
Environmental impact	No or low toxicity in use, no or low impact when spilled
Lifetime	Durability of properties for long use

Source: Pathmaster Marketing Ltd.

- Micro-dieseling: This is a process in which air bubbles implode as they move from a low-pressure zone to a high-pressure zone, most often in a pump or bearing. This causes adiabatic compression and temperatures of 1,000°C to create sub-micron carbonaceous deposits. The oil often turns black from the soot in the fluid. Micro-dieseling also creates a black patch with a dark brown varnish background in the membrane patch calorimetry (MPC) test.
- Electrostatic spark discharge: This is a phenomenon in which the lubricant accumulates static electricity from the Helmholtz double layer created in high velocity, low clearance lubricant zones. The static electricity accumulates to the point of discharge, creating sparks up to 10,000°C and thermally degrading the hydraulic oil. They can also cause spark erosion of servo-valves, significantly impairing their functioning.
- Extreme temperature zones: This can occasionally occur in an hydraulic system. Malfunctioning valves can generate very high, localised temperatures. Improperly sized reservoir heaters may also have high surface temperatures. Extreme temperature zones can crack hydrocarbon molecules, creating both light ends and heavy ends that can impact a fluid's viscosity, flash point and other physical characteristics. Under extreme conditions of temperature, coke products can be formed, and as coke forms, additional fluid degradation products also form.

Although there are multiple ways in which hydraulic oils can degrade, the result of oil degradation is typically the generation of deposits. These deposits are most often the link between the oil degradation and machine performance and reliability.

Varnish can cause a wide range of operational challenges in lubricated systems. Increased wear rates are experienced due to the "sandpaper" effect of particles trapped by the varnish. Heat exchanger efficiency declines as varnish creates an insulating layer on the tubes. Oil flow is impaired causing starvation issues. Filters blind

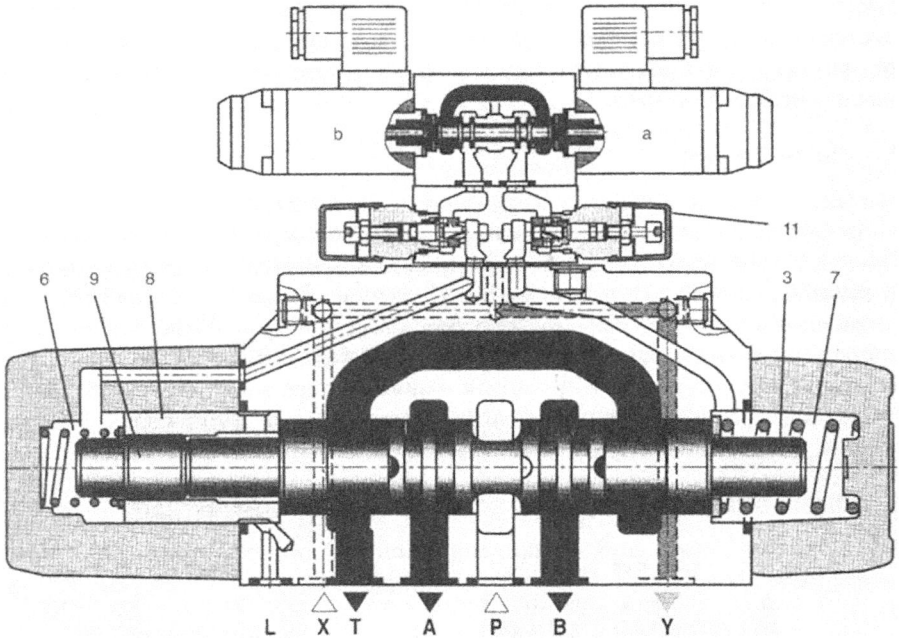

FIGURE 12.5   Electro-hydraulic servo-valve.

Source: Pathmaster Marketing Ltd.

off as their pore sizes are reduced. The fluid life itself is shortened due to the reactive nature of these deposits.

Servo-valves are the most sensitive component impacted by varnish. A leading cause of valve hysteresis and chatter is caused by the increased static friction required to move varnish-coated components. A diagram of an electro-hydraulic servo valve is shown in Figure 12.5, illustrating the fine clearances involved. Some of the operational impacts in an hydraulic system are:

- Loss of clamp speed or cycle time, which can impact production.
- Reduced repeatability and precision, which can increase the scrap rate.
- Increased set-up and interchangeability times, due to additional maintenance and variance adjustments, which can also impact production.

Contamination is a major source of lubricant component wear. Oil analysis programmes generally include particle counts, elemental spectroscopy and tests for water content to measure these contaminants. However, contaminants also can speed up the rate of degradation or cause entirely new failure pathways to occur. Some of these contaminants include:

- Dirt and hard particles are well known to cause wear. But they also can act as catalysts to increase the rate of oxidation. Iron and copper are good examples of wear metals that are catalysts in the oxidation process.

- Incompatible lubricants or other liquids, such as cleaning fluids or solvents. All these can react with the hydraulic oil to form precipitates. Even if precipitates are not formed, incompatible liquids can impair the hydraulic oil through its interaction with other contaminants. An hydraulic oil's interaction with air, for example, can be seen by an increase in foam or air release times. Its interaction with water can be impaired as measured by demulsibility tests.
- Water contamination will cause a multitude of fluid film, corrosion and viscosity issues. Water can degrade the hydraulic oil causing it to fail. Hydrolysis reactions with hydraulic base oil are an example of water's impact on fluid degradation. Water also can react with various additives, particularly zinc dialkyl dithiophosphate (ZDDP) antioxidant and anti-wear additives, the degradation products from which can block filters and servo-valves.

Numerous methods exist with which to identify and monitor these issues.

12.3 HYDRAULIC OIL PERFORMANCE

Hydraulic oil formulations are becoming more advanced so they can provide greater performance under the demanding conditions described earlier. Older industrial petroleum-based hydraulic oil formulations used an API Group I solvent refined base oil and ZDDP additive as a combined anti-wear and antioxidant additive system, together with a corrosion inhibitor and a pour point depressant.

Newer hydraulic oil formulations have begun to use higher-quality API Group II and Group III base oils. When combined with appropriate oxidation inhibitors, these oils are can provide improved oxidation stability and longer drain intervals. However, higher purity hydrocracked base oils have lower solubility properties for oxidation and thermal degradation products, particularly varnish, often resulting in poorer deposit control. This can be counter-acted using appropriate polar components.

Hydraulic oils can have a direct impact on the energy efficiency of the overall hydraulic system, so new formulations are being designed to provide improved energy efficiency. Shear-stable, high viscosity index (HVI) hydraulic oils have shown energy efficiency improvements across multiple applications. This is accomplished through improved viscometrics, using Group III base oils and/or shear-stable VI improvers. Studies have shown energy reduction may be up to 20% in some applications.

Hydraulic oils are now commonly formulated with primary antioxidants to improve the fluid's oxidative resistance and to protect the anti-wear additives. High-performance ashless hydraulic oils, using combined phenolic and amine antioxidants, sulphur-phosphorous anti-wear additives and succinimide corrosion inhibitors are increasingly replacing the older ZDDP-based oils.

Although synthetic fluids, particularly polyalphaolefins and diesters, are being used as base fluids for high-performance or biodegradable hydraulic fluids, similar performance requirements and monitoring and control methods apply to these fluids as well. The choice between mineral oil, Groups I, II or III, and synthetic oils is likely to depend primarily on application, cost, performance requirements and environmental factors. For example, hydraulic fluids used in many forestry and agricultural applications must be biodegradable, by law.

12.4   MONITORING HYDRAULIC OILS AND SYSTEMS

Measuring hydraulic oil degradation is essential in a predictive maintenance strategy. Determining how the fluid is failing is a central part of understanding how to increase the life and performance of hydraulic oils. Many of the modes of degradation have inexpensive solutions when properly identified. For example, spark discharge may be eliminated by changing the type of filter. This provides great payback to hydraulic oil users.

Oil analysis is an essential step in determining the mode of hydraulic oil failure. Another essential step is analysis of deposits when possible. Several tests are used to monitor hydraulic oils.

Increases in viscosity may be due to oxidation. However, by the time a noticeable change has occurred, many other problems are well advanced. To adequately predict hydraulic oil failure due to oxidation, the following tests are recommended:

- Voltammetry (ASTM D6971, RULER test): Measures the antioxidant performance of both primary and secondary antioxidants used in hydraulic oils.
- FTIR: ASTM D7414 measures oxidation, ASTM D7412 measures phosphate anti-wear additives, an ASTM draft method is currently being developed to measure acid content by FTIR, and FTIR is also effective at measuring phenolic antioxidants.
- Membrane patch colorimetry (ASTM D7843): This test method is effective in determining an hydraulic oil's propensity to form varnish.
- Ultracentrifuge: Extracting and qualitatively measuring the amount of degradation products through high centrifugal forces have been successful in predicting the onset of varnish.

Determining acid number (AN) has been an integral part of hydraulic oil analysis for decades. The final component in an oxidation reaction is acid, so this test has shown value in measuring the results of oxidation. However, because Group II and III base oils have lower solubility properties, the degradation products may not remain in the oil long enough to be converted to acidic molecules. Also, acids will only start to develop in the oil once all the antioxidants have been depleted, often beyond fluid failure and too late to prevent operational problems.

As noted in Section 6.2.5, the revised ASTM D7843 test for MPC can be used to account for the effects of the solvent used, by measuring the iMPC value and comparing it with the original MPC value. Although this analysis takes more time, important additional information may be gained about the potential of the hydraulic oil to form varnish.

Elemental spectroscopy is a powerful oil analysis tool and is effective for measuring metallic additive elements and wear metal contents. Where necessary, it can be supplemented and extended using ferrography.

Almost all hydraulic systems are sensitive to the presence of dust, dirt and/or wear debris, so the cleanliness of the hydraulic oil can be critically important.

Unfortunately, many lubricating oils are not very clean when they are delivered to customers, even though they look "clear and bright". While most lubricant manufacturers claim to make quality products, not every manufacturing process is of the highest quality. It is these processes, or a lack thereof, that are at least partially responsible for the poor cleanliness of new lubricants.

Lubricating oils are manufactured in a blending plant, using one or more base oils and, most usually, a number of additives or additive packages. The manufacturing processes for both base oils and additives can introduce dust, dirt, rust particles and/or water into these fluids. The methods of delivering these components to a blending plant can also introduce airborne dust or dirt.

In the blending plant, the storage tanks, blend vessels and pipework may not be completely clean and free of dust, dirt or rust particles. Many modern blending plants maintain clean blending vessels and pipework and filter base oils and additives into each blend. The blending vessels also have breathers fitted, to keep out airborne dust.

Finished product storage tanks also need to be kept clean, as must filling lines and small product plastic bottles, cans and drums. For some products, new drums are less desirable, as they are made in a metalworking environment with grinding and welding. The slag and grinding dust is likely to end up inside the drums. In some cases, lubricant blenders just fill the drums without inspecting them for cleanliness. Reused drums can be cleaner than new drums, as the washing, rinsing and drying processes used to recondition them can remove most of the debris inside them.

In the case of bulk deliveries, lubricants should be filtered prior to being transferred into the tank. The way in which the tank was cleaned can also affect its cleanliness. Steam cleaning or flushing with diesel fuel can cause cleanliness or purity issues with the finished lubricants. Tanks that are dedicated to specific products will eliminate the possibility of cross-contamination.

Unfortunately, there is no universal standard for container cleanliness. Examples of the cleanliness of base oils, additives and blended hydraulic oils are shown in Table 12.3.

Aeration of hydraulic oils can be a serious problem if not dealt with quickly and properly. It can lead to cavitation of the pump, spongy or slow hydraulic response, as well as the loss of control of the hydraulic system. There are four states of air in oil: dissolved, entrained, free and foam.

Aeration can be caused by a variety of reasons. The reservoir design could be creating problems. If the hydraulic oil return line to the reservoir is dumping straight in from the top, a diffuser should be added to reduce the amount of turbulence. Air leaks in an hydraulic system can also lead to aeration problems. The suction side of the pump should be checked for leaks. Pump seals, pipe fittings and unions are all possible areas for air leaks that allow the system to pull in air and pressurise it. These leaks can be found using an ultrasonic gun or by simply placing a small drop of grease around the suspected leak site. If there is a leak, a hole will appear in the grease, revealing the area of air ingression.

Poor air release can be caused by oxidation and contamination or anything that results in low surface tension within the fluid. A lower surface tension allows the

TABLE 12.3

Example ISO Cleanliness Codes for Base Oils, Additives and Hydraulic Oils

Fluid	ISO Cleanliness	
	Optical	Pore Blockage
Group I base oil	17/14/11	14/13/10
Group II base oil	17/17/13	16/15/13
Polyalphaolefin base oil	18/16/13	19/18/15
Diester base oil	14/14/13	19/16/14
Polyol ester base oil	16/15/13	13/12/10
ZDDP-based hydraulic oil additive package	16/14/12	18/17/14
Phenolic antioxidant	15/14/11	16/15/12
Aminic antioxidant	16/14/12	15/14/12
Calcium sulphonate detergent	18/16/13	16/15/12
Group I-based hydraulic oil	17/15/12	16/16/13
Group II-based hydraulic oil	19/17/13	16/15/12
Polyalphaolefin-based hydraulic oil	18/15/12	15/14/12
Polyol ester-based hydraulic oil	21/18/14	21/20/19

Source: Noria Corporation, with permission.

Note that some additives produce "ghosts" that register as particles when using optical particle counting.

bubbles to break apart below the surface. The larger the bubble, the quicker it rises to the surface. When bubbles begin to break apart below the surface, the amount of time it takes them to rise increases. This keeps them in suspension longer, increasing the risk of cavitation and also leading to problems with foam.

Often the only tests and actions performed on an hydraulic system involve changing the filters, sampling the oil and checking the oil level. As long as the system is operating, the mentality of "if it isn't broke, don't fix it" frequently prevails. However, on any given hydraulic system, 15 to 20 regular reliability tests should be performed while the system is operating. There are also several checks and procedures that should be completed during shutdowns or downtime.

The oil reservoir should be checked when the system is down. An illustration of a typical oil reservoir is shown in Figure 12.6. The reservoir should be cleaned at least once a year. In addition to storing the hydraulic oil, two other main purposes of the reservoir are to dissipate heat and to allow contaminants to settle. If the reservoir is not cleaned, not only will its ability to dissipate heat be diminished, but it will act as a heat sink. Temperatures can easily rise well above the maximum recommended level of 60°C. The higher the temperature, the faster will be the oxidation, leading to sludge and varnish in the system. If the contaminants are not removed from the reservoir, they will be drawn into the pump, causing premature failure of the system components.

Many reservoirs contain a suction strainer to keep large particles from entering the pump. Most suction strainers have a 74 μm rating, whereas the tolerances

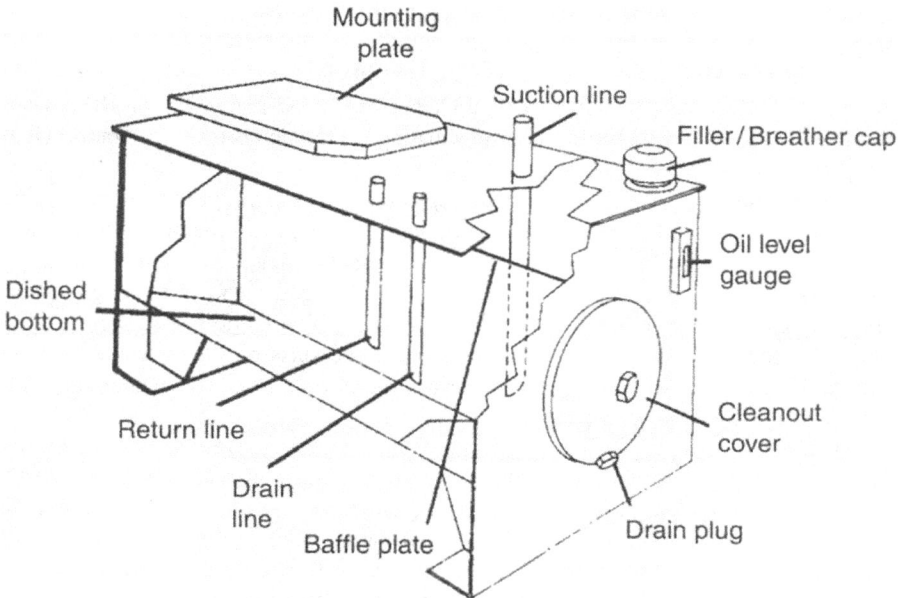

FIGURE 12.6   Hydraulic system reservoir.

Source: Pathmaster Marketing Ltd.

inside pumps and valves are typically 3 to 8 μm. A lint-free cloth should be used when cleaning a reservoir. If a solvent is used, it should be one recommended for hydraulic systems. Even small amounts of the wrong solvent can impair certain additives.

When hydraulic oil is removed from the reservoir, it should be filtered going into a storage tank with a flushing and filtering unit, which can remove solid contaminants and water. A quality, high-capture-efficiency filter (ISO 16889) that matches the target cleanliness level of the system should be used. Unless the oil is severely degraded, it is not necessary or even desirable to change it. After the reservoir is cleaned, the oil can be re-filtered during filling. The entire system should then be flushed to clean the oil in the lines to the valves and actuators.

System flushing is done by connecting the inlet and outlet lines of the cylinders and motors. If possible, electrical or manual actuation of the directional valves will allow the fluid to re-circulate through the piping. If this is not possible, the directional valves can be bypassed by connecting the pressure and tank lines to the outlet lines of the actuators. The machine's existing pump can be used to re-circulate oil through the lines. A high-velocity flushing unit will enable re-circulation of the oil in the reservoir through the filters during the flushing process. The system should be run for as long as possible.

In some hydraulic systems, the heater in the reservoir is disconnected during the summer or may not have been fitted when the system was built initially. The heater thermostat should be checked and set to turn the heater on when the temperature drops to 21°C. If the pump is mounted on top of the reservoir and the oil temperature drops below about 15°C, then some cavitation of the pump may occur.

Most hydraulic system reservoirs have two oil level switch settings: warning and shutdown. The problem is that the difference between these two levels may be several hundred litres of oil. By eliminating the warning switch and setting the shutdown at a higher level, oil loss will be minimal if a hose ruptures.

The breather cap is usually the most neglected component on the reservoir. The breather cap filter should have a capture efficiency that matches the target fluid cleanliness. This is the first line of defence for contaminants entering the tank. Depending on the location, the breather cap may need to be changed a couple of times a year. Many breathers have a mechanical indicator that will provide a visual indication when the element is dirty. Other options include pressurising the reservoir with an internal bladder or using a moisture-removal type of breather. Money spent upgrading a breather cap is never wasted.

Mineral oil hydraulic oils will oxidise faster at temperatures above 60°C, but many systems will not shut down the unit until the oil temperature reaches 70°C. Hydraulic systems are designed to operate below 60°C. For every 10°C increase in temperature, the oxidation rate is approximately double, meaning that the oil lifetime is halved. If the oil temperature rises above 60°C, then a problem exists in the system. This could be caused by a cooler malfunction or excessive bypassing at the pump. The high-temperature switch should be set at 60°C. The recommended operating temperatures for different types of hydraulic fluids are shown in Table 12.4.

TABLE 12.4
Recommended Operating Temperatures for Alternative Hydraulic Fluids

Fluid	Temperature, °C	
	Long Service	Short Durations[a]
Group I mineral oils	50–80	100–120
Group II mineral oils	50–80	100–120
Group III mineral oils	50–90	110–120
Polyalphaolefins	50–120	<200
Diesters	50–120	<200
Polyol esters	50–140	<220
Polyalkylene glycols[b]	60–110	130–140
Phosphate esters	60–140	<270
Silicones	180–220	250–280

Source: Pathmaster Marketing Ltd., compiled from numerous published sources.

• Notes

[a] Short durations means less than 15 minutes.

[b] Neat PAGs only, not Water-Glycol fluids.

In a shell-and-tube type of heat exchanger, oil flows over the tubes. Water flow is ported through the tubes in the opposite direction. The heat in the oil is transferred from the oil to the water. To achieve the most efficient heat transfer, the water flow should be 25% of the oil flow. The water flow can be controlled by manual valves, a water-modulating valve or an electrical solenoid valve. Circulating hot wash oil or light distillate through the tube or shell side can effectively remove sludge or similar soft deposits. Soft salt deposits may be washed out by circulating hot, fresh water. A mild alkaline solution, such as a 1.5% solution of sodium hydroxide, can be used. The tubes should be flushed in the opposite direction to which the oil normally flows.

If the system is fitted with an air cooler, the cooler fan should turn on at about 50°C and turn off at about 40°C. The fins should be cleaned so daylight can be seen through them. If necessary, combs should be used to straighten the fins on the unit. When cleaning the fins with an air hose, care should be taken so as not to damage them.

On variable-volume hydraulic pumps, the flow out of the case drain line can be checked by porting the line into a container and timing it. This test should be made with the outlet pressure at the maximum level. It is not recommended that the line be held during this test. Also, the line should be secured to the container prior to starting the pump. The normal case flow is 1% to 5% of the maximum pump volume. Vane pumps usually bypass more than piston-type pumps. If 10% of the maximum volume flows out of the case drain line, then the pump should be changed. An excellent method of monitoring the case drain flow while operating is to permanently install a flow meter in the case drain line.

Fixed-displacement pumps can be tested by checking the flow through the relief valve. The pump can be turned on and the flow out of the relief valve tank line can be recorded for 1 minute. With the setting of the relief valve to its minimum, the test can be repeated. There should be less than a 10% difference in flow rates between the two tests. If a pump is badly worn, the flow will be considerably less at the highest pressure.

An accumulator that is used for volume should be pre-charged with dry nitrogen to one-half to two-thirds the pump's compensator setting. When the hydraulic system is turned off, a charging rig with a gauge can be used to check the pre-charge level. To confirm an accumulator is operating properly, the side of the shell with a temperature probe or infrared camera can be checked. The bottom half should be hotter than the top half. If heat is only indicated at the bottom, the accumulator may be overcharged. If there is no heat, the bladder may have ruptured, the piston seals may be bad, the pre-charge may be above the compensator setting or all the nitrogen may have leaked out. If heat is felt all the way to the top, the accumulator is undercharged. Another check that can be made is to watch the system pressure gauge while the system is operating. The pressure should not normally drop more than 100 to 500 psi when the accumulator is properly pre-charged.

If piston accumulators are used, the charging rig should be installed when the system is down and the oil bled off the top of the piston. With the pump on and the bleed valve open, there should be little or no flow out of the bleed valve. Care should be taken so all personnel are away from the bleed valve prior to turning on the pump. If there is continuous flow, the piston seals or barrel may be worn. If no flow exists, the accumulator should be recharged to the proper dry nitrogen level.

All hydraulic system hoses should be checked for the proper length and wear. Hoses rarely burst due to the rated working pressure being exceeded but rather because of a poor crimp or rubbing on a beam, another hose or other item. Hose sleeves are available from a variety of manufacturers if rubbing cannot be avoided. Hoses generally should not exceed 4 feet in length unless they move with the machine.

The system piping should also be examined to verify that a hose is installed prior to connecting to a valve bank or cylinder. The hose will absorb the hydraulic shock generated when the oil is rapidly deadheaded. One exception to this rule is that hard piping should be used when connecting to a vertical or suspended type of load. Pilot-operated check valves and counterbalance valves can be employed to hold the load in the raised position.

System clamps should be checked to confirm they are the correct type for hydraulic lines. Beam and conduit clamps are not acceptable, as they will not absorb the shock generated in the piping or tubing. Clamps should be spaced approximately 5 feet apart and installed within 6 inches of the pipe or tubing termination point.

On any system, one or more valves will be closed while the system is operating. These include relief valves used with pressure-compensating pumps, air bleed valves and accumulator dump valves. The tank lines of these valves should be checked regularly with a temperature probe or infrared camera to verify that the valves are closed and no oil is being lost back to the reservoir.

ASTM D6224 "Standard Practice for In-Service Monitoring of Lubricating Oil for Auxiliary Power Plant Equipment" complements ASTM D4378 (discussed in the next chapter) in that it has a similar format and addresses important equipment other than the major turbines. Lubricating oils in equipment such as gears, large diesel engines, pumps, compressors, hydraulic systems and electrohydraulic control systems are included. As with all ASTM standards, this is a consensus document and represents the expertise of oil producers, oil users and those with a general interest (such as commercial testing laboratories).

The practice seeks to educate lubricant users by describing the general properties of the various types of lubricating oils and by describing the various tests which can be used to monitor the lubricating oil or equipment in which it is used. It also describes the factors affecting the service life of oils and defines proper sampling techniques. Users of the practice should reduce operating costs by changing oil only when needed as determined by oil testing. Users should also reduce maintenance costs by relying more on oil analysis results to determine when to make equipment repairs instead of doing so on a scheduled basis.

The five tables in the practice provide guidance on sampling and testing. The first table provides guidelines on the types of testing that would be appropriate for characterising or determining the acceptability of new oils. The second table contains guidelines for sampling and testing in-service oils. In both tables, each type of test is described as being recommended, optional, as-needed or not normally relevant for individual oil types. A recommended test is one which provides the most information about the oil quality or equipment condition and which should be included in a good oil-monitoring programme. An optional test is one that may provide additional information but could be omitted with less impact if the sample amount or analysis time is limited. A test that is deemed as-needed may be beneficial if there is an operational problem (for example, foaming) or if another test result indicates the need. If a test is not generally informative for a given type of oil, then there will be no entry in either table. Each user is able to customise their oil-monitoring programme by using these guidelines and considering their own operation and resources. While the practice references mostly ASTM test methods, it does not preclude the use of alternative instrumentation or test methods.

The third table includes warning levels of in-service oil test results. These warning levels are either an absolute value, a percent of the new oil or a statistical deviation from trend data. The fourth table provides an interpretation of the oil test results and gives recommended action steps. The last table lists possible sources of inorganic elements (wear metals and additive elements) in oil in order to help interpret elemental analysis results.

An abstract of the second table in ASTM D6224 is reproduced in Table 12.5, for hydraulic, gear, diesel engine and air compressor oils.

TABLE 12.5
Guidelines for Testing In-Service Oils

Test	Test Method	Hydraulic Oils	Gear Oils	Diesel Engine Oils	Air Compressor Oils
Frequency		Every three months			
Appearance	Visual	R	R	–	R
Viscosity at 40°C	D445	R	R	O	R
Viscosity at 100°C	D445	–	–	R	–
Acid number	D975	R	R	–	R
Water content	D1744	R	R	R	R
Antioxidants	D6971	–	–	–	O
RVPOT	D2272	–	–	–	O
Colour	D1500	O	–	–	O
Flash point	D92	–	–	O	O
Insolubles	D893	–	–	O	–
	D2273	O	O	–	O
Demulsibility	D1401	–	–		O
Anti-corrosion	D665A	O	O		O
Foaming	D892	AN	AN	AN	AN
Air release	D3427	AN	AN	AN	AN
Base number	D4739	–		R	–
Glycol content	D2982	–		O	–
Fuel dilution	D3524	–		O	–
Particle counts	OEM	R	R	–	O
Wear metals	OEM	O	O	O	O
Wear debris	OEM	AN	AN	AN	AN
Elemental analysis	D5185	R	R	R	R

Source: ASTM D6224.

Notes

R = recommended, O = optional, AN = as needed.

Tests for foaming characteristics, air release and wear debris analysis are as-needed for all four types of oils.

The frequency of sampling for large diesel engine oils is 250 to 500 hours for continuously operated equipment or every six months for standby equipment. ASTM D1744 (water content) or ISO 6296 is not recommended for certain lubricants (notably diesel engine oils) due to interferences by additives. Water content in diesel engine oils should be measured using ASTM D95, IP 74, ISO 3733, DIN 51582 or AFNOR T60-113.

Recommended warning levels for all these tests for all four types of oils are reproduced in Table 12.6. ASTM D6224 is a particularly useful guide to setting up an oil condition monitoring programme.

12.5 HYDRAULIC FLUID FILTRATION AND CONTAMINATION CONTROL

To achieve a target cleanliness level in an hydraulic system requires the successful capture and removal of contaminants. Filtration must be done in a way that does not disrupt the flow of oil or unduly increase the pressure drop within the system. However, it also requires minimising the ingress of contaminants into the hydraulic system in the first place and minimising the wear of metallic components. Particle filtration of an hydraulic system is usually sufficient for controlling the level of oil contaminants.

Removing contaminants from the hydraulic oil is generally considered to be secondary to the primary functions of the hydraulic system and thus is often neglected and under-emphasised as a vital part of the design. This can result in filters performing poorly due to lacking structural integrity, dirt-holding capacity or particle-capture efficiency. There are many ways in which the system could be inadequate for performing filtration, including its flow rate, the presence of vibration or if the overall differential pressure is lost because of the filters' locations within the system. Each of these factors must be considered for the filtration system to work effectively.

If more particles are coming into the system from ingression points such as breather points, hatches and seals than are being removed, the system will never be clean.

If an alternative method is needed to remove contaminants from the hydraulic oil, centrifugation, magnetic filtration or a combination of periodic, portable filtration systems may be used. Although filters can be very efficient, sometimes they are not enough and must be complemented with other methods to remove finer particles. For example, with magnetic filtration, ferromagnetic particles can be captured regardless of their size. These types of particles can make up nearly 90% of particles suspended in hydraulic oil. Centrifuges also offer advantages over traditional filter media. The efficiency of a centrifuge is consistent throughout its service, and the lower limit of its particle size trapping capability can be much less than 1 μm.

Filtration efficiency is normally expressed as the ratio of dirt entering a filter compared with the dirt exiting the filter, of a specified micron (μm) size. Filter testing is completed in a laboratory so the results can be compared and statistically verified. Filtration tests attempt to accurately control the quality of the test dust, the flow rates,

TABLE 12.6
Warning Levels of In-Service Oil Test Data

Test	Hydraulic Oils	Gear Oils	Diesel Engine Oils	Air Compressor Oils
Appearance	Not clear and bright, hazy, visible debris			
Viscosity	±5% of new oil, max	±5% of new oil, max	±10% of new oil, max	±5% of new oil, max
Acid number	Increase of 0.2	Increase of 0.5	–	Increase of 0.2
Water content (%wt)	>0.05	>0.1	>0.2	>0.6
RULER phenols (%)	<25	–	–	<10
RULER amines (%)	<25	–	–	<25
RVPOT	–	–	–	<25% of new oil
Colour	Unusual or rapid darkening			
Flash point	–	–	Decrease of 30°C	Decrease of 20°C
Insolubles	>0.1 %wt	>0.5 %wt	>2.5 %wt, pentane	–
Demulsibility	>30 minutes	–	–	>30 minutes
Anti-corrosion	Light fail	Light fail	–	Light fail
Foaming, Seq I	>450/>10	>450/>10	–	>450/>10
Air Release[a] (minutes)	>5 to 20	>5 to 20	–	>5 to 20
Base number	–	–	<20% of new oil	–
Glycol content	–	–	Any detected	–
Fuel dilution	–	–	>1% to 5%[b]	–
Particle counts	–/17/14 max	[c]	–	–/17/14 max
Wear metals	[d]	[d]	[d]	[d]
Elemental analysis	±25% of new oil			

Source: ASTM D6224, adapted by Pathmaster Marketing.

• *Notes*

[a] The maximum air release time depends on the oil viscosity.

[b] The maximum fuel dilution for diesel engines depends on the type of engine.

[c] The warning level ranges from –/18/15 to –/22/18 depending on the application. Longer component life is generally correlated with better cleanliness.

[d] A recommended warning level is a concentration of two standard deviations above the mean of six or more prior results. In cases when there are insufficient analyses to provide a good baseline or where there the increase in concentration is steady but gradual, the OEM should be consulted for guidance.

temperatures, measuring equipment and many other variables to ensure the repeatability of the test. The test, called a multi-pass filter test, produces a Beta (β) rating for the filter at a given micron size. For example, a $\beta_{10} = 75$ filter removes 74 of 75 of the particles greater than 10 μm entering the filter.

It is disappointing, but understandable, to find that filters do not always perform in the service exactly as the laboratory tests would suggest. Issues contributing to poor filter performance can include pulsating flows, flow rate, start-up contamination and overall design flaws within the hydraulic system. In the laboratory, filters are tested under steady-state flow conditions. The pulsating flow of real-life applications can significantly reduce performance. For this reason, it is advisable to avoid conditions of pulsating flow when installing a filter into a circuit design.

Different configurations of media can also cause different efficiencies at increasing differential pressures. For example, two filter elements can have the same micron rating (average efficiency) but differing maximum and minimum efficiencies and, consequently, different dirt-holding capacities.

Contamination levels in a system are a function of many factors, such as the amount of contaminant in the oil at start-up and the rate of contaminant ingression. For most hydraulic systems, the prevention of contamination is more cost-effective than the cure. To control contamination, troubleshooting the system design will yield dramatic results. Typical levels of contamination in different hydraulic systems are listed in Table 12.7. Some systems, such as those in farm tractors and excavators, are designed to tolerate relatively high levels of contaminants, because of where they are used. Other systems, including aircraft landing gear and spacecraft antenna, are required to be very clean, so that they will not fail to function when required.

Many hydraulic system breathers consist of an open cover or tube or at best a filler cap without a proper filter element inside. No unfiltered air should be allowed to enter an hydraulic system. Wherever possible, oil reservoirs should have adequate breather filters installed; ambient air carries considerable quantities of contaminants. A quality breather filter with an absolute efficiency of $\beta_{10} = 75$ or better will be satisfactory for most circumstances. A quality 10 μm filter for liquids will typically be more efficient at capturing fine dirt particles when applied in air filtration applications. In humid conditions, breathers should remove water, which means they should incorporate a desiccant.

Good reservoir design will ensure that any water or heavy dirt settles to a small area or standpipe at the base of the reservoir, which can be drained periodically. Water left in the oil leads to bacteria growth and chemical degradation. Good reservoir design will provide return line diffusers, adequate baffles and sufficient volume

TABLE 12.7
Typical Hydraulic System Contamination

Hydraulic System	NAS Class
New oil (drum)	8–9
Farm tractors	11–12
Excavators	8–12
Mobile cranes	7–12
Machine tools	7–12
Factory robots	5–12
Aircraft landing gear	5
Missile control systems	4
Spacecraft antenna	2

Source: Pathmaster Marketing Ltd.

to settle out heavy dirt particles, water and any entrained air. Oil should be regularly recycled or used to prevent long-term degradation. Ideally, oil should be filtered going into and out of the reservoir.

Three mechanisms of filtration are described:

- Edge: Particles stick to the edges of surfaces. This occurs in the fine gaps in servo-control valves.
- Surface: Particles sit on the surface of filter media. This is the main mechanism in cake filtration.
- Depth: Particles accumulate throughout the filter medium.

With depth filtration, particle transport forces particles through the filter medium and to approach media surfaces. Here, attachment occurs when Van der Waal attractive forces cause particles to stick to media surfaces and attached particles increase the attractive forces. Then, hydrodynamic forces cause particles to detach from the media and continue until they reach an unfilled media surface. The filling of pores with particles increases the hydrodynamic forces.

Examples of edge filters include stacked metal discs and stacked paper discs. Surface filters include filter paper, pleated paper, woven fibres, woven wire, polymer membranes and flat ceramic filters. Depth filters include wound fibres, felt, wood pulp, mineral wool, diatomaceous earth, sintered metal and deep ceramic filters.

Cartridge filters are the most commonly used types in hydraulic systems. They have removable elements of cylindrical shape incorporating a filter medium surrounding a drainage core with top and bottom seals. They are generally used to clarify liquids where the concentration of solids is usually below 0.01 %wt. They may be disposable or cleanable and may filter through the depth, surface or edge. Particle size removal is 0.02 to 100 µm. A typical cartridge filter is shown in Figure 12.7.

The filter cartridges are generally one of four types:

- Pleated paper, fibre mat or wire mesh.
- Wound fibres.
- Stacked disc.
- Hollow disc.

Examples of a pleated paper cartridge and wound fibres are shown in Figures 12.8 and 12.9.

The performance of a cartridge filter is measured by particle retention, described as either absolute or nominal filter rating, flow rate maintenance, defined by cartridge media permeability and pressure drop and solids holding capacity, which directly affects filtration run length and filter life. With flow rate maintenance, when a cartridge filter is correctly matched to the filtration application, flow rate will be maintained with only a gradual increase in pressure drop across the filter, until the filter is fully loaded, when the pressure drop will rise rapidly. In terms of the solids holding capacity of a depth cartridge filter, solids must build up throughout the depth of the

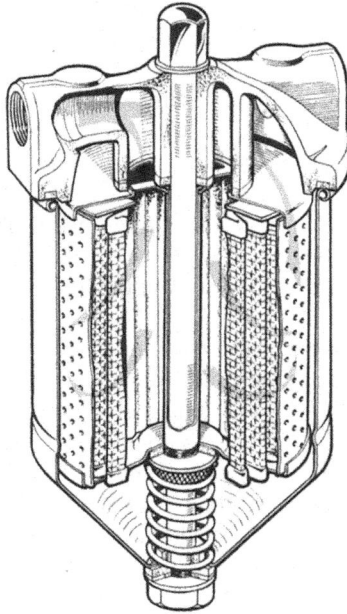

FIGURE 12.7   Cartridge filter housing.

Source: Pathmaster Marketing Ltd.

FIGURE 12.8   Pleated cartridge filter.

Source: Pathmaster Marketing Ltd.

FIGURE 12.9   Wound fibre filter.

Source: Pathmaster Marketing Ltd.

filter at a rate that is about equal in percentage of full load at all depths. If this does not happen, some depth will be limiting and cake filtration will start. The inefficient use of depth filters is caused most commonly by surface filtration blocking.

In most hydraulic systems, filtration can be either full-flow or by-pass, depending on the location of the filter. Diagrammatic illustrations of both full-flow and by-pass filtration (taken from a diesel engine for simplicity) are shown in Figure 12.10.

Different components in an hydraulic system require different levels of cleanliness. Target ISO 4406 cleanliness codes for illustrative components are shown in Table 12.8. It is worth noting that, at higher system pressures, cleanliness levels need to be lower. Also, servo-valves and cylinders require higher levels of cleanliness than pumps or motors.

It is important to note that different types of filters will remove more than contaminant particles. This is illustrated in Table 12.9. For example, surface and edge filters will remove large particles, but may not remove all small particles and will not

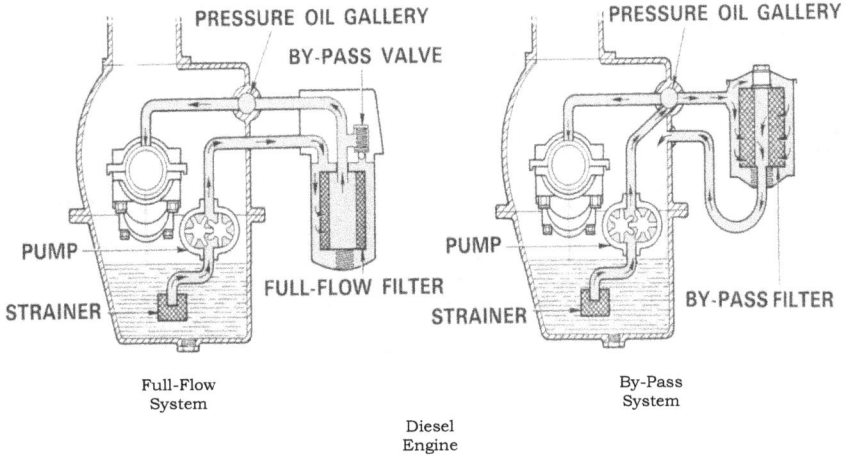

FIGURE 12.10   Full-flow versus by-pass filtration.

Source: Pathmaster Marketing Ltd.

TABLE 12.8
Target ISO 4406 Cleanliness Codes for Hydraulic System Components

Component	<140 bar	212 bar	>212 bar
Fixed gear pump	20/18/15	19/17/15	–
Fixed piston pump	19/17/14	18/16/13	17/15/12
Fixed vane pump	20/18/15	19/17/14	18/16/13
Variable piston pump	18/16/13	17/15/13	16/14/12
Variable vane pump	18/16/13	17/15/12	–
Flow control valve	19/17/14	18/16/13	18/16/13
Proportional flow control	17/15/12	17/15/12	16/14/11
Servo-valve	16/14/11	16/14/11	15/13/10
Cylinder	17/15/12	16/14/11	15/13/10
Vane motor	20/18/15	19/17/14	18/16/13
Axial piston motor	19/17/14	18/16/13	17/15/12
Radial piston motor	20/18/15	19/17/14	18/16/13

Source: Pathmaster Marketing Ltd, compiled from numerous published sources.

TABLE 12.9

Contamination Removal by Filters

Contaminant	Filter Type		
		Depth	
	Surface and Edge	Inactive	Active
Large particles	Yes	Yes	Yes
Small particles	Some	Yes	Yes
Dispersed oil oxidation products	No	Some	Yes
Coagulated oxidation products	Yes	Yes	Yes
Oil soluble contaminants	No	No	Most
Additives	No	No	Some
Water	No	Some	Some
Full-flow filtration	Yes	Rarely	Rarely
By-pass filtration	Yes	Yes	Yes

Source: Pathmaster Marketing Ltd.

remove oil soluble contaminants or water. Conversely, active depth filters (containing diatomaceous earth, for example) will remove all particles as well as oil soluble contaminants, some additives and water. Inactive depth filters (pleated paper, for example) will not remove oil soluble contaminants or additives, but may remove some water and dispersed oil oxidation products.

Analysis of the particulates either found on or recovered from hydraulic system filters can be a useful part of a lubricant condition monitoring programme. ASTM D7898 "Lubrication and Hydraulic Filter Debris Analysis (FDA) for Condition Monitoring of Machinery" was developed as a guide to such analysis. The objective of filter debris analysis (FDA) is to diagnose the operational condition of oil-wetted machinery systems to identify abnormal wear or incipient component failures. Oil system filters (typically lubrication system or hydraulic systems) capture the vast majority of metallic and non-metallic debris generated or contained within a system. The ASTM practice enables a consistent approach to the analysis of in-service debris captured in filters and is intended primarily for lubrication or hydraulic systems. According to the ASTM, lubrication and hydraulic system filters are a rich source of information about system health that are seldom exploited for machinery condition monitoring purposes. The practice includes components of debris analysis, including the size, quantity, morphology and composition of debris trapped by the system filter. System filters have an additional advantage over traditional sample-based techniques in that they capture a high percentage of total system debris (metallic, non-metallic and organic particulate contamination) within the size range useful for machinery condition monitoring.

ASTM D7919 "Filter Debris Analysis (FDA) Using Manual or Automatic Processes" compliments ASTM D7898, providing practical guidance on FDA of in-service lubricant filters. Various techniques for debris removal, collection and analysis are presented with their associated benefits and limitations. Debris removal techniques include manual and automatic methods. Analytical techniques range from visual, particle counting, microscopic, X-ray fluorescence (XRF), atomic emission spectrometry (AES) and scanning electron microscopy with energy dispersive X-rays (SEM-EDXR).

12.6 MONITORING FIRE-RESISTANT HYDRAULIC FLUIDS

12.6.1 HIGH WATER-BASED HYDRAULIC FLUIDS

Two types of high water-based hydraulic fluids (HWBF) are used in low-pressure hydraulic systems where there is a risk of fires or explosions. HFAE fluids are typically 5% soluble oil emulsions (oil-in-water) which are cloudy, milky white in appearance. HFAS fluids are typically 5% solutions of synthetic, water-soluble components in water, which are clear or translucent in appearance.

Both types of fluids are extremely fire-resistant and also have very good cooling properties. Unfortunately, they are not very good lubricants and can tend to promote corrosion (rusting) in some hydraulic systems unless they are formulated properly with high-performance corrosion inhibitors. Because of their high water content, and hence high density, compared with mineral oil, hydraulic pumps need to be de-rated in order to obtain service life.

The operating temperature of systems using HWBFs must be limited to 50°C, to reduce the possibility of evaporation of the water and deterioration of the fluid's properties and performance. If these fluids are exposed to temperatures below freezing (0°C), separation of the phases may occur, particularly for HFAE fluids. Hydraulic systems that use these fluids in colder climates are likely to need effective lagging, heating or winterising. Additionally, if these fluids are not properly formulated or maintained in-service, there is a risk of bacterial and/or fungal growth.

A number of tests are used to determine the fire-resistance of fire-resistant hydraulic fluids. These include:

- Spray flammability: A spray of fluid is injected through an orifice under pressure and is contacted with a source of ignition, usually a flame.
- Hot metal flammability: The fluid is contacted with a hot metal surface at a prescribed temperature.
- Wick test: A specified wick material is soaked with test fluid and is then passed in and out of a flame until fluid ignition is maintained. The test is designed to evaluate the effect of evaporation of volatile components on fluid flammability, particularly with regard to fibrous lagging round hydraulic pipes.
- Auto-ignition temperature: The temperature at which the fluid will burn in air without an additional ignition source.

All of these tests are used to develop and evaluate the properties and performance of fire-resistant fluids. None of them is used to monitor fluids in use.

It is worth noting that HWBF fluids do not have a meaningful auto-ignition temperature.

Tests used to monitor HWBF fluids in service are water content, which must be maintained at close to 95% volume, corrosion inhibition, appearance and microbial growth. The tests involved are the same as those used for water-mix metalworking fluids, which are discussed in Chapter 14. Monitoring of hydraulic systems that use HWBFs involves temperature and pressure control.

12.6.2 WATER GLYCOL HYDRAULIC FLUIDS

Water glycol fire-resistant fluids (HFCs) are typically formulated with water (usually 35% to 45% weight), ethylene glycol and a high molecular weight water-soluble polymer, to provide a degree of viscosity, together with anti-wear, corrosion inhibiting and anti-fatigue additives. They are solutions, clear or translucent, and not emulsions.

These fluids are, consequently, better lubricants than HWBFs, but have less fire resistance. They can be used at higher pressures than HWBFs, but not as high as with mineral oil or synthetic hydraulic oils or fire-resistant phosphate esters. Because they have higher densities than mineral oils or synthetic oils, the vacuums created in pump inlet ports can be higher. Some metals, particularly zinc, magnesium and cadmium, may have an adverse reaction to HFCs. These reactions are likely to result in the production of sticky residues that can block pipes, orifices and filters and cause servo-valves to stick.

As with HWBFs, the two most important criteria for monitoring these fluids in use are water content and system operating temperature. The fluid's water content must not be allowed to drop below a limit set by the fluid manufacturer. If more water needs to be added to correct the water content, it must be distilled or deionised, not tap water. The hydraulic system's operating temperature should not be allowed to exceed 60°C, to minimise the possibility of water evaporation. HFCs tend to be stable to temperatures a little below freezing, but not too low. Monitoring of hydraulic systems that use HFCs also involves temperature and pressure control.

12.6.3 PHOSPHATE ESTER HYDRAULIC FLUIDS

Phosphate esters are used in higher-pressure hydraulic systems where there is a risk of fires or explosions. These include the electro-hydraulic control (EHC) systems in steam turbines for electrical power generation. They have a balance of desirable and undesirable properties that may create conflicts for some users, especially if they are not maintained properly.

Phosphate esters have good thermal stability, excellent boundary lubrication properties, low volatility and fair hydrolytic stability. Unfortunately, phosphate esters can soften some elastomers, such as Buna-N or nitrile, PVC coatings and paints. However, they are more compatible with butyl, nylon, PTFE, EPR, Viton and epoxy-based paints. Because phosphate ester degradation products can catalyse further degradation, they require vigilant condition monitoring and specialised reconditioning.

Triaryl phosphate esters are relatively unique among hydraulic fluids because they are self-extinguishing. The fluid doesn't create enough energy to support its own combustion in a fire. Other lubricants can burn readily once they reach their "fire" points. In addition, phosphate esters tend to have higher flash and fire points, higher auto-ignition temperatures and perform better in spray flammability and wick-type fire propagation tests.

Although they have reasonable thermal and acceptable hydrolytic stability properties, phosphate ester lubricants require routine attention to maintain stability throughout the hydraulic system's service life. Fluids can remain in service for 15 years or more unless they are exposed to conditions that contribute to premature degradation. Examples of such conditions include:

- The purification media is not changed often enough, or it is changed incorrectly.
- The purification system flow rate is wrong (too high or too low), and/or it is not in-service continuously.
- The facility is using inferior or dated parts and/or following bad procedures, especially on fluid top-ups and media changes.
- The facility is using the wrong type of phosphate ester and/or one that does not meet the specifications for that machine.
- The facility is not using suitable condition-monitoring devices, for example, working media and filter housing pressure gauges.
- The facility is not using current or correct OEM maintenance and operating procedures.
- The facility is following inappropriate OEM documentation or procedures.

Determining the failure root cause (and there may be more than one) requires proper fluid testing and a careful review of the fluid and equipment history. Many of these systems have been in service for 20 years or more, during which numerous improvements are likely to have been made.

Most of the tests recommended for phosphate esters are similar to those recommended for mineral oil condition monitoring. However, because of differences in chemistry, experience with mineral oil testing does not necessarily mean competence with phosphate ester analysis. For example, there are limits on the chlorine content and on resistivity for these fluids when used in servo-controlled hydraulic systems. Tests for chlorine and resistivity are seldom conducted for mineral oil but are required to prevent electrokinetic erosion of the servo-valve metering edges.

Most fluid suppliers offer some tests as a free service to accompany the use of their fluid. While this certainly simplifies the challenge of initiating an oil analysis programme, it does not promise success. The full slate of tests should be correctly selected even if the test is not offered as a free service.

Stabilising the acid number of phosphate esters is the most basic fluid control requirement. Most turbine manufacturers set a maximum limit of 0.20 mg KOH/g. With good control, less than 0.10 mg KOH/g should be achievable.

The concern about water in phosphate esters is two-fold. First, because the base fluid is a product of a reaction between an acid and an alcohol, and because the reaction can be reversed, too much water creates a risk of hydrolysis and fluid degradation.

Fluid oxidation and contamination from aggregate-type purification media (fuller's earth, activated alumina and others) can increase this risk. Secondly, steel components in the system are prone to rusting in the presence of too much moisture. It is important to gauge the sensitivity of each EHC system and to maintain the moisture levels as low as is reasonably achievable. Some systems can tolerate higher water concentrations without apparent difficulty.

A high chlorine content can cause servo-valve electrokinetic wear. The source can be chlorinated cleaning solvents. The fluid's electrical resistivity needs to be kept high to prevent electrokinetic wear of servo-valve internal components. Excessive foaming can affect the pumps and level controls. It can also impair air and/or water release. A high air release value can indicate fluid degradation or contamination.

Most turbine manufacturers recommend ISO particle count targets between about 18/15/13 and 17/14/12. However, caution should be exercised when developing oil analysis and contamination control programmes to limit or eliminate as much variability as possible. Development of effective sampling, collection and analysis procedures, including installation of sample points at the correct location, is critical. It is also necessary to verify the laboratories' sample preparation procedure and method of particle counting when interpreting questionable or problematic data. Poor preparation of the sample can lead to erroneous results and misguided operational decisions in response.

Fluid contamination by purification media can often be identified spectrometrically through high magnesium and calcium content in the case of Fuller's earth and high sodium in the case of activated alumina, zeolites, Selexsorb and ion exchange resins. However, some elemental spectrometers do not pick up sodium very well and the alternate use of X-ray fluorescence spectrometers (XRF) can be poor sensitivity for magnesium.

Such media migration contaminants can contribute to deposit formation, foaming and/or decreased air release properties. These conditions can lead to adiabatic compression, fluid darkening, varnish formation, plugging of filters/servo valve screens and/or sticking servo valves and solenoids.

An illustrative oil monitoring programme for phosphate ester hydraulic fluids is shown in Table 12.10.

Fluid maintenance issues have also been complicated by formulation changes in the fluid, the purification media and the filter elements. The workplace environment and access to in-house resources have changed as well. These systems can run trouble-free without expensive modifications, but it is important not to substitute one problem for another.

12.7 SUMMARY

Hydraulic systems are among the most complex and sensitive items of equipment in terms of their lubrication and protection. Monitoring and controlling hydraulic fluids and systems require more attention to more parameters than many other lubricated applications.

TABLE 12.10

Illustrative Fluid Monitoring Programme for Phosphate Esters

Property	Test Method	Frequency
Appearance	–	Every 2 months
Colour	ASTM D1500	Every 2 months
Kinematic viscosity at 40°C, cSt	ASTM D445	Every 2 months
Acid number, mg KOH/g	ASTM D974	Every 2 months
Water content, mg/kg	ASTM D6304	Every 2 months
Chlorine content, mg/kg	GE E50A345	Every 2 months
Mineral oil content, ml/l	–	Every 2 months
Particle count	ISO 11500	Every 2 months
Resistivity at 20°C, giga-ohm cm	ASTM D1169	Every 2 months
Elemental contents, ppm	ASTM D6595	Every 6 months
Flash point, °C	ASTM D92	Every 12 months
Relative density at 15°C	ASTM D1298	Every 12 months
Foaming properties, sequences I, II and III	ASTM D892	Every 12 months
Air release value at 50°C, minutes	ASTM D3427	Every 12 months
Other tests[a]	–	Every 12 months

Source: Noria Corporation, with permission.

Notes

[a] Other tests could include Dissolved Gas Analysis (DGA), FTIR, Heptane Insolubles or Ferrography.

Because hydraulic systems are used in a very wide range of applications, both inside and outdoors, each type of hydraulic fluid used in each different application requires a different set of monitoring tests to establish its suitability to continue in use. Consequently, trend analysis and flagging limits need to be identified for each type of fluid in each application.

As with many other industrial lubricants, time, practical experience and attention to detail are required from maintenance engineers and supervisors. While general guidance can be very helpful, accumulated practical knowledge is even more valuable.

13 Condition Monitoring of Compressors and Turbines

13.1 INTRODUCTION

Compressors and vacuum pumps are vitally important mechanical devices that are used to pressurise and circulate gases through processes, facilitate chemical reactions, provide inert gas for safety or control systems, recover and recompress process gases and maintain correct pressure levels by adding or removing gases or vapours from process systems. Compressors are also used to provide pneumatic (compressed air) power for construction and manufacturing operations. Gas compressors are used in almost all industries, including automotive, steel, chemical, mining, food, natural gas and petroleum production and processing and storage and energy conservation. Refrigerator compressors are used in refrigeration and air conditioning.

A large number of types of compressors have been designed and used, as illustrated in Figure 13.1. They are used for a variety of applications. In addition to being used to compress gas, many compressors serve as blowers or can be used as vacuum pumps. Compressors are classified as either positive displacement or dynamic. The positive displacement class includes reciprocating (piston) types, several rotary types and diaphragm types. Dynamic compressors are either of the centrifugal or axial flow type, although mixed flow machines that combine some elements of both types are used.

The most commonly used compressors for industrial applications are reciprocating and rotary screw types. Crosshead reciprocating compressors tend to be used for very high-pressure applications. There are two types of trunk reciprocating compressors, single acting and double acting, as shown in Figure 13.2. Reciprocating compressors are lubricated in the same way as a gasoline or diesel engine in a car or truck.

Most reciprocating compressors are either single-stage or two-stage, with smaller numbers of multi-stage (three, four or more) machines. Lubricating single-stage and two-stage machines is generally similar, while multi-stage units may have somewhat different requirements, depending on pressures, temperatures, gas conditions and the size and speeds of the pistons.

The principal parts common to all reciprocating compressors are pistons, piston rings, cylinders, valves, crankshafts, connecting rods, main and connecting rod bearings (crankpin bearings) and suitable frames that generally contain the lubrication system. Double-acting machines (which compress on both faces of the pistons as shown in Figure 13.2) require piston rods, packing glands, crossheads and crosshead

DOI: 10.1201/9781003245254-13

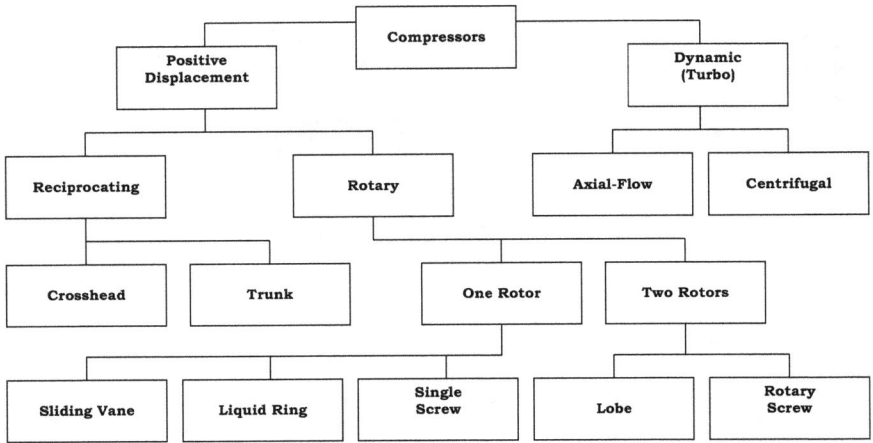

FIGURE 13.1   Types of compressors.

Source: Pathmaster Marketing Ltd.

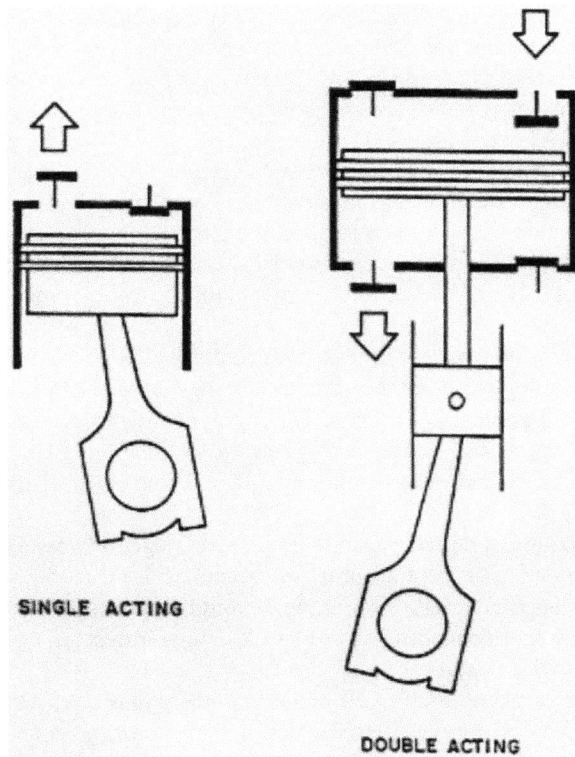

SINGLE ACTING

DOUBLE ACTING

FIGURE 13.2   Reciprocating compressors.

Source: Pathmaster Marketing Ltd.

guides. The connecting rods are connected to the crossheads by crosshead pins. Crossheads and associated parts are also used in some mult-istage, single-acting compressors, but the majority of single-acting compressors are of the trunk piston type, with the connecting rods connected directly to the pistons by piston pins (wrist pins). When lubricating these compressors, all parts associated with the cylinders, including pistons, rings, valves and rod packing (on double-acting machines) are considered to be cylinder parts. All parts associated with the driving end, including main, connecting rod, crosshead pin or wrist pin bearings and crankshaft and crosshead guides, are considered to be running parts or running gear. In many applications, lubricant requirements differ so substantially that there are two lubricating systems to separate the cylinder lubrication from the running gear lubrication.

Reciprocating compressors are provided with cooling facilities to limit the final discharge temperature to a reasonable value and to minimise power requirements. The cylinder walls and heads are cooled, and in the case of two-stage and multi-stage machines additional cooling is often used between stages.

Positive displacement rotary compressor types are straight lobe, rotary lobe, helical lobe (more commonly called rotary screw compressors), rotary vane and liquid piston. Many design variations are available for each of these types, based on application requirements. They can be single-stage or multiple-stage units that are designed for low-pressure/high-flow applications or for relatively high pressure requirements. Rotary screw and rotary lobe compressor designs can be either dry or flooded with respect to their lubrication. In the flooded types, oil is used to help to remove heat from the compressed gas and to seal the spaces between the rotors, so the lubricant comes into contact with the gas being compressed, In the dry types, only the gears and bearings of the compressor are lubricated and the lubricant does not come into contact with the gas being compressed. Rotary vane compressors are almost always flooded, while liquid piston compressors are almost always non-lubricated.

Rotary screw compressors are available in single-impeller or the more common two-impeller (rotor) designs. In the two-impeller types, a diagram of which is shown in Figure 13.3, one common design uses a four-lobed male rotor meshing with a six-lobed female rotor. Timing gears may individually drive the rotors, or the male rotor may drive the female rotor. Gas is compressed by the action of the two meshing rotors. The machines come in single-stage and multiple-stage units. With the oil flooded compressors, since oil is available in the cylinder (casing) to lubricate the rotors, these machines are now usually built without timing gears. They require an external circulation system to control the temperature of the oil and a system to remove most of the oil from the compressed gas, as shown in Figure 13.4.

Turbo compressors, shown in Figure 13.5, which are usually relatively small in comparison with other compressors, tend to be used in lower-pressure applications. Centrifugal and axial flow compressors deliver oil-free gas, as only their bearings are lubricated and the oil does not usually come into contact with the gas. Positive displacement compressors are also manufactured in non-lubricated arrangements, although these are not recommended for severe operating conditions such as high pressures (oil helps to seal), high temperatures, with wet gas or when there are corrosive compounds in the gas. Many operators of non-lubricated positive displacement compressors use some lubricant, particularly in reciprocating cylinders, to provide

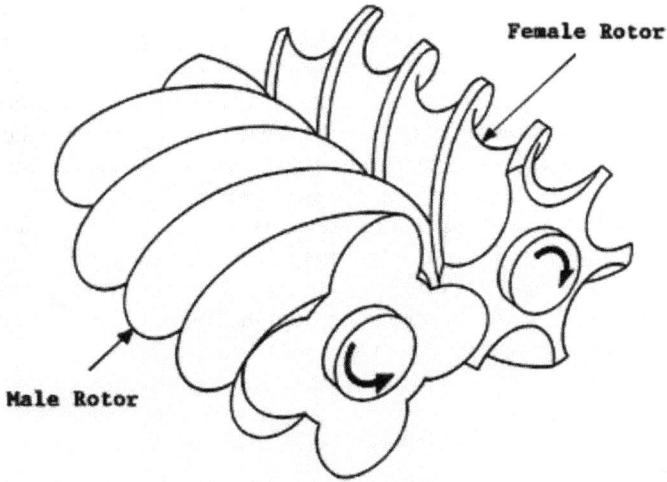

FIGURE 13.3   Rotary screw compressors.

Source: Pathmaster Marketing Ltd.

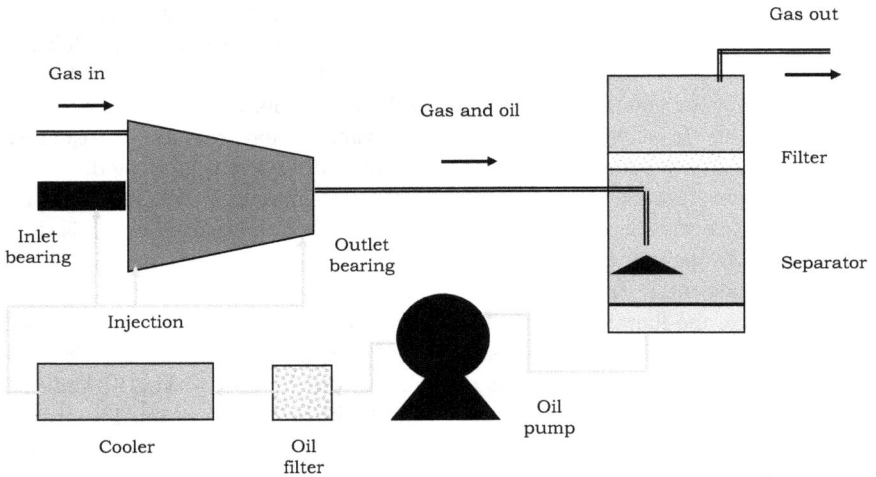

FIGURE 13.4   Lubrication system for a rotary screw compressor.

Source: Pathmaster Marketing Ltd.

longer life of components and improved sealing. A few reactive gases, such as oxygen, require special consideration, and hydrocarbon-based lubricants should not be used for gases of these types.

Improvements in the design of helical lobe (screw) compressors have resulted in higher-pressure capabilities, with efficiencies approaching those of reciprocating

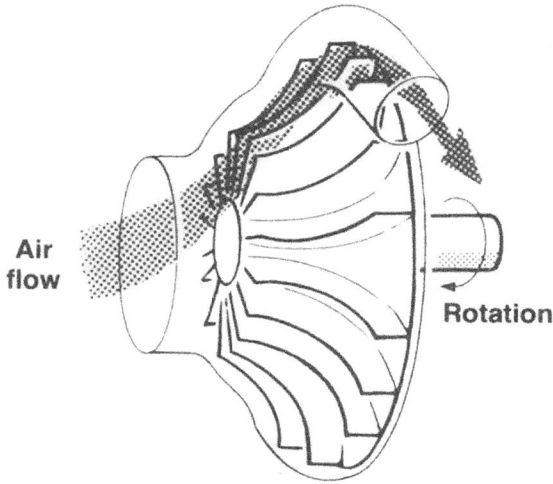

FIGURE 13.5 Turbo-compressor.

Source: Pathmaster Marketing Ltd.

TABLE 13.1
Compressed Gases

Reactive Gases	Hydrocarbon Gases	Inert Gases
Air	Natural gas	Carbon dioxide
Oxygen	Methane	Carbon monoxide
Chlorine	Propane	Nitrogen
Synthesis gas	Butane	Hydrogen
Hydrogen chloride	Ethylene	Helium
Vinyl chloride	Propylene	Ammonia
Sulphur dioxide	Acetylene	Neon
	Butadiene	

Source: Pathmaster Marketing Ltd.

compressors. This advantage, combined with the lower noise levels associated with rotary machines, has also increased the use of rotary screw compressors.

Numerous types of gases need to be compressed for a wide range of domestic, commercial and industrial applications. They are generally grouped into three classes, listed in Table 13.1. The reactive and hydrocarbon gases are obvious, but the class of inert gases contains some gases that are not truly chemically inert.

Lubrication requirements vary considerably, depending not only on the type of compressor but also on the gas (including any contaminants) being compressed. Air and gas compressors are generally mechanically similar. The main difference is in the effect of the gas on the lubricant and the compressor components. Lubricants

TABLE 13.2

Effect of Compressor Staging on Gas Discharge Temperature

Discharge Pressure		Discharge Temperature[a], °C		
psi	kPa	Single Stage	Two Stages	Three Stages
70	483	203	98	–
80	552	219	104	–
90	621	233	109	–
100	689	247	114	–
110	758	256	119	–
120	827	271	122	83
250	1,724	–	163	108
500	3,447	–	207	132

Source: Pathmaster Marketing Ltd.

[a] The calculated discharge temperatures are based on adiabatic compression of air. Actual temperatures will be lower due to heat losses within the compressor.

play roles in preventing wear, achieving sealing, minimising viscosity dilution and additive reactions with the gas and preventing corrosion. Refrigeration and air conditioning compressors require special consideration because of the recirculation of the refrigerant and mixing of the lubricant with it.

Compressing a gas causes its temperature to rise due to adiabatic compression. When a gas is compressed to a higher pressure, its discharge temperature will be higher, as shown in Table 13.2. When high discharge pressures are required, compression can be achieved in two or more stages, with the gas being cooled between stages to limit temperatures to reasonable levels. This also improves compressor efficiency and reduces power consumption for the range of temperatures that can be reached. The temperatures shown in Table 13.2 are based on adiabatic compression of air with an intake pressure of 14.7 psia (101.3 kPa abs), a temperature of 15.6°C and intercooling between stages to the same temperature. Adiabatic compression assumes that all the mechanical work done during compression is converted to heat in the gas. The cylinder is assumed to be perfectly insulated, ensuring that no heat is lost from it. However, although the temperatures indicated in Table 13.2 are higher than the discharge temperatures reached in practice, they must still be considered when considering the thermal and oxidative stabilities of the lubricant being used.

Steam and industrial gas turbines are used extensively in electricity generation, as prime movers for generators. They are also used for mechanical drive applications in many industries, to power centrifugal pumps, compressors, blowers and other machines. They continue to be used for shipboard propulsion. Gas turbines are used to power civil and military jet aircraft.

In a steam turbine, medium- to high-pressure steam is expanded in the nozzles, where part of its heat energy is converted to kinetic energy. This kinetic energy is then converted to mechanical energy in the turbine runner, either by the impulse principle or by the reaction principle. If the nozzles are fixed and the jets directed towards moveable blades, the jets' impulse force pushes the blades forward. If the nozzles are free to move, the reaction of the jets pushes against the nozzles, causing them to move in the opposite direction.

In small, purely impulse turbines, steam is expanded to exhaust pressure in a single set of stationary nozzles. As a result of this single expansion, the steam issues from the nozzles in jets of extremely high velocity. To obtain maximum power from the force of the jets' impact on a single row of moving blades, the blades must move at about half the velocity of the jets. As a result, single-stage impulse turbines operate at very high rotational speeds. To reduce rotatational speed while maintaining efficiency, the high velocity can be absorbed in more than one step, which is called velocity compounding.

With velocity compounding, steam is first expanded in the stationary nozzles to high velocity, which is then reduced in two steps through a first two rows of moving blades. The steam is then expanded again in a set of stationary nozzles and delivered to the first pressure stages. In each case, velocity is increased and pressure is decreased in the stationary nozzles. In the moving blades, velocity decreases but the pressure remains constant.

Another method of reducing rotor speed while maintaining efficiency is to decrease the velocity of the jets by dividing the drop in steam pressure into a number of stages. This is called pressure compounding. The steam pressure is reduced by somewhat less than half in the velocity-compounded stage and the steam then passes to the pressure-compounded stages, where it is reduced in four steps to the final exhaust pressure. Each pressure stage consists of a row of stationary nozzles and a row of moveable blades, so that the whole assembly is equivalent to mounting four single-stage impulse turbines on a common shaft.

Gas turbines consist of an axial compressor to compress the intake air, a combustor section and a power turbine. The compressed air is mixed with fuel (liquid or gas) and burned in combustion chambers (combustors). The hot gases expand through a turbine or turbines to drive the load. There may be a single shaft with a single turbine to drive both the compressor and the load or two shafts with a high-pressure turbine to drive the compressor and a low-pressure turbine to drive the load.

Industrial gas turbines in power generation will operate at speeds that range from 3,000 to 12,000 rpm and drive generators that rotate at 1,800 or 3,600 rpm for 60 Hz electricity or 1,500 or 3,000 rpm for 50 Hz electricity. Some turbines are connected directly to the generators, while others go through gear reduction units to operate the generators at the appropriate speeds. Aviation gas turbines have a similar arrangement, with the exhaust gas providing thrust for the aircraft. Newer designs of aviation gas turbines additionally power a large turbo fan at the front of the compressor, the exhaust air from which flows around the body of the turbine to provide additional thrust, thereby maximising the energy efficiency of the turbine.

Water turbines (also known as hydraulic turbines) are used for generating electricity in hydroelectric power stations. They are either impulse (Pelton) or reaction (pressure) types. Reaction turbines include the inward flow (Francis), diagonal flow (Deriaz) or propeller types. Bulb turbines are reaction turbines using a propeller-type runner. The choice of which type of unit to use in a particular application is a function of the pressure head and the quantity of water flow available.

In an impulse turbine, usually called a Pelton turbine, jets of water are directed by nozzles against shaped buckets on the rim of a wheel. The impulse force of the jets pushes the buckets on the rim of a wheel and causes the wheel to revolve, so the buckets move in the same direction as the jets of water. Optimum efficiency is achieved by having a high-pressure head of water and maximum water velocity. Pelton turbines are built with the shaft either horizontal or vertical. Horizontal shaft machines are built with either one or two nozzles per runner, for small to medium installations. Vertical shaft Pelton turbines with four to six nozzles are now being used for larger installations.

The flow of water in a reaction turbine impinges on a set of curved blades which, in effect, are the nozzles. The reaction of the water on the blades causes them to rotate in the opposite direction. In a Francis turbine, the water flows radially inward from a volute casing and is turned through 90° in the blades before flowing to the tail race outlet. In a Deriaz turbine, the direction of water flow is partially turned in the volute casing so that the flow is diagonally through the blades. The shaped boss then turns the water through the remaining angle to direct it into the tail race outlet. In propeller turbines, the direction of flow is controlled by a volute casing or a flume so that the water flows axially through the turbine. Francis and Deriaz turbines are intermediate to high head machines. The various propeller turbines (fixed blade, Kaplan and bulb) are low head machines. Bulb turbines (also called tubular turbines) are low head machines for what are referred to as "flow-of-stream" river applications. They can be used for relatively small, low cost installations and have permitted the development of hydroelectric power in locations where installation of the older types of turbine would not be practical or economic.

Lubricating, monitoring and maintaining compressors and turbines are obviously vitally important for a huge number of users of these machines. The large number of different types and applications for compressors and turbines mean that monitoring requirements vary widely.

13.2   AIR COMPRESSORS

In terms of the number of compressors in use, more compressors are used to compress air for utility use than for any other purpose, excluding refrigeration and air conditioning applications. Although many air compression applications require high pressures, the large majority of pneumatic equipment is designed for pressures between 90 and 100 psig, so most compressed air systems are designed to operate between 100 and 125 psig. These requirements are satisfied by both portable compressors used on construction projects, in mining and in other outdoor applications and by stationary compressors used to provide plant air in applications ranging from service stations to industrial plants. Previously, most air compression was done by reciprocating compressors, but large numbers of compressors of other types are now being used for reasons involving various factors such as design and metallurgical improvements, the need to increase speed capabilities and the need to achieve size reduction.

Condition monitoring of reciprocating air compressors is relatively straightforward. Operating data, particularly discharge pressures, should be monitored continually. Vibration should also be monitored and any deviation from the baseline is an indicator of an anomaly that should be investigated. Monitoring also involves those critical components that are unique to each compressor depending on its type, stages, oil type and other characteristics.

The discharge temperature on each discharge valve is probably the most important diagnostic indicator. They indicate the condition of the discharge valve, as well as how the piston rings on a double-acting cylinder are sealing. When ring wear starts to become a factor, the discharge temperature starts to climb due to internal by-passing of the rings and subsequent re-compression of hot gas in the cylinder. To identify high temperatures on discharge valves, the valve covers can be painted with paint that discolours or chars at temperatures above a specified maximum, depending on the compressor type.

Monitoring the suction valve temperature indicates whether there is any leakage of the seats on the valve. Some operators choose not to monitor this because of the cost, but it would immediately alert an operator of a leaking suction valve. Infrared thermography can be used to do this, as it can with discharge valves. Monitoring compressor frame vibration is an attempt to detect any bearing wear and knocking. Depending on the compressor design, operators may need to contend with babbitt metal bearings and crosshead tolerances that are subject to normal mechanical wear.

Monitoring rotary screw air compressors is a little more difficult. As with reciprocating air compressors, contaminated air contaminates the compressor oil, resulting in equipment failures and downtime. Ingestion of air from sources containing acidic gases, such as boiler exhaust, diesel exhaust and any operations discharging acidic gases, should be avoided.

A proper discharge air temperature is important to maximising oil and compressor life. As a general rule for a humid environment, the discharge temperature should be 55°C higher than the ambient temperature of the inlet air to prevent accumulation of water in the compressor oil. Air-cooled compressors typically have coolers sized to prevent the compressor from running below this temperature and thus to automatically avoid this problem. A water-cooled compressor may cool too efficiently and condense water in the oil. If the water separates from the oil and collects in the receiver tank, the free water may not be detected in oil samples, because it may not be truly representative. One solution is to observe the temperatures and make adjustments as necessary. Depending on fluid type, temperature and humidity, water levels of up to 0.7% in a rotary screw compressor are normal. Levels above that amount indicate free water in the system and require intervention.

High operating temperatures will increase the rate of oxidation. Every 10°C increase in temperature will approximately double the oxidation rate, reducing oil life by approximately half. The temperature at which a fluid is rated for its nominal life, typically 8,000 hours for most polyalphaolefin (PAO), polyalkylene glycol (PAG) or ester compressor oils, will vary. A fluid rated for 8,000 hours at 95°C would be expected to significantly outlast a fluid rated for 8,000 hours at 80°C.

Load and unload performance affects carryover and energy savings. A compressor, which is allowed to run unloaded with a minimal air demand, will typically experience high oil consumption and more internal condensation and corrosion. It will also use much more electricity, resulting in significant energy waste and increased cost. This cycle and loading should be observed and the compressors' usage should be adjusted to match the air demand. Compressor original equipment manufacturers (OEMs) have developed computerised control systems to continually monitor and adjust to maximise savings.

Minor oil leaks may be a warning of an impending failure of a coupling, gasket or seal. They should not be ignored. The volume of makeup oil added to each compressor should be logged for frequency and amount. A "material balance" or log of oil addition should be done. A sudden increase in oil makeup rate might correspond with the change to an inefficient, defective or improperly installed air/coolant separator. Finding this quickly can result in fluid savings and avoid the contamination of downstream components. OEM separators are typically closely matched to the air flows and other characteristics of specific compressors. Also, the materials of construction are compatible with the OEM fluids, to prevent failures due to incompatibilities.

It is important to track separator and bearing filter differential pressures and frequency of changes. These are useful trends for several reasons. The depletion of corrosion protection in the oil may result in shortened filter element life, due to blinding with corrosion particles. Progressively shorter intervals between filter changes, while on the same charge of oil, are strong indicators of this condition.

The ingestion of particulates in the air may also result in short separator life. Particulate levels and element life may relate to the quality of the inlet filter element, an air leak or improper installation. Also, changing separator elements at the OEM-prescribed differential pressure (usually about 10 psi) results in energy savings and avoids the potential collapse of a separator, which results in massive amounts of fluid being discharged downstream into the air system.

Several key elements must be addressed to ensure an effective air compressor oil analysis programme, as errors or omissions related to these elements can lead to unnecessary costs and lower reliability.

The significant air quality issues for centrifugal compressors are particulate removal through filtration and the effect of acid gases on inter-cooler corrosion. In a centrifugal compressor, there is only minimal contact between the air and the lubricant and reservoir sizes are typically quite large, resulting in dilution of any contaminants. It is therefore rare for ambient air conditions to significantly affect oil life or to have a detrimental effect on the running gear of a centrifugal compressor.

Double-acting reciprocating compressors are not totally immune to air contaminants, but are less subject to inlet air quality issues because of a continuous infusion of fresh cylinder lubricant. The fresh oil serves to flush contaminants through the system with a protective effect, even though these compressors have a high lubricant consumption.

The ambient air is a more serious concern for rotary screw compressors, where the entire flow of air through the compressor contacts the oil, which is effectively acting as a scrubber to absorb the oxidation products (acids) and contaminants. Even a low concentration of acid is significant, when the volume of air being handled is considered. Some of this acid will be absorbed by the fluid, which will show up analytically as a lower pH and higher acid number (AN). Oils for rotary screw compressors are formulated with good corrosion protection, but eventually even that is overwhelmed. Once this occurs, filters may plug more frequently due to corrosion particulates. This effect results in significantly shortened fluid life. It is not unusual in a contaminated environment to see the life of a nominal 8,000-hour fluid reduced to 2,000 hours.

In addition, the service life of downstream components, such as after-coolers and dryers, is often compromised by corrosion caused by acid gases which pass through the compressor from the environment. These gases then condense with water in the coolers and dryers and increase corrosion rates significantly.

To extend compressor oil life, remote air inlets may be installed to obtain inlet air from a source away from the contamination. Air can be tested by suspending corrosivity coupons of copper and silver in the air near the compressor. After a specified period, laboratory analysis of the resulting compounds on the surface of the coupons will reveal the type and extent of contaminant in the air. Inlet air scrubbers may then be used to remove these contaminants from the inlet air.

The key oil analysis parameters vary with the type of compressor oil being used. These can be mineral oil, PAO, diester, polyol ester or PAG-based. Most of the oil analysis parameters will be common to all types of oil.

AN is an indication of remaining useful oil life. AN may increase with either oil oxidation or accumulation of contaminants from the environment. Accumulated acid reflects the depletion of the corrosion inhibitor(s). Suggested flagging limits vary, typically from 1.0 to 2.0 mg KOH/g. The oil life from the time the AN reaches 1.0 until the time it will reach 2.0 mg KOH/g is only 10% to 20% of the total oil life.

The viscosities of some OEM compressor oils are specifically designed for the needs of that compressor application. In general, reciprocating air compressors require ISO 100 or ISO 150 viscosity grades, depending on ambient temperature, while rotary screw air compressors require ISO 32 or ISO 46 viscosity grades. With PAG compressor oils, it is unusual for the fluid to fail due to viscosity change, because they are resistant to varnish and sludge formation, and while the viscosity will normally increase initially by about 10%, it then stabilises and is unlikely to increase further.

Mineral oil and PAO-based compressor oils will oxidise in use and the remaining lifetime can be determined using either Fourier transform infra-red (FTIR) spectroscopy or the rotating pressure vessel oxidation test (RPVOT) test. This is not necessary with PAGs, because AN is a reliable indicator of fluid condition. Monitoring the oxidation level is useful in preventing varnish and deposit formation. The degree of oxidation depends of the type of compressor and compressor oil. Rotary screw compressor temperatures typically range from 80°C to 115°C, while vane compressors operate at temperatures from 80°C to 150°C, and multi-stage reciprocating compressors have valve discharge temperatures between 160°C and 210°C. At the upper ends of these temperature ranges, the risk of forming varnish on bearings, deposits on filters or deposits on discharge valves increases.

The water content of air compressor oils ranges from 0.4% to 0.6%, whereas for other oils in manufacturing plants, the water content should be kept below 0.05%. When the air is compressed, water vapour is condensed. Higher water contents are normal for rotary screw compressor oils, which are formulated to function in this environment. PAG-based compressor oils will tolerate about 0.8% water before free water becomes a problem. With mineral oils, PAOs and diesters, free water will typically become an issue at lower levels.

The analysis of compressor condensate is a useful tool in detecting some corrosive or acid gases in the air that may not be effectively absorbed by the compressor oil. A low pH or high AN in the condensate may reveal a corrosive condition, which if left unchecked, will lead to short after-cooler and refrigerated dryer life.

Metallic elements and contaminants should always be monitored in compressor oils. A rotary compressor is unique in that the metals and particulates in the oil can originate from several sources. Primary sources include:

- Ingestion with the inlet air, either through or by-passing the inlet filter.
- Corrosion particles, primarily from the upper portion of the receiver tank.
- Wear debris from rotors, housing, gears and/or bearings.

Analytical ferrography is one simple and useful technique for differentiating between these three sources. Once determined, any of these problems can be readily resolved. Elements of particular interest include iron (wear of rotors, gears, pistons or rings), copper, lead and/or tin (wear of bearings), calcium, sodium or magnesium (coolant water leakage), silicon (dust in inlet air or anti-foam additive) and aluminium (corrosion of the cooler).

An example oil analysis report, compiled over several years, for a rotary screw air compressor is shown in Table 13.3. Note that test results from one of the samples contain highlighted alarm limits. Example flagging limits for reciprocating and rotary screw air compressor oils are shown in Table 13.4.

13.3 GAS COMPRESSORS

Inert gases, such as nitrogen, carbon dioxide, carbon monoxide, helium, hydrogen and neon, do not react with lubricating oils and do not condense on cylinder walls at the highest pressures reached during compression. Ammonia is relatively inert, but some special considerations need to be used.

Inert gases generally do not introduce any special problems, and they can be handled satisfactorily by the oils used for air compressors. However, carbon dioxide is slightly soluble in mineral oil and tends to reduce its viscosity. If moisture is present, carbonic acid, which is slightly corrosive, will form. To minimise the formation of carbonic acid, the system should be kept as dry as is practical. To counteract the dilution effect, higher-viscosity oils than those normally used in air compressors are recommended.

Ammonia is usually compressed in dynamic compressors, but occasionally it may be compressed in positive displacement compressors. In the presence of moisture, it can react with some oil additives and oxidation products to form soaps. Ammonia is not compatible with anti-wear compounds such as zinc dialkyldithiophosphate (ZDDP), and oils containing additives of these types should not be used. Automotive engine oils and many anti-wear-type hydraulic oils contain ZDDP. Ammonia may also dissolve in the oil to some extent, resulting in viscosity reduction. Highly refined

TABLE 13.3
Example Oil Analysis Report for a Rotary Screw Air Compressor Oil

Date sampled	18/5/15	10/5/16	23/5/17	15/5/18	14/5/19
Date received	19/5/15	12/5/16	24/5/17	17/5/13	16/5/14
Date reported	21/5/15	13/5/16	25/5/17	18/5/13	17/5/14
Lab number	nd	nd	nd	nd	nd
Oil brand	nd	nd	nd	nd	nd
Oil type	nd	nd	nd	nd	nd
Oil grade	ISO 46	ISO 46	ISO 46	ISO 46	ISO 46
Oil added (litres)					
Oil changed	Unknown	Yes	Yes	Yes	Yes
Time in compressor (hours)	Unknown	8,064	8,184	8,040	8,040
Ruler (ASTM D6971)					
Amine %	98	**35**	91	100	99
Phenol %	94	**<25**	85	75	94
Physical/Chemical					
VP pentane insolubles		**920**	52	12	114
Colour (ASTM D1500)	2.5	7.5	3.0	1.5	1.5
UC sediment (MM1169)	–	7	1	1	1
Blotter test	–	2	2	1	1
Membrane patch calorimetry	10	24	17	5	16
Varnish potential rating	Low	**High**	Moderate	Low	Moderate
Particle counts					
ISO 4406 rating	19/17/12	18/16/11	18/15/11	18/16/12	19/17/13
>4 μm (particles/ml)	2,798	2,467	1,832	1,624	3,730
>6 μm (particles/ml)	695	346	292	358	980
>14 μm (particles/ml)	32	18	19	30	78
>23 μm (particles/ml)	6	4	5	7	25
>50 μm (particles/ml)	2	1	1	1	3

Notes
nd = not disclosed.
Source: Pathmaster Marketing.

straight mineral oils are usually used. PAO-based compressor oils are also used because of their low solubility for ammonia.

When gases are compressed for human consumption, such as carbon dioxide for use in carbonated beverages, carryover of conventional lubricating oils is undesirable. Generally, medicinal white oils are required for cylinder lubrication in these circumstances.

Conventional mineral oils cannot be used in inert gas compressors for some applications. For example, in some chemical processes, traces of hydrocarbons cannot be tolerated in the process gas, or some constituents of lubricating oils might poison catalysts used in later processes. Compressors similar to those used to produce oil-free air or systems equipped with sophisticated filtration and conditioning equipment are used

TABLE 13.4

Example Flagging Limits for Air Compressor Oils

Property	Reciprocating Compressor	Rotary Screw Compressor
Viscosity	±10% of new oil	±10% of new oil
AN, mg KOH/g	2.0 max	1.0 max
Water content, %wt	0.6% max	0.8% max
Ruler amine, %	>40%	>40%
Ruler phenol, %	>25%	>25%
Varnish potential, max	Moderate	Moderate
Discharge temperature, °C	220 max	120 max

Source: Pathmaster Marketing Ltd.

where hydrocarbon carryover cannot be tolerated. Polyisobutene (PIB)-based compressor oils are used where carryover of conventional mineral oils might poison catalysts.

More energy is consumed in compressing natural gas than any other gas except air. When the volumes of other hydrocarbons that are compressed for the chemical and process industries are considered, the total energy consumed in compressing hydrocarbons is extremely large. Dynamic compressors are usually used if the hydrocarbons must be kept free of lubricating oil contamination, but if high pressures are required, reciprocating compressors are used. With improved technology and the ability of some rotary compressors to achieve higher pressure and volume capacities, there is also a trend towards the use of rotary screw compressors in hydrocarbon compression.

While natural gas is mainly methane, other gases usually are present in small amounts. These include ethane, carbon dioxide, nitrogen and heavier hydrocarbon gases. The heavier hydrocarbon gases are similar in many respects to the hydrocarbons that are compressed for process purposes. Occasionally these heavier hydrocarbons are in liquid form, which complicates the lubricant selection process.

The temperature at which a material will condense from the gaseous state to the liquid state (also the temperature at which it will pass from its liquid state to the gaseous state, that is, its boiling point) increases with increasing pressure. With the higher boiling point, heavier hydrocarbons, the condensation temperature may be above the cylinder wall temperature at the pressure in the cylinders. The condensate formed under this condition will tend to wash the lubricant from the cylinder walls and dissolve in the lubricating oil, resulting in viscosity reduction. Using an oil that is somewhat higher in viscosity than would be used for air under the same operating conditions can generally compensate for the dilution effect. Generally, compounded oils help to resist washing where condensed liquids are present in the cylinders. It is usually advisable also to operate with somewhat higher than normal cooling jacket temperatures, to minimise condensation. This also requires the use of higher-viscosity oils.

Natural gas that contains sulphur compounds as it comes from the production well is referred to as "sour" gas. Compressors handling sour gas are usually lubricated with detergent-dispersant engine oils, such as automotive engine oils or natural gas engine oils. These oils provide better protection against the corrosive effects of sulphur. The viscosities used most frequently are ISO VG 100 and 150, but if the gas is wet (that is, carrying entrained liquids), heavier oils may be used. The compressor of integral engine-compressor units is usually lubricated with the same oil used in the engine. However, depending on the contaminants contained in the gas, compressor cylinders may require a lubricant different from that used in the engine crankcase.

Natural gas is considerably less soluble in water-soluble PAGs than it is in either mineral oils or PAOs. This is due to the higher polarity of PAGs. As a result, an increasing number of natural gas compressors are lubricated using PAG-based compressor oils, particularly for reciprocating compressors.

Among the chemically active gases that must be considered most frequently are oxygen, chlorine, hydrogen chloride, sulphur dioxide and hydrogen sulphide. Mineral oils should not be used with oxygen because they form explosive combinations. Oxygen compressors with metallic rings have been lubricated with soap solutions. Compressors with composition rings of some types have been lubricated with water. Compressors designed to run without lubrication are also being used. Some of the inert synthetic lubricants, such as the chlorofluorocarbons or fluorinated oil, can be used safely and provide good lubrication. Dry-type solid lubricants such as Teflon or graphite can be used to minimise metal-to-metal contact in this service.

Mineral oils should not be used for the lubrication of chlorine and hydrogen chloride compressors. These gases react with the oil to form gummy sludges and deposits. If the cylinders are opened to remove these deposits, rapid corrosion takes place. Compressors designed to run without lubrication are used. Diaphragm and non-lubricated rotary compressors are also used for these corrosive and reactive gases.

Sulphur dioxide dissolves in mineral oils, reducing the viscosity. It may also form sludges by reacting with the additives in the presence of moisture or by selective solvent action. The system must be kept dry to prevent the formation of acids. Highly refined straight mineral oils or white oils from which the sludge-forming materials have been removed, either by acid treating or severe hydroprocessing of the base oils, are often chosen. Oil feed rates should be kept to a minimum.

Hydrogen sulphide compressors must be kept as dry as possible because hydrogen sulphide is corrosive in the presence of moisture. Compounded oils are usually used, and corrosion and oxidation inhibitors are considered to be desirable.

Monitoring the compressors and compressor oils for all these gases is essentially similar to air compressors, particularly with respect to operating temperatures and pressures, vibration, contaminants, viscosity, water and wear metals. Most of the differences resulting from the reactivity or otherwise of the respective gases have been highlighted above.

13.4   REFRIGERATION AND AIR CONDITIONING COMPRESSORS

The five essential parts basic to every refrigeration or air conditioning compression cycle are evaporator, compressor, condenser, receiver and expansion valve (or capillary). Liquid refrigerant flows from the receiver under pressure through the expansion valve to the evaporator coils, where it evaporates, absorbing heat and resulting in a cooling action. The vapour is then drawn into the compressor where its pressure and temperature are raised.

At the higher pressure in the discharge of the compressor, the condensing temperature of the refrigerant is higher than it would be at atmospheric pressure. When the hot, high-pressure vapour flows from the compressor to the condenser, the cooling water (air in some applications) removes enough heat from it to condense it. The heat removed from the refrigerant in the condenser is equal to the amount of heat removed from the cold room (cooling action) plus the heat resulting from the mechanical work done on the refrigerant in the compressor that is not removed by the jacket cooling of the compressor. In many commercial installations, the evaporator cools a heat transfer fluid such as brine, which is then pumped through the area to be cooled. Smaller units, such as home refrigerators and freezers, room air conditioners and automotive air conditioners, have air-cooled rather than water-cooled condensers. A diagram of a typical commercial refrigeration system is shown in Figure 13.6.

In commercial installations, two or three stages of compression may also be used. If system pressures or cooling capacities dictate the use of two stages of compression, two-stage compressors are used, or a combination of separate single-stage compressors. Rotary sliding vane, scroll or rotary screw compressors are sometimes used at low to moderate pressures or for booster purposes. Multi-stage reciprocating compressors are used for large air conditioning installations, with a trend towards the use of more scroll compressors. Reciprocating compressors are commonly used for refrigeration systems, with a trend towards the use of rotary vane compressors. Centrifugal compressors are also used on some commercial refrigeration systems

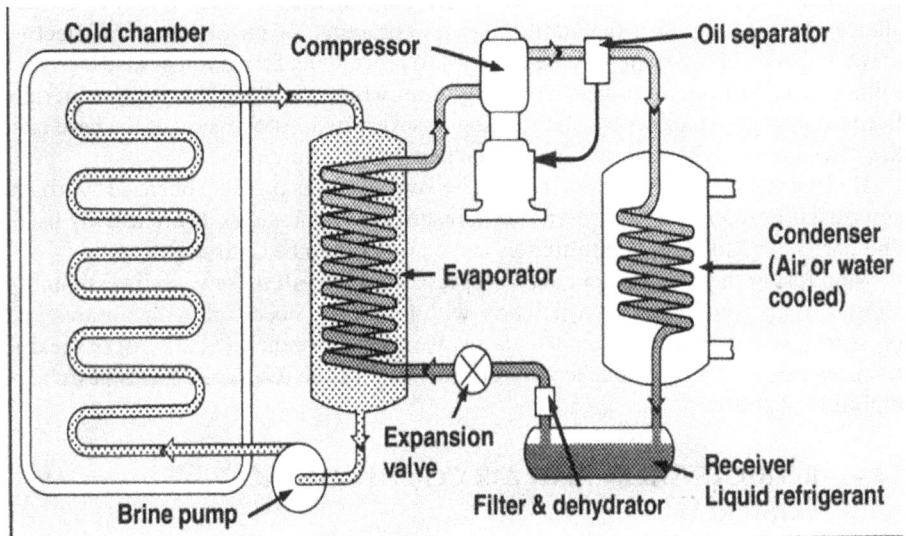

FIGURE 13.6   Refrigeration system.

Source: Pathmaster Marketing Ltd.

as well as in chillers. Reciprocating, sliding vane and scroll compressors are used for automotive air conditioning systems, with some screw and axial piston compressors also used. Some very small units such as dehumidifiers may be equipped with diaphragm-type compressors. Reciprocating compressors are used in most other applications.

Most reciprocating compressors for commercial installations are of the single-acting, trunk piston-type and have closed crankcases. As a result of refrigerant leakage past the pistons, the crankcases are filled with a refrigerant atmosphere. The same is true of axial piston units used for automobile air conditioning. Crosshead and double-acting compressors have open crankcases. The majority of small to medium-sized electric motor driven refrigeration and air conditioning units are hermatically sealed, with the operating parts, including the electric motor, inside the sealed unit.

Selecting the oil to be used as the lubricant for a refrigeration or air conditioning compressor depends on the refrigerant. Refrigerants, of which there are several types, must be able to absorb and transfer heat. They must be able to readily change states from a liquid to a gas. Depending on the refrigerant used, very low temperature refrigeration or simply basic cooling capacity can be achieved. Refrigerant selection is based on a number of criteria:

- Application requirements: The nature and amount of cooling, evaporator temperature range and system size.
- Refrigerant thermodynamic characteristics.
- Safety considerations: Flammability and toxicity.
- Cost: Refrigerant and operational.
- Regulatory compliance.

Refrigerant types are generally referred to by their ASHRAE classification (ANSI-ASHRAE Standard 34-2001):

- R717: Ammonia.
- R11, R12, R113, R114, R500, R502: Chlorofluorocarbon (CFC).
- R22, R123, R125: Hydrochlorofluorocarbon (HCFC).
- R600a: Iso-Butane.
- R744: Carbon dioxide.
- R134a, R 143a, R404a, R507: Hydrofluorocarbon (HFC).

CFCs were banned in 1989 under the Montreal Protocol, due to ozone depletion in the upper atmosphere. HCFCs are being phased out, due to the potential to add to global climate change. This leaves just ammonia, iso-butane, carbon dioxide and HFCs.

Lubricants perform several functions in a compressor system. Of course, they must be able to lubricate the machine, but they also need to be compatible with the refrigerant. In some systems, the lubricant is required to act as a cooling fluid as well as a sealant. Compressor lubricants are often a specialised blend of base oils and additives in order to provide the necessary lubricating properties while still being compatible with the refrigerant. Any incompatibility of the base oil and the refrigerant could have disastrous results for the equipment. Moisture contamination can be very detrimental to some synthetic base oils that are hydrolytically unstable. Moisture reacts with the base oil to form acids, change the viscosity and impairs the oil's lubricating properties. This can lead to premature compressor failure as well as improper system cooling. Moisture in the system can also form ice crystals, which may block the expansion valve.

The majority of refrigerator compressor lubricants are synthetic. This allows them to have a longer service life and handle the demands of the system better than mineral oils. Even so, naphthenic mineral oils, dewaxed paraffinic mineral oils, PAOs or alkyl benzenes are used with ammonia. PAOs, esters and PAGs are used with CO_2, PAGs or naphthenic mineral oils are used with iso-butane and polyol esters are used with HFCs.

Oil sampling and analysis are impractical for domestic refrigeration and air conditioning, automotive air conditioning and small industrial refrigerators and freezers. These systems are typically "sealed for life", and if the unit fails, it is simply replaced.

In industrial plants, refrigeration compressor systems tend to be among the most critical machines. It is therefore important to take oil samples regularly to check the health of the lubricant and the machine. The oil analysis tests performed on these fluids include elemental analysis, viscosity analysis and wear debris analysis. The viscosity must be monitored because refrigerant dilution can lead to a decrease in viscosity and an increase in machine wear.

In some cases, oil samples must be degassed before they can be shipped to the laboratory or analysed. Since the gas expands with temperature, it can result in a pressure increase in the bottle, causing a leak or the oil to erupt upon opening the bottle. While pressure-relieving caps can be used with these bottles, every time the bottle is opened, the oil sample is exposed to contamination, which can affect the particle count results.

13.5 STEAM AND GAS TURBINES FOR ELECTRICITY GENERATION

Modern steam and gas turbines subject the turbine oil to ever greater demands. Higher temperatures are encountered in bearings, smaller reservoirs reduce residence times, and issues with varnish deposits have become critical concerns. Since the oil is essential to reliable turbine operation, a sound oil condition monitoring programme is needed to ensure long trouble-free operations.

A picture of a large steam turbine is shown in Figure 13.7, to illustrate the sizes of these machines in comparison with the person in the top right-hand corner of the picture. The internal arrangement of a steam turbine is shown in Figure 13.8, and a schematic of the lubrication circuits of a steam turbine is shown in Figure 13.9.

Steam and gas turbine oils used for power generation are increasingly using API Group II base oils, rather than Group I base oils. This has required a shift to more complex and effective oxidation inhibitors and corrosion inhibitors, as well as additional polar components to correct the poorer solvency properties of Group II base oils.

FIGURE 13.7 Large steam turbine.

Source: Pathmaster Marketing Ltd.

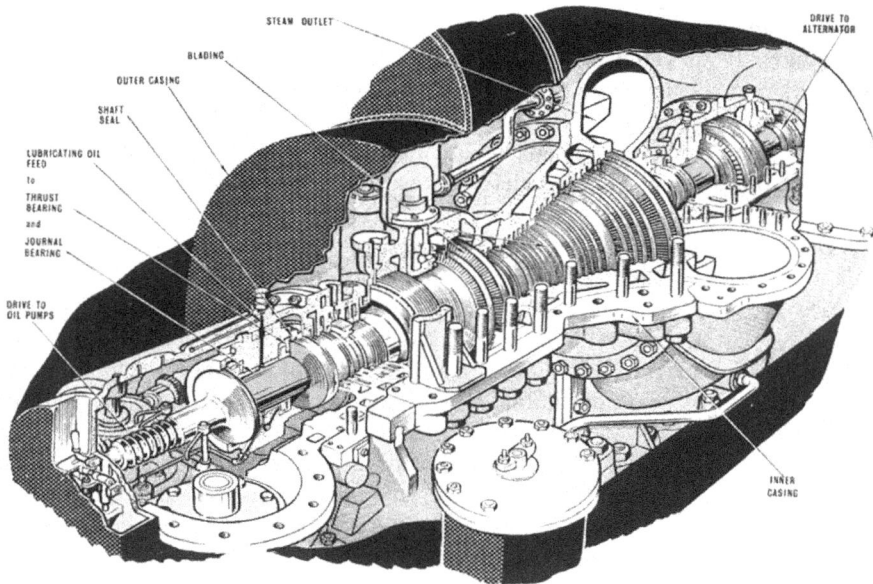

FIGURE 13.8 Steam turbine arrangement.

Source: Pathmaster Marketing Ltd.

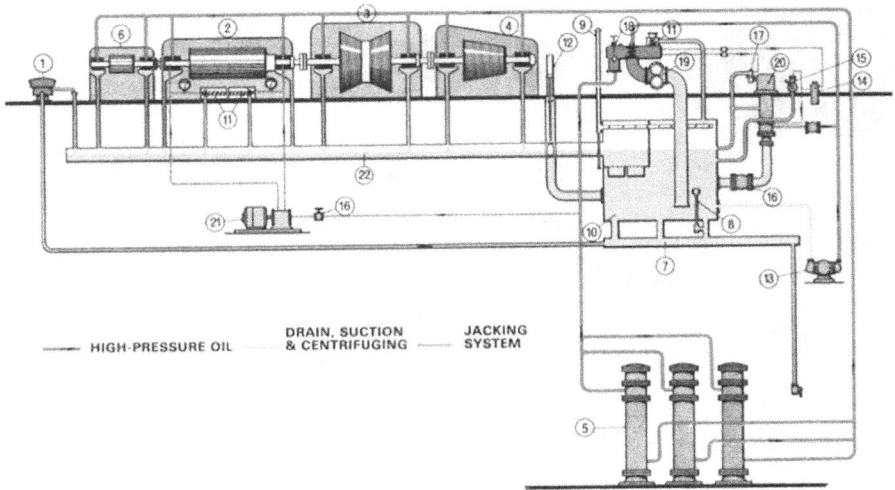

FIGURE 13.9   Steam turbine lubrication diagram.

Source: Pathmaster Marketing Ltd.

There are four primary reasons why turbine oils degrade in service. All oils oxidise in service when exposed to oxygen in the atmosphere. Oxidation is not limited to the reservoir, since air is dissolved in the oil. With increasing temperatures found in turbines, increasing flow rates and shorter reservoir residence times, oxygen and oil have more opportunities to interact.

Secondly, the oil can be exposed to temperatures in a turbine that cause base oil and additive molecules to thermally degrade. The result of this degradation is the formation of materials that are not readily soluble in the oil. These materials then form deposits within the oil system, which can sometimes lead to equipment failures.

Turbine oils are subject to a variety of contaminants such as water (especially in steam turbines), dust and other ingress materials, wash down chemicals and internally derived contamination, such as wear metals. While none of these are a direct result of oil degradation, they often contribute to other degradation issues. Wear metals, such as copper, iron and lead, catalyse oxidation reactions. Water (especially chemically treated water) can have very adverse effects on the ability to dissipate foam and separate from water. Excess foaming can lead to sluggish response from hydraulic control systems, cavitation in pumps and bearings and safety issues if the foam over-fills the reservoir and spills on the floor.

Additive depletion is the fourth major cause of turbine oil deterioration. Some additive depletion is normal and expected. Antioxidant additives are consumed as they perform their function. Demulsifiers help the oil shed water, but if exposed to large amounts of water contamination, the demulsifiers can be removed. Anti-foam additives can be removed from ultra-fine filtration or can agglomerate when the oil is not circulated for extended periods of time.

All of these factors should be consistently monitored throughout the life of the turbine oil.

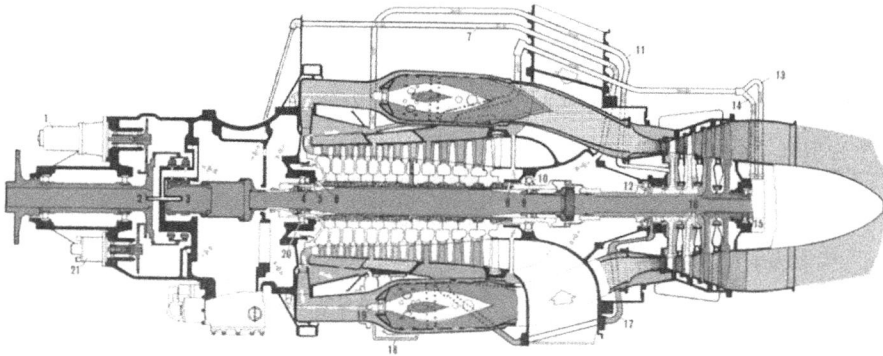

FIGURE 13.10   Industrial gas turbine.

Source: Pathmaster Marketing Ltd.

Gas turbines are the most demanding application for turbine oils, which also has been one of the fastest-growing markets for power generation. A diagram of an industrial gas turbine used for power generation is shown in Figure 13.10. The efficiency and firing temperature of gas turbines are continually rising as more advanced metallurgies are developed. In general, the stress on the turbine oil also increases with more efficient engines, and most turbine OEMs are demanding increased turbine oil performance. The stress on turbine oils, however, may be more complex than just looking at the lubricant's residence time, high-temperature zones and reservoir temperatures. Data suggests that "Class E" turbine systems may impart more stress to its turbine oil when compared with "Class F" systems. In reality, turbine oils lubricating "Class F" units typically degrade faster and have more performance challenges. Additional factors to consider when assessing the stresses on the turbine oil in a system include the environmental conditions of where the unit is located, its duty cycle and the plant's maintenance practices. Some "Class F" units are known to exert further stress on the turbine oil due to a phenomenon known as spark discharge, which occurs in the unit's main turbine oil filters.

For any power generation facility, the turbine is considered to be central to the operation. Any problem requiring an unexpected shutdown of the main turbine is likely to cause a significant unplanned outage, potentially resulting in millions of dollars of downtime costs. Due to the volumes of oils used in large steam turbines and the rate of oil top-up, turbine oils are expected to last between 10 and 20 years. For this reason, careful monitoring of in-service turbine oil physical and chemical properties is required, together with common contaminants such as water and particulates. This is also true for new turbine oils, which must meet rigorous performance specifications prior to selection and use in a new application.

The testing of turbine oils is of such significance that ASTM has developed a standard specifically for these oils; ASTM D4378 "Standard Practice for In-Service Monitoring of Mineral Turbine Oils for Steam and Gas Turbines". Oil parameters that need to be tested regularly include viscosity, AN, RPVOT, water content, cleanliness and corrosion inhibition.

Viscosity is the most important characteristic of a turbine oil because the oil film thickness under hydrodynamic lubrication conditions is critically dependent on the oil's viscosity characteristics. Turbine blade clearances are critical to power plant efficiency and reliability. These blade clearances are directly impacted by lubricant viscosity. Changes in oil viscosity can result in unwanted rotor positioning, both axially and radially. Axial movements directly impact turbine blade efficiency and in extreme cases can lead to blade damage. Radial movements caused by changes in viscosity can result in oil whip, where the rotor does not settle into one radial position. Oil whip can often be identified from vibration analysis, but is often a direct result of high viscosity.

For in-service turbine oils, the viscosity should remain consistent over years of service, unless the oil has become contaminated or severely oxidised. ASTM D4378 identifies a 5% change from the initial viscosity as a warning limit. It is important to note that this is a change with respect to a new oil baseline, not the typical value reported on the turbine oil supplier's specification sheet.

RPVOT (ASTM D2272) was developed for the monitoring of in-service oils to warn of a loss in oxidation resistance. As a turbine oil degrades, it forms weak organic acids and insoluble oxidation products that adhere to governor parts, bearing surfaces and oil coolers. After a period of time, these oxidation by-products and carbonaceous insolubles adhere to surfaces causing a significant change in critical clearances and in some instances preventing the oil from providing adequate cooling to the bearings and fouling turbine control elements and heat exchangers.

ASTM D4378 identifies an RPVOT drop to 25% of the new oil value with a concurrent increase in AN as a warning limit. Many turbine OEMs simplify this by using the 25% of initial RPVOT without reference to AN, while others list a 100-minute RPVOT minimum. If the RPVOT test result is approaching 25% of the new oil value, the testing frequency should be increased from quarterly to monthly. It should be noted that the RPVOT test is designed to determine a lubricant's suitability for continued use, not to compare competitive oils. Competitive oil comparisons should be evaluated on the basis of RPVOT longevity, rather than the absolute RPVOT value.

In gas turbines that use a common oil reservoir for bearings and system hydraulics, ultra-centrifuge testing should be used in conjunction with RPVOT as a way to determine varnish formation.

Typically, an oil that has reached its minimum allowable RPVOT values needs to be changed. However, as a short-term measure, the "bleed and feed" method of turbine oil rejuvenation is suitable to extend the life of the turbine oil for a limited time. Addition of supplemental antioxidants to an oxidised turbine oil in service can put equipment at risk. An oil that has an RPVOT value below 100 minutes is likely to have diminished its inherent base oil oxidation stability. The addition of supplemental oxidation inhibitors (known as readditization) may temporarily boost the RPVOT value but, given the diminished nature of the base oil, may sharply reduce the time frame before heavy varnishes and sludges are formed. Without the use of special filters such as Fuller's Earth, to strip all polar materials, contaminants and additives, followed by complete readditization, the rejuvenation of a degraded turbine oil is inadvisable.

The other method of monitoring the potential of a steam or gas turbine oil to form varnish is the membrane patch colorimetry (MPC) (ASTM D7843) test. As noted in

Section 6.2.5, the revised ASTM D7843 test for MPC can be used to account for the effects of the solvent used, by measuring the iMPC value and comparing it with the original MPC value. Although this analysis takes more time, important additional information may be gained about the potential of the turbine oil to form varnish.

Testing for water, using Karl Fischer titration (ASTM D6304), particularly in steam turbines, is important because water is a precursor to oil oxidation and rust formation. Excessive water will also alter an oil's viscosity, which may reduce its load-carrying capacity. Studies also warn that water levels above 250 ppm in hydrogen-cooled generator windings may lead to stress corrosion cracking of generator rotor retainer rings. Water in a turbine oil in warm storage tanks, where the oil is typically stagnant, can promote the spread of microbial growth that will foul system filters and small-diameter gauge and transducer line extensions.

ASTM D4378 identifies 1,000 ppm or 0.1% of water as a warning level, while some gas and steam turbine OEMs have identified 500 ppm. In hydrogen-cooled generators, an upper limit of 250 ppm should be maintained. Because free and emulsified water are the most harmful, it is advisable to keep water levels below saturation, typically 100 to 200 ppm depending on base oil types, additive formulation and turbine oil age.

Sharp increases in AN (ASTM D664, IP 177 or ISO 6619) may indicate contamination or a severely oxidised oil. Organic acids formed by oxidation can corrode bearing surfaces and should be corrected as soon as practicable. ASTM D4378 offers guidelines of 0.3 to 0.4 mg KOH/g above the initial value as an upper warning level. However, many oil analysts view an upward movement in AN as small as 0.1 mg KOH/g as concerning.

Turbine journal bearing clearances (10 to 20 μm) and hydraulic servo-valve clearances (3 to 5 μm) dictate the need for clean oil. Excessive bearing wear and servo-valve sticking can result if good cleanliness standards are not maintained. Typical OEM recommended turbine oil cleanliness levels are ISO 4406 18/16/13 or NAS 1638 level 7, although significant component life extension can be achieved by keeping cleanliness levels significantly lower than these limits. Of course, this means higher costs of filtration to remove particles.

Rust particles act as oxidation catalysts and can cause abrasive wear in journal bearings. Rust inhibitors are normally kept at satisfactory levels through the addition of top-up oil. Rust inhibitors can impact water separation so field readditization is generally not recommended. In-service oil testing should be conducted with distilled water as identified in ASTM D665A, IP 135A, ISO 7120, DIN 51585 or AFNOR T60-151. ASTM D4378 considers a light fail as a warning limit.

Demulsibility (ASTM D1401, IP 412, ISO 6614 or AFNOR T60-125) is important to turbine oil systems that have direct contact with water. This is particularly true for steam turbines where gland seal leakage is difficult to avoid. The ability of the oil to shed water will have a direct impact on its long-term oxidation stability. Demulsibility can be compromised by excessive water contamination or the presence of polar contaminants and impurities. The impact of demulsibility depends on the system residence time and anticipated levels of water contamination. Demulsibility testing can show failure in the laboratory, but with sufficient residence time, the turbine oil may shed water in the reservoir at an acceptable rate that does not impact

TABLE 13.5

Suggested Oil Analysis Program for Steam and Gas Turbines

Property	Turbine Type Steam	Gas	Frequency	Limit
Viscosity	✓	✓	Monthly	±5% of new oil value
Acid number	✓	✓	Monthly	0.3 to 0.4 mg KOH/g above new oil value
RPVOT	✓	✓	Quarterly	25% of new oil value; monthly tests if approaching 25%
Visual water	✓	✓	Daily	Check for haziness
Water content	✓	✓	Monthly	<0.1% in steam turbines <0.05% in gas turbines
Cleanliness	✓	✓	Monthly	Target 18/16/13 or better
Corrosion	✓	✓	Monthly	Pass ASTM D665A
Demulsibility	✓	✓	Monthly	<15 ml emulsion after 30 minutes
Varnish potential	–	✓	Quarterly	50 or more

Source: ASTM D4378 "Standard Practice for In-Service Monitoring of Mineral Turbine Oils for Steam and Gas Turbines".

turbine oil performance. Small reservoirs with lower residence times will require better demulsibility performance than larger reservoirs.

ASTM D4378 does not offer warning limits for demulsibility although some turbine OEMs identify levels of 3 millilitre (ml) emulsion after 30 minutes on new oils. In-service oil test results of 15 ml or greater of emulsion after 30 minutes should serve as a warning limit.

A suggested oil analysis programme for steam and industrial gas turbines, abstracted from ASTM D4378, is shown in Table 13.5. The frequency of testing varies from monthly to quarterly, although any abnormal test results shown by trend analysis may indicate the need for more frequent sampling and testing.

Turbine oil properties that do not require regular testing, but are used in new oil specifications, include viscosity index (VI), TOST, foaming, air release, FZG and flash point.

VI is an indication of the oil's change in viscosity with a change in temperature. Most gas and steam turbine OEMs require a turbine oil with a VI of at least 90, which is met by most turbine oil suppliers. The VI for turbine oils should not vary in service, because turbine oils typically do not contain VI improvers and therefore do not need to be tested routinely.

The TOST (ASTM D943, ISO 4263) test attempts to determine the expected turbine oil life by subjecting the test oil to oxidative stress using oxygen, high temperatures, water and metal catalysts, all of which increase sludge and acid formation. The test was developed to evaluate the expected performance of new turbine oils. Most turbine OEMs use TOST in their specifications to screen out high-risk turbine oils. Current gas turbine OEM specifications for TOST range from 2,000 to 4,000 hours with new gas turbine technology specifications at 7,000 hours. All TOST reporting above 10,000 hours is done through non-ASTM test modifications that may not correctly represent a turbine oil's performance. Because a TOST test can take up to a year or more to complete, it is impractical as an in-service oil test and is rarely performed for this reason.

An in-service test for a turbine oil sample for foam (ASTM D892, IP 146, ISO 6247, DIN 51566 or AFNOR T60-129) will often give a higher result than new oil levels suggested by a turbine OEM, but the oil does not typically present a field foaming issue because of the position of the suction line relative to the oil surface, where foam accumulates. If the foam level in the turbine reservoir is 6 inches or less and does not overflow the reservoir or cause level-monitoring issues, then turbine oil foaming is not usually a major cause for concern. A sudden increase in foaming may indicate a more serious problem. Oil at the reservoir surface should show at least one clear area (no bubbles) and larger breaking bubbles should be seen at this interface.

ASTM D4378 offers warning limits for Sequence I of the foam test of 450/10 ml for foaming tendency and stability. A foam stability of less than 5 ml is a good indication that foam bubbles are breaking and the turbine should not experience operating problems from foam. When tackling a foam problem, cleanliness, contamination or mechanical causes should be investigated before field addition of an anti-foam additive should be considered. Addition of too much anti-foam additive is likely to cause an even greater problem with increased air entrainment. Dirt is a leading cause of foam, so ISO cleanliness should be tested as a likely cause. Testing for foaming properties should be conducted only when foam presents an operating problem and for product compatibility testing.

Some steam and gas turbine OEMs specify air release limits (ASTM D3427) in their new oil specification requirements. These limits can be as low as 4 minutes, which is typically not a problem for most ISO VG 32 turbine oils, but can be an obstacle for ISO VG 46 oils, due to the higher viscosity. In turbines with small reservoirs and minimal residence time, entrained air mixtures could be sent to bearings and critical hydraulic control elements causing film strength failure problems, loss of system control, particularly in EHC systems, and an increased rate of oxidation. Air release of turbine oils should not vary with in-service time and therefore may not need to be tested for condition assessments routinely, unless a specific problem is suspected.

Turbines with geared shaft connections to the generator often require anti-wear or extreme pressure additives to support gear tooth loading. Industry standard testing for gear load performance is the FZG Gear Test. Typical rust and oxidation inhibited (R&O) ISO 32 turbine oils have an FZG failure load stage of 6 or 7. ISO VG 32 oils with anti-wear or extreme pressure additives, as well as rust and oxidation inhibitors, can have an FZG failure load stage of 10, which meets all major turbine OEM specifications. FZG gear tests on turbine oils should not vary with in-service time and do not need to be tested for condition routinely unless a specific wear related problem is encountered.

Flash point (ASTM D92, IP 36, ISO 2592, DIN 51356 or AFNOR T60-118) testing is done primarily to confirm product integrity from contamination. ASTM D4378 identifies a drop in 17°C from the new oil flash point as a warning limit. Flash point testing is required only if product contamination from a different oil or solvent is suspected.

13.6 AVIATION GAS TURBINES

Oils used in all modern aviation gas turbines are polyol ester-based, with high-performance oxidation inhibitors, corrosion inhibitors, load-carrying additives and deposit control additives. The specifications for these oils are very onerous and, consequently, their performance properties are exceptional. In the civil aviation market, there are only four suppliers of aviation gas turbine oils, worldwide. Many engines used in civil aircraft do not have major overhauls until 20,000 to 40,000 hours of operation have elapsed. As a consequence, any deposit formation in the engine, particularly in the plain bearings, oil injectors and oil filters, may cause lubrication problems. A diagram of a typical lubrication system for an aviation gas turbine is shown in Figure 13.11.

Aviation gas turbine oils are not routinely monitored, as there is almost always a loss of oil during flight and significant oil top-up between flights. Instead, the engines are very closely monitored during flight. Any issues that may be due to oil oxidation, thermal degradation, deposit formation or wear metal generation will quickly show up in engine operation and will be corrected shortly after the plane lands and before the next flight.

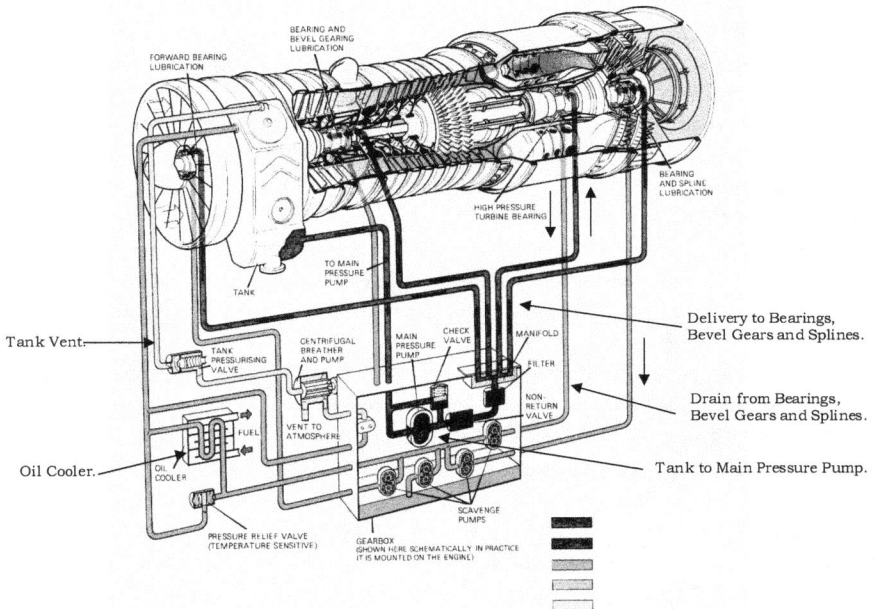

FIGURE 13.11 Rolls-Royce Olympus 593 gas turbine.

Source: Pathmaster Marketing Ltd.

Rolls-Royce uses Engine Health Management (EHM) to track the health of thousands of engines operating worldwide, using onboard sensors and live satellite feeds. A corporate EHM team covers all the business sectors, which enables Rolls-Royce to develop technologies and best practice. In the civil aviation market, for example, the Trent family of engines is supported by a comprehensive Rolls-Royce EHM capability and accessible as appropriate by the airlines involved. The in-flight monitoring system for a Rolls-Royce Trent 1000 gas turbine is shown in Figure 13.12.

EHM is a pro-active technique for predicting when something might go wrong and averting a potential threat before it has a chance to develop into a real problem. EHM covers the assessment of an engine's state of health in real time or post-flight and how the data is used reflects the nature of the relevant service contracts. The evolution of EHM and the subsequent revolution in its use has significantly reduced costs by preventing or delaying maintenance, as well as flagging potentially costly technical problems.

Broader engineering disciplines can benefit from the data that is collected. As operational profiles of technical performance are revealed in ever more detail, from individual components to whole engines, so engineers can develop more thorough and cost-effective maintenance schedules, and designers can feed higher reliability features into the engine products of the future.

EHM uses a range of sensors strategically positioned throughout the engine to record key technical parameters several times each flight. The EHM sensors in aero engines monitor numerous critical engine characteristics such as temperatures, pressures, speeds, flows and vibration levels to ensure they are within known tolerances and to highlight when they are not. In the most extreme cases, air crew could be

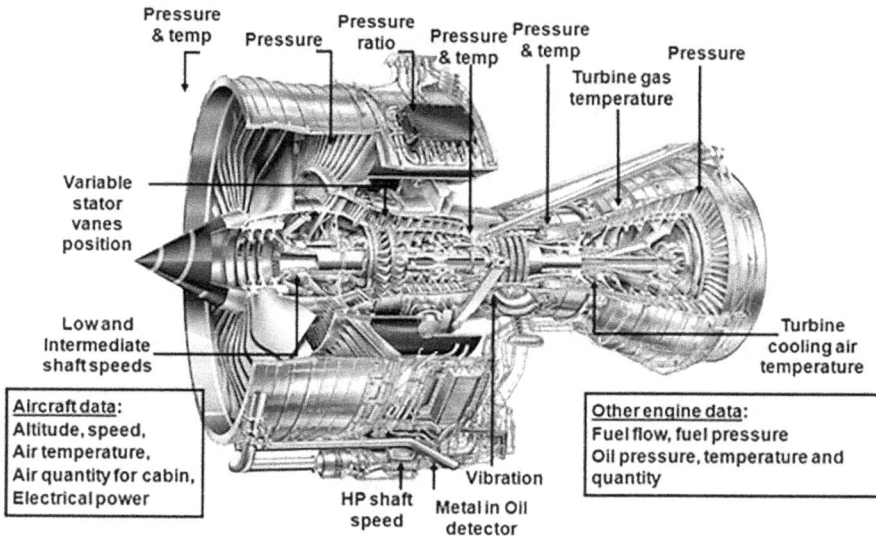

FIGURE 13.12   Rolls-Royce Trent 1000 gas turbine monitoring.

Source: Pathmaster Marketing Ltd.

contacted, but far more often the action will lie with the operator's own maintenance personnel or a Rolls-Royce service representative in the field to manage a special service inspection.

Rolls-Royce Trent engines can be fitted permanently with about 25 sensors. Figure 13.12 shows the locations of the sensors in a Trent 1000 gas turbine. Many of the sensors are multi-purpose as they are used to control the engine and provide indication of engine operation to the pilot as well as being used by the EHM system. These are selected to make the system as flexible as possible.

The main engine parameters, shaft speeds and turbine gas temperature (TGT) are used to give a clear view of the overall health of the engine. A number of pressure and temperature sensors are fitted through the gas path of the engine to enable the performance of each of the main modules (including the fan, the intermediate and high-pressure compressors and the high-, intermediate- and low-pressure turbines) to be calculated. These sensors are fitted between each module, except where the temperature is too high for reliable measurements to be made.

Vibration sensors provide valuable information on the condition of all the rotating components. An electric magnetic chip detector is fitted to trap any debris in the oil system that may be caused by unusual wear to bearings or gears. Other sensors are used to assess the health of the fuel system (pump, metering valve, filter); the oil system (pump and filter); the cooling air system; and the nacelle ventilation (nacelle is the cover housing, separate from the fuselage that holds engines, fuel or equipment on aircraft). As engine operation can vary significantly between flights (due to day temperature or pilot selection of reduced thrust), data from the aircraft to provide thrust setting, ambient conditions and bleed extraction status is also used.

As soon as the individual reports arrive at the specialist EHM analysts, they are processed automatically. The data is checked for validity and corrections applied to normalise them. The snapshot data is always "trended", so that subtle changes in condition from one flight to another can be detected. Automated algorithms based on neural networks are used to do this and multiple sensor information is fused to provide the most sensitive detection capability.

When abnormal behaviour is detected, this is confirmed by an analyst based in the Operations Centre, before being sent to the aircraft operator and logged by the Rolls-Royce Technical Help Desk. Manual oversight is still an important part of the process, as false alerts can cause unnecessary maintenance actions to be taken by airlines and these need to be avoided. Trended data, and data from the other types of ACMS report, are also uploaded onto the Rolls-Royce customer website, so that plane operators can easily view the health of their fleet of engines.

Engine rpm is not given in an absolute value but in a percentage of maximum and also for each spool in the engine. This is indicated in N1, N2 and so on. Most systems use an AC alternator where rpm is related to the frequency generated and not influenced by voltage drops due to long wiring. Thrust is measured by reading the inlet and exhaust pressure and indicating this as a ratio to the crew. Engine pressure ratio (EPR) is taken at a central inlet (P1) and at another engine station, for example 7 (P7).

Oil pressure and temperature are measured in each engine, with displays in the cockpit. A low-pressure warning light is usually included and should not be lit during engine operation. Exhaust gas temperature is an indirect indication of turbine inlet temperature (TIT, TET or ITT). For a long service life of the turbine blades and burner cans, these temperatures may not exceed manufacturer limits under all circumstances.

Fuel flow is measured in weight over time, indicating the fuel consumed by the engine. It will also include an indication for total fuel used, so that this can be compared with the amount of fuel taken on board. A sensor is installed in the low-pressure fuel line, and it consists of a small vane which is rotated by fuel flowing past it. Fuel pressure and temperature give an overall idea to the crew of the health of the fuel system and possible pending blockages due to water in fuel.

Multiple vibration detectors are installed at several locations in the engine to monitor vibration. An increase in any type of vibration deviating from normal could be an indication of a pending failure.

13.7 WATER (HYDROELECTRIC) TURBINES

Water turbines, used in hydroelectric power stations, range in size from less than 1 megawatt (MW) to over 750 MW. They operate at speeds as low as 40 rpm to as high as 2,200 rpm, but typically 100 to 200 rpm. There are several types of water turbines, generally classified as either impulse (Pelton type) or reaction (pressure; Francis, Deriaz, Kaplan and Bulb) type, as described in Section 13.1.

The main parts of these turbines that require lubrication are the turbine and generator bearings, the guide vane bearings, the control valve, the governor and control system and the compressors.

The bearings of hydroelectric sets are either self-lubricated or supplied with lubricant by a central circulation system. Circulation systems may be either unit systems (a separate system is used for each unit in a station) or station systems (all the units in a station are supplied from one system). In many cases, one or more of the bearings may be of the self-lubricating type, with a unit system supplying the other bearings.

In self-lubricated bearings, the oil is contained in a tank surrounding the shaft. Oil is lifted by grooves in the bearings or by a ring pump on the shaft. Cooling coils can be located in the tank, or with a cylindrical shell bearing, a cooling jacket may be located around the bearing shell. External cooling coils may also be used, but these are generally suitable only for relatively high-speed machines, which generate sufficient pumping force to circulate the oil through the external circuit.

Older hydroelectric units were equipped with mechanical hydraulic governor control systems with a mechanical speed-sensitive device and a hydraulic system to actuate the guide vanes, together with the runner blades if a Deriaz or Kaplan turbine was used. Newer machines are often equipped with electrical speed-sensitive devices and electronic systems.

Older turbine governor hydraulic systems generally operated at 150 psi and used the same oil as the bearing oil system. Current units operate with pressures around 1,000 psi, but can go as high as 2,000 psi. The hydraulic systems are now usually separate systems and require anti-wear hydraulic fluids for the higher-pressure systems. There is also a trend towards the use of environmentally acceptable fluids, particularly synthetic or vegetable ester-based types, for these applications. Hydraulic pumps are driven by electric motors. Air-charged accumulators (air over oil) are used to maintain system pressure and supply the large fluid flow necessary to adjust rapidly to meet sudden changes in turbine load. They also provide a source of fluid under pressure to shut the turbine down in the event of a failure in the system. Emergency shutdown may also be assisted by the use of closing weights on the guide vane operating mechanism or by designing the vanes that will be closed by water pressure if the hydraulic system fails.

The guide vanes, or wicket gates, are manufactured with an integral stem at each end that serves as the bearing journal. One bearing is used at the bottom and one or two bearings at the top. A thrust bearing may also be required. These bearings, as well as the bearings of the operating mechanism, are grease-lubricated. Centralised application systems are now usually used to supply these bearings.

In some hydroelectric turbines, the guide vanes are arranged to close tightly and act as the shutoff valve for the turbine. In most installations, however, separate closing devices on the water inlet are used. In the case of pump turbines, closing devices on both the inlet and outlet are used. Closing devices include sluice valves, rotary valves, butterfly valves and spherical valves. All are designed for hydraulic operation. Closing weights may be used for emergency shutdown. Bearings are grease-lubricated.

In most hydroelectric power stations, compressed air is required to maintain the pressure in the hydraulic accumulators. Compressed air is also used to blow out the draft tube and turbine casing when maintenance is to be performed. Compressed air also blows out the pump or turbine when the changeover from pump to turbine operation, or vice versa, is made in pump/turbine installations. Compressed air is also used in some impulse turbine installations to keep the tail water out of the turbine when the tail water level is high. Compressors operated in hydroelectric plants are critical pieces of equipment. Air compressors can be four-stage units and can operate with discharge pressures up to 1,000 psi.

The need for extreme reliability and long service life of hydroelectric plants generally dictates that premium, long-life lubricants are used. Rust- and oxidation-inhibited premium oils are usually used for oil applications. Viscosities usually are of ISO viscosity grade 32, 46, 68, or 100, depending on bearing design, speeds and operating temperatures. Oils with excellent water-separating characteristics are desirable. While start-up temperatures are rarely below freezing, the oils used must have adequate fluidity for proper circulation at those temperatures. Where oil lifts are not used for starting, oils with enhanced film strength may be desirable to provide additional protection during starting and stopping.

Greases used in grease-lubricated bearings require good water resistance and corrosion protection properties. They should be suitable for use in centralised lubrication systems and should have good pumpability at the lowest water temperatures, which could be as low as 1°C. Calcium and lithium complex greases are used, either NLGI 1 or 2 grades. Air compressors used in hydroelectric power stations are lubricated as described earlier in this chapter.

Monitoring hydroelectric turbines and lubricants involves the same parameters as for bearings, hydraulic systems and compressors, except for one significant property. Because hydroelectric turbines operate at relatively low temperatures, usually closer to ambient temperature, oxidation is much less of a problem compared with other lubricant applications.

13.8 SUMMARY

Because compressors and turbines are vital pieces of equipment in most industrial, civil and military applications, effective condition monitoring programmes for machines and lubricants are also vitally important.

A large number of types of compressors have been designed and used for a wide variety of applications and to compress many different types of gases. In addition to being used to compress gas, many compressors serve as blowers or can be used as vacuum pumps. Compressors are classified as either positive displacement or dynamic. As a result, their lubrication requirements can be very different.

Lubricants for gas compressors are generally grouped into those for air, those for inert gases, those for reactive gases and those for hydrocarbon gases. Lubricants for refrigeration and air conditioning systems are different again, as are condition monitoring programmes.

There are fewer types of turbines and their lubrication requirements are more similar, although there are some significant differences. Steam and gas turbines used for generating electricity require lubricants with different properties to gas turbines used in civil and military aircraft. Water turbines for generating electricity have slightly different requirements, again with slightly different approaches to condition monitoring.

Many of the measurements and tests used in condition monitoring programmes are similar for compressors and turbines, but some are very specific, depending on the applications and operating conditions.

14 Condition Monitoring of Metalworking and Production Engineering Fluids and Pastes

14.1 INTRODUCTION

Production engineering fluids are usually grouped into three main classes: metalworking fluids and pastes, heat treatment fluids and temporary corrosion protectives.

Metalworking fluids and pastes are engineering materials that optimise metalworking processes. Metalworking encompasses metal removal and metal deformation. Metalworking fluids used for metal removal are known as cutting and grinding fluids, while fluids and pastes used for drawing, rolling, bending or stamping processes are known as metal forming fluids or pastes. The outcome of the two types of processes differs. The processes by which the machines make the products, the mechanics of the operations and the requirements for the fluids used in each process are different.

Heat treatment, also known as quenching, is one or more operations involving the controlled heating and cooling of a metal in the solid state for the purpose of obtaining specific properties. Many types of heat-treating processes can be used to fulfill a wide variety of hardness and mechanical properties that may be required for metal components.

Temporary corrosion protectives provide corrosion protection for relatively short durations, during the storage or transportation of manufactured metal components or assemblies. "Temporary" refers to their ease of removal, not to the duration of the protection. They provide a water and oxygen-resistant barrier by reason of their blanketing effect and natural or added corrosion inhibitors, which form an adsorbed layer on the metal surface.

14.2 PRODUCTION ENGINEERING PROCESSES

14.2.1 METALWORKING AND METAL FORMING

The properties and performance required of metalworking and metal forming fluids and pastes are governed by the mechanics of the processes involved. The fluid or paste must provide a layer of lubricant to act as a cushion between the workpiece and the tool in order to reduce friction. Fluids may also need to function as a coolant to reduce the heat produced during machining or forming. Otherwise, distortion of the

DOI: 10.1201/9781003245254-14

workpiece and changed dimensions could result. In addition, the fluid or paste must prevent metal pick-up on both the tool and the workpiece by flushing away the chips as they are produced or distortion that could arise during the process. All of these attributes function to prevent wear on the tools and to reduce energy demands. Just as importantly, the metalworking fluid or paste is expected to produce the desired finish on an accurate component part. Production engineering must focus on producing components at the highest rate, with maximum tool life, minimum downtime and the fewest possible part rejects (scrap), while maintaining component accuracy and finish.

A very large range of metalworking and metal forming processes are used to make component parts:

- Grinding.
- Drilling, tapping and boring.
- Turning and facing.
- Milling.
- Broaching and reaming.
- Hobbing.
- Rolling.
- Forming and forging.
- Pressing, drawing and wall ironing.
- Stamping and blanking.
- Electrical discharge machining (also called electro-discharge machining).

Grinding is high-speed, multi-point metal cutting. Metal is removed by an abrasive grinding wheel rotating in contact with the workpiece. Grinding has two basic formats. With surface grinding, the workpiece is secured to a flat table that rotates or reciprocates horizontally beneath a vertical or horizontal rotating grinding wheel. With cylindrical grinding, the workpiece rotates between centres mounted on a reciprocating bed and the grinding wheel rotates on a fixed head and the workpiece moves back and forth across the face of the wheel.

Centreless grinding involves two grinding wheels running on almost parallel axes. The workpiece rests on a narrow rail between the grinding wheels and the gap between the wheels determines the finished diameter of the workpiece. One of the wheels performs the grinding, while the other is a control wheel. Plain shafts feed themselves continuously through the gap, due to the screwing action of the slightly tilted axes of the grinding wheels. The workpiece diameter is progressively reduced in size until it reaches the designed size at exit.

Honing and lapping are variations of grinding. Honing is performed by shaped abrasive stones pressed lightly against pre-machined surfaces. It is used to give a precisely controlled surface finish or a fine dimensional accuracy. Lapping is used for the production of flat surfaces on light or easily distorted surfaces. The component is placed flat on a rotating flat abrasive surface and is randomly moved over the surface. Lapping is also used to rub two components together with an abrasive paste between them, to produce accurately matching faces and fine clearances.

Drilling uses a spiral fluted tool with two symmetrical cutting edges, in the form of tapered blades (the drill point), to remove metal in a circular motion. The resulting chips are transported out of the resulting hole by way of the spiral flutes. Drill geometry is complex. The cutting speed is a maximum at the drill periphery and is zero at the centre of the drill point. The angle of the drill point is key to successful drilling. Angles between 118° and 135° are required for steels, 90° to 140° for aluminium and soft alloys and 80° for plastics.

Boring and deep-hole drilling are variants of conventional drilling and cutting. Boring uses a cutting tool in place of a drill and is designed for wide and deep holes. A hole is considered to be "deep" if its depth is more than ten times its diameter, so standard twist drills cannot be used to drill such holes. Deep-hole drilling uses a cutter at the end of a hollow tube, through which cutting fluid is fed. The tube is centralised in the hole by guide pads. Gun drilling uses a tool configuration that cuts to one side. The cutting point is not symmetrical, and the drill uses the wall of the hole as its support for cutting and maintaining straightness.

Tapping is a variant of drilling, done at a slower speed. A thread is cut or formed into the walls of a previously drilled hole. Careful attention must be paid to the axial feed so that it follows precisely the lead of the thread. Chip clearance is provided by the flutes of the tap, but chips are broken by sporadic reversal and withdrawal of the tap to clear the flutes. Cut taps make threads by cutting the metal, thus generating chips, while form taps make threads by pushing the metal aside, making no chips.

With turning, a stationary tool is moved axially along a rotating workpiece. Straight, tapered or complex-shaped shafts can be produced using combinations of tool shape, geometry and tool path in two dimensions. Chip formation is continuous, as the tool is always in contact with the workpiece. Long, stringy chips can be difficult to handle, so chip breakers are placed behind the point of the cut. Chips can also be broken using a high-pressure jet of cutting fluid onto the back of the chips. Turning may be performed on the outside of a shaft, as well as on the inside of a tubular shaft.

Milling uses a cutter with multiple teeth which is rotated and moved across the workpiece. Using the face of the cutter produces a flat surface. Using the periphery of the cutter produces a form or slot. Milling involves intermittent cutting, so is fundamentally different from drilling or turning. Milling cutters are susceptible to thermal fatigue, resulting in chipping and poor tool life. When milling slots or forms, up-cutting requires very sturdy fixturing to avoid lifting the workpiece off the table, while down-cutting can result in cutter snatching, causing poor surface finish and short tool life.

Broaching is analogous to milling, but using a flat cutter. Each tooth of the broach takes the same depth of cut, but the broach tool is designed to have a set "rise" per tooth and so removes all the required metal in one pass. For example, if the rise per tooth is 0.05 mm and the required depth of cut is 5 mm, the broach tool will have 100 teeth and may be 1 metre long. Broach tool design is complex. Tooth geometry and gullet spacing for chip clearance depend on the metal to be cut and whether the cutting is on an external or internal surface. Broaching is a high-force, slow-speed cutting process, so chips can get jammed in the broach during internal broaching.

Gear shaping generates gear teeth using a cutter with a similar form to the shape of the gears being cut. A gear shaping machine has a two-phase cutting action, the cutting stroke and the relief stroke. The cutter and the blank rotate in phase during the shaping motion, as if they were gears in mesh. Shaped gear teeth can either remain as cut or be further machined by planning or grinding.

Gear hobbing generates gear teeth using a cutting tool (hob) with an axis of rotation nearly 90° to the helix angle of the gear teeth to be formed on the gear blank. Longitudinal flukes in the hob cut across its thread and form the cutting edges. The hob is fed parallel to the axis of the gear and along the path of the gear teeth. Both the gear blank and the hob rotate at steady speeds, and the form of the cutter is designed to take into account its feed from one end of the blank to the other. Gear hobbing is a continuous process with no indexing and no return stroke and is used for high-speed production of finished gears.

With rolling, a metal ingot, rectangular or circular in cross section, is rolled into a thinner sheet or rod, between two rolls. One roll, usually the top one, is loaded against the other roll, usually by very high hydraulic pressure. Each pair of rolls is called a "mill stand". Successive mill stands are used to form thinner and thinner sheets. A "rolling mill" (production plant) may have up to ten mill stands. Any sizing error in one mill stand could over-stress and damage the following mill stand. The accuracy of a rolling mill depends on the strength of its component parts. Rolling of metals can be done "hot" (~900°C) or "cold" (~400°C).

Forging is the process of shaping metal and increasing its strength by hammering or pressing. Most usually, an upper die is forced against a heated workpiece positioned on a lower, stationary die. If the upper die or hammer is dropped onto the workpiece, the process is known as drop forging. To increase the force of the blow, power is sometimes applied to augment gravity. The number of blows is carefully gauged to provide maximum effect, with minimum wear on the die.

Near net-shape forging uses powdered or sintered metal placed in the lower die. The upper die is then forced into the lower die under very high pressure, to force the powdered metal to bond into a shape that is as close as possible to the required finished component. This is different from conventional forging, which is done using heated solid metal workpieces. The near net-shaped component may then need some finishing processes, such as grinding, drilling or tapping.

A newer process, called "additive manufacturing", is being used increasingly to manufacture metallic components. Additive manufacturing, also known as 3D printing, is a transformative approach to industrial production that enables the creation of lighter, stronger parts and systems. It is another technical advance made possible by the transition from analogue to digital processes. Additive manufacturing uses computer-aided design (CAD) software or 3D object scanners to direct hardware to deposit material, superfine layer upon superfine layer, in precise geometric shapes. As its name implies, additive manufacturing adds material to create an object. Each successive layer bonds to the preceding layer of melted or partially melted material. Objects are digitally defined by the CAD software that essentially "slice" the object into ultra-thin layers. This information guides the path of a nozzle or print head as it precisely deposits material on the preceding layer.

In the 1980s, 3D printing techniques were considered suitable only for the production of functional or aesthetic prototypes. 3D printing focused initially on polymers, due to the ease of manufacturing, handling and bonding polymeric materials. Currently, the precision, repeatability and material range of 3D printing have improved such that 3D printing processes are being used to manufacture components with very complex shapes or geometries that would be otherwise impossible to construct by hand, including hollow parts or parts with internal truss structures to reduce weight. Metallic and ceramic components can now be made using additive manufacturing. A laser or electron beam selectively melts or partially melts in a bed of powdered material, and as the material cools or is cured, it fuses together to form a three-dimensional object.

However, because the process can be fraught with complications, due to isolated and monolithic algorithms, additive manufacturing is only being used at present to manufacture components that are very difficult to make using other production engineering methods. It is obvious that additive manufacturing does not use any metalworking or metal forming lubricants.

Pressing or drawing involves a punch, a die and a flat blank. The punch forces the blank into the die, reshaping it into a three-dimensional shape. Pressing involves much less force than drawing. Drawing has a number of degrees of severity. The draw ratio is the depth of draw divided by the diameter of the blank. With a shallow draw, the draw ratio is less than 1.5, a moderate draw has a draw ratio between 1.5 and 2.0, while a deep draw has a draw ratio greater than 2.0.

Wall ironing involves thinning the side walls of a previously deep-drawn cup. The die is specially shaped so that the clearance at the bottom of the die is less than the thickness of the metal being drawn. For example, manufacturing aluminium beverage cans involves a three- or four-stage process in which a shallow drawn cup is then deep drawn, followed by wall ironing.

Wire is made by pulling metal bar stock through a series of reduction dies, until the correct shape and size are reached. Most wires are cylindrical in shape, but they can be drawn to oval, flat or rectangular shapes. Wires can be made of steel, aluminium, copper or any ductile metal. It is exceedingly easy to break wires while drawing them because the forces required to draw wires are very large.

With stamping and blanking, a metal blank is pressed out from a sheet or strip. The remaining scrap is in the form of a skeleton, which can be reused by re-melting into a solid bar which can be rolled into another sheet or strip. Blanks can be round, square or any shape required for subsequent operations. The metal sheet or strip is clamped tightly and a punch, whose edges function like a moving knife edge, cuts the metal much like a cookie-cutter through dough. The stationary die, below the punch, also serves as a cutting edge. With rough blanking, only 50% to 70% of the depth of the sheet is cut, and the rest is torn off. In fine blanking, the punch cuts all the way through the sheet or strip.

Variations of stamping include piercing (punching a hole in the metal sheet, leaving a round slug of metal as the scrap), notching (forming a slit rather than a hole), coining (making the blanks that will later be stamped (pressed) with the "head" and "tail" surfaces of coins), shearing (cutting strip sheet into rectangular or square blanks) and trimming (removing excess metal from a formed workpiece to give the final shape).

Electrical discharge machining (EDM), also called electro-discharge machining, spark machining, spark erosion, die sinking, wire burning or wire erosion, is a metal fabrication process in which a desired shape is obtained by using electrical discharges (sparks). It is used to shape materials which are difficult to machine, have high-strength temperature resistance and are electrically conductive. Material is removed from the workpiece (the anode) by a series of rapidly recurring current discharges between it and a shaped die or wire (the cathode), separated by a dielectric fluid. The die (also called the tool) is generally graphite or copper, and the dielectric fluid is generally kerosine or deionised water. Copper wire is used for wire erosion. The wire is moved in three dimensions, under computer control, to shape the workpiece as required.

The voltage is applied across the tool and the workpiece in pulse form, while a small gap is maintained constantly between them. The dielectric fluid is circulated through the gap. When the voltage between the two electrodes increases, the intensity of the electric field in the gap between the electrodes becomes greater, causing dielectric breakdown of the fluid and producing an electric arc, which results in localised temperatures of around 10,000°C. This melts material from the workpiece. Once the current stops, dielectric fluid is conveyed into the inter-electrode space, enabling the solid particles (debris) to be carried away and the insulating properties of the dielectric to be restored. After a current pulse, the voltage between the electrodes is restored to what it was before the breakdown, so that a new liquid dielectric breakdown can occur to repeat the cycle. The process is continued until the required shape of the workpiece is achieved.

14.2.2 HEAT TREATMENT (QUENCHING)

Heat treatment, also known as quenching, is one or more operations involving the controlled heating and cooling of a metal in the solid state for the purpose of obtaining specific properties. Many types of heat-treating processes can be used to fulfill a wide variety of hardness and mechanical properties that may be required for metal components.

For example, heat treatment of steel involves the rapid cooling of the austenite form to transform it into the hard form, martensite. This is generally achieved by cooling at a sufficiently fast rate to avoid the formation of soft constituents in the steel, such as pearlite and bainite. For a given steel composition and heat treatment condition, there is a critical cooling rate for full hardening at which all the high-temperature austenite is transformed into martensite without the formation of either pearlite or bainite. The steel begins to harden at a specific temperature, called "Ms", which depends on its carbon content and is fully hardened at a lower temperature, called "Mf". For steels with a carbon content of 0.2%, Ms is 430°C, for 0.4% carbon Ms is 360°C and for 1.0% carbon Ms is only 250°C.

In practice, however, when a steel component is quenched, the surface cools more rapidly than the centre. This means that the surface could cool at the critical cooling rate and hence be fully hardened, whereas the centre cools more slowly and forms soft pearlitic or bainitic structures. Lack of through hardening can be overcome by one of two methods. Increasing the hardenability of the component by using a steel with higher alloy content usually causes a delay in transformation. This reduces the critical cooling rate for martensitic transformation producing fully hardened components with maximum hardness and mechanical properties. However, alloying elements can be expensive and may not be beneficial to other metalworking processes, such as machining or forging.

Increasing the quenching speed means that the cooling rate at the centre of the component is higher than the critical cooling rate. This can be achieved by changing from a normal speed heat treatment oil to a high-speed accelerated quenching oil or, if using an aqueous polymer quenching fluid, by reducing the concentration of the solution. As a result, the steel composition, component section thickness and type of heat treatment fluid all have significant impact of the properties obtained by heat treatment.

Whichever type of quenchant is used (see Section 14.4), cooling generally occurs in three distinct stages, each of which has very different characteristics. The first stage of cooling is characterised by the formation of a vapour film around the component. This is a period of relatively slow cooling during which heat transfer occurs by radiation and conduction through the vapour blanket. The stable vapour film eventually collapses and cool quenchant comes into contact with the hot metal surface resulting in nucleate boiling and high heat extraction rates. In the third stage, boiling ceases and heat is removed by convection into the liquid. Heat is removed very slowly during this stage.

The duration of the vapour phase and the temperature at which the maximum cooling rate occurs have a critical influence on the ability of the steel to harden fully. The rate of cooling in the convection phase is also important since it is generally within this temperature range that martensitic transformation occurs, and it can, therefore, influence residual stress, distortion and cracking.

However, cooling curves produced under laboratory conditions must be interpreted carefully and should not be considered in isolation. Results on used quenchants should be compared with reference curves for the same fluid. Quenching characteristics are also influenced significantly by the degree of agitation for a normal-speed quenching oil under varying degrees of propeller agitation.

Again, in practice, the performance obtained from a quenchant depends upon actual conditions in the quench tank. For this reason, it is highly desirable to measure quenching characteristics on site. Specialised, fully portable and self-contained equipment has been developed for this purpose by IVF (The Swedish Institute of Production Engineering Research).

14.2.3 TEMPORARY CORROSION PROTECTION

Temporary corrosion protectives provide corrosion protection for relatively short durations, during the storage or transportation of manufactured metal components or assemblies. "Temporary" refers to their ease of removal, not to the duration of the protection. They provide a water and oxygen-resistant barrier by reason of their blanketing effect and natural or added corrosion inhibitors, which form an adsorbed layer on the metal surface.

Most of the metallic components made using production engineering processes are likely to have bare metal surfaces which may require protection until they are either assembled into a machine or system or until they are further processed. The protection may be required during shipping or storage. The protection is unlikely to be required in use, because parts in the machine or system should be protected from corrosion either by lubricants or by more permanent coatings, such as paint or lacquers. Typical examples are automotive parts, fasteners, steel strip, steel sections, aluminium alloys or yellow metals, which require protection for a few days, weeks or months, during a delay in the manufacturing and supply schedule.

In some cases, the parts may need to be protected after machining, particularly if this has involved the use of a water-mix metalworking fluid (see also the following section). The parts may need to be dewatered or dried as part of the overall protection. Each of the metals to be protected may require a specific approach to the formulation of the protective. The metal may be part of an assembly together with other metals, plastics or rubber seals, and contact compatibility with other parts of the assembly will need to be considered.

14.3 TYPES OF METALWORKING FLUIDS

There are three distinct types of metalworking fluids:

- Soluble-oil emulsions: Milky soluble-oil emulsions, clear soluble-oil emulsions, micro emulsions and extreme pressure (EP) soluble-oil emulsions. All are emulsions of oils and additives in water.
- Aqueous (water) solutions: Grinding fluids and synthetic cutting fluids.
- Neat oils: Mineral oils, fatty oils, blends of mineral and fatty oils, sulphur-based EP oils, chlorine-based EP oils and sulphur-chlorinated oils.

Their primary functions are to provide cooling and lubrication of workpieces and metalworking tools. Their secondary functions are to remove chips from the tool/workpiece interface(s), to minimise "built-up-edge" (damage to tool surfaces) and to protect workpieces and tools from corrosion.

At the same time, high-performance lubrication and/or cooling will enable increased machining speeds, reduced power (energy) consumption and optimum accuracy and surface finish of machined components.

The numerous metalworking processes described earlier require very different metalworking fluids. As the degree of lubrication increases, so does the need to use oil-based fluids. Conversely, when more cooling is required, water-based solutions or emulsions need to be used. The gradations are illustrated in Figure 14.1. High operational severity requires more lubrication, less cooling and higher metalworking fluid activity. High cutting speed generates more friction and heat, so requires more cooling and lower cutting fluid activity.

Water-mix metalworking fluids provide several advantages, but have some disadvantages. Because they are sold as concentrates and then diluted in water, they almost always have low in-service costs and the concentrations are quite flexible. They obviously provide good cooling, but also good chip removal, and there is no fire risk. However, they tend to have high operating costs, as their monitoring and maintenance costs are greater than for oil-based fluids. Water-mix metalworking fluids,

	Operation severity	Cutting speed	Cutting fluid activity
	V High	Low	V High
Broaching			
Tapping			
Threading			
Gear shaping	↑		↑
Reaming			
Drilling		↓	
Milling			
Turning			
Grinding			
	Low	V High	Low

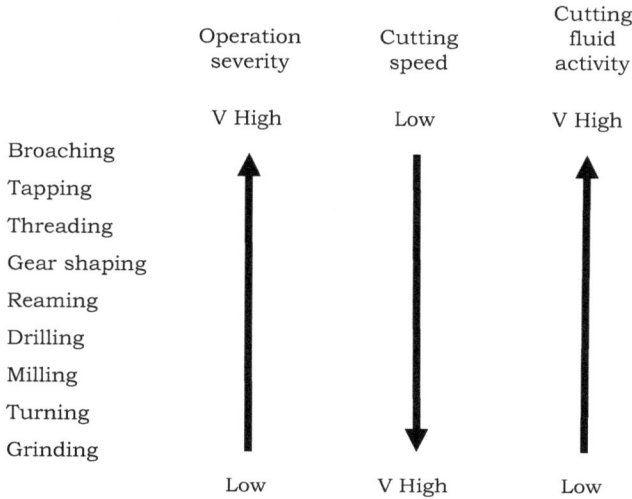

FIGURE 14.1 Machining operation versus metalworking fluid activity.

Source: Pathmaster Marketing Ltd.

because of the combination of water and organic compounds (oils and additives), are prone to suffer from microbial (bacterial and/or fungal) degradation. In addition, their compatibility with other fluids used in metalworking machines, such as hydraulic oils, gear oils and slideway lubricants, can be problematic.

Neat metalworking fluids also have advantages and disadvantages. Although they have higher initial purchase costs, their monitoring and maintenance costs are lower than water-mix metalworking fluids. They enable higher rates of metal removal, improved tool life, fewer corrosion problems and compatibility with gear oils, hydraulic oils and slideway oils. However, their cooling properties are much lower than water (even with low-viscosity neat oils), they have higher drag-out rates, and there is some risk of fire if friction forces are too high. Neat oils are also easy to regenerate and reuse, lowering their environmental impact, and there are no requirements for biocides to control microbial growth, lessening their health and safety risks.

The relationships between cooling and lubrication for the different types of metalworking fluids are shown in Figure 14.2. Although the base fluids are either water or oil(s), additives are used to improve lubrication and control other performance properties. Unfortunately, none of the additives used improve cooling performance very much.

Soluble-oil emulsion metalworking fluid concentrates contain mineral oil(s), emulsifiers, coupling agents, machining performance additives, corrosion inhibitors, metal passivators, anti-foam additives and (sometimes) biocides. Many of the newer emulsion-type products do not contain biocides, but the other additives are selected to be less susceptible to microbial degradation.

Aqueous solution metalworking fluid concentrates contain many of the same classes of additives, although no mineral oil(s). All of the additives though must be water-soluble or water-miscible and there are higher amounts of corrosion inhibitors.

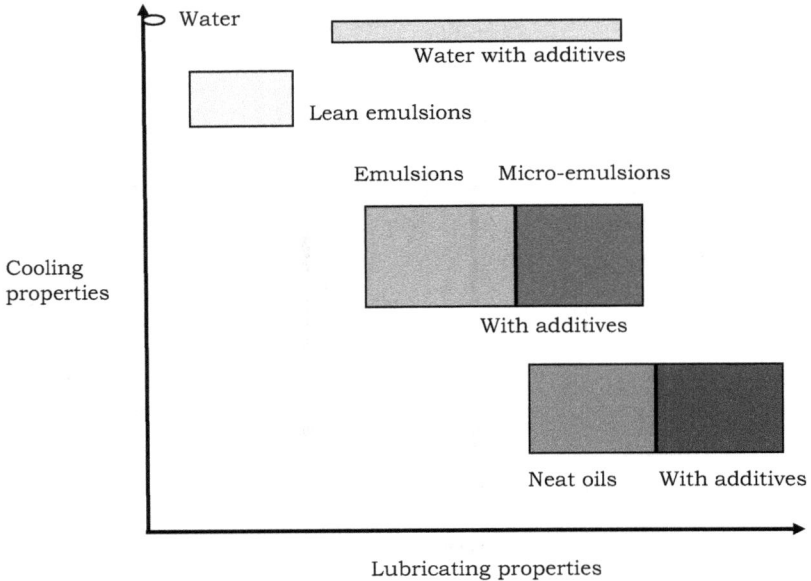

FIGURE 14.2   Cooling versus lubrication.

Source: Pathmaster Marketing Ltd.

Neat metalworking fluids contain mineral oil(s) together with some or all of the following additives: lubricity additives, load-carrying additives, extreme-pressure (EP) additives, dispersants, viscosity index (VI) improvers, anti-mist additives, removability additives (for water or detergent wash-off) and antioxidants.

The selection of which type of metalworking fluid to use depends on:

- The material to be machined.
- The process engineering operation.
- The tool type and material.
- The water type (for water-mix fluids).
- The level of filtration.
- The machining equipment.

The characteristics of different types of metals to be machined are summarised in Figure 14.3. An illustration of the different types of metalworking fluid required for different types of tool material is shown in Figure 14.4.

Particular attention needs to be paid to the quality of the water used to dilute water-mix metalworking fluids in preparation for use. Factors that affect the properties and machining performance of water-mix metalworking fluids include:

- Total hardness.
- Temporary hardness.
- Chloride ion content.
- Sulphate ion content.

- Nitrate ion content.
- Phosphate ion content.
- Bacterial content.
- Fungal content.

The impacts of water quality on water-mix metalworking fluids are discussed in depth in Section 14.8.1.

14.4   TYPES OF METAL FORMING FLUIDS AND PASTES

Each type of metal forming process has its own type of metal forming oil or paste. Rolling, drawing, forging, stamping and casting lubricants are all different and require different properties.

The factors that affecting the lubricating regime in metal forming include:

- Contact macro-geometry: This varies with the type of process and the die design.
- Contact micro-geometry: The micro-topography or surface finish of the tool, die and workpiece.

Type	Characteristics	Metals
Normal	Surface has a natural affinity to retain lubricant films	Carbon and low-alloy steels
Active	Surface has high attractive energy and reacts readily with chemicals	Brass, copper, tinplate, magnesium alloys
Inactive	Surface has low attractive energy and requires special wetting agents	Stainless and high-alloy steels, aluminium alloys

FIGURE 14.3   Types of metal surfaces.

Source: Pathmaster Marketing Ltd.

Tool Material	Metalworking Fluid
High speed steel	Neat oils Water-miscible fluids
High carbon steel	Water-miscible fluids
Non-ferrous cast alloys	Neat oils
Cemented or sintered carbides	Water-miscible fluids Low viscosity neat oils
Cubic boron nitride	Water-miscible fluids
Coated ceramics	Water-miscible fluids Dry machining
Polycrystalline diamond	Water-miscible fluids Dry machining

FIGURE 14.4   Metalworking fluids and tool materials.

Source: Pathmaster Marketing Ltd.

- Load: The contact force, from the tool to the workpiece.
- Speed: The surface speed of the tool relative to the workpiece and of the workpiece relative to the die.
- Environment: The atmosphere and moisture.
- Lubricant properties: Viscosity, load carrying, friction modifying and cooling.

For example, the functions of a pressing lubricant are to control friction between the workpiece and dies/tools, to control the rate of wear of dies and tools, to prevent metal pickup, welding or spot seizures, to control metal heating and to minimise corrosion of workpiece, dies, tools and machines. By comparison, the functions of a drawing lubricant are to cool the die and work material, to lubricate the die and work material, to prevent the pickup of work metal onto the die, by avoiding adhesion or welding, to cushion dies when drawing is being initiated and to minimise corrosion of work material, dies and machines.

Metal forming lubricants, unlike metalworking fluids, are required to prevent metal-to-metal contact. To do this, the fluid or paste must provide either a physical or chemical barrier, to prevent the contact of punch or die with the metal workpiece. The separation is necessary to protect the tooling, the cost of which in most forming operations is quite significant in comparison with all other components in the process. A lubricant that improves tool life can pay for itself through reduced tooling changes.

A metal forming fluid or paste must provide sufficient lubrication to make the component. Good lubrication is usually defined as a reduction in friction at the interface between the die and the workpiece. Lower friction results in lower energy and less drag on the metal, which yields a more uniform flow of metal through the forming dies. Lower friction also benefits the operation by reducing the amount of wear, which in turn reduces the amount of metal fines and debris. Fewer metal fines means that metal pickup or deposition on the punch is less likely.

As with other production engineering lubricants, an important function of metal forming oils and pastes is to prevent corrosion. Some products are required to be temporary corrosion protectives following production of the parts, while they are stored before the next stage of the process. In addition to controlling corrosion of the finished components, metal forming lubricants must prevent corrosion of punches, dies and machines.

Metal forming fluids and pastes must be easy to apply. Application methods include drip applicators, roll coaters, electrodeposition, airless spray and mops or sponges. Selection of a method should be based on optimised lubricant delivery for performance and on fluid conservation. Over-application of a lubricant will not increase lubrication performance, but may even make it worse and may make the workplace so messy that operators will dislike the fluid or paste.

Following forming, the lubricant must be easy to remove and should be compatible with the selected cleaning equipment, whether it be vapour degreasing or alkaline wash.

Operator health and safety issues are very important. Operators using reasonable safety practices must be able to work with the fluids or pastes, without the risk of unpleasant odours, dermatitis, respiratory distress or other health problems.

Metal forming fluids and pastes are required to function either as boundary lubricants or hydrodynamic lubricants. In many cases, this is achived by using solid lubricants, such as graphite, talc, silica, polytetrafluoroethylene (PTFE) or several others. However, the solids will sometimes attach onto the surfaces of the tooling or the component. Removal of solid films from finished parts can be a major problem.

Metal forming fluids or pastes contain mineral oils, polar additives (fats, soaps or esters), EP additives, emulsifiers, corrosion inhibitors, metal passivators, oxidation inhibitors and solid additives. The types of metal forming lubricants used in the different production engineering processes are summarised in Figures 14.5 and 14.6.

Metal	Pressing	Blanking	Stamping	Punching
Low-alloy steels	Fatty oils and pastes	Fatty oils and emulsions	Fatty oils and emulsions	Fatty oils
Stainless and high-alloy steels	High viscosity chlorinated oils	Sulphurised oils	Chlorinated oils	Chlorinated oils
Aluminium	Fatty oils and emulsions	Fatty oils	Fatty oils and emulsions	Fatty emulsions
Copper and brass	Fatty oils and pastes	Fatty oils	Fatty oils	Fatty pastes

FIGURE 14.5   Lubricants used in pressing, blanking, stamping and punching.

Source: Pathmaster Marketing Ltd.

Metal	Stretch Forming	Deep Drawing	Bar and tube Drawing	Rod and wire Drawing
Low-alloy steels	Fatty pastes + EP additives	Solid filled fatty EP pastes	Heavy oils, fatty pastes	Fatty oils and emulsions
Stainless and high-alloy steels	Sulphurised oils	High viscosity sulphur-chlorinated oils	High viscosity chlorinated oils and pastes	Soaps, EP pastes
Aluminium	Fatty emulsions + EP additives	Concentrated EP emulsions	High viscosity fatty oils	Fatty oils
Copper and brass	Fatty pastes	Solid filled fatty pastes	Fatty oils and emulsions	Fatty emulsions

FIGURE 14.6   Lubricants used in forming and drawing.

Source: Pathmaster Marketing Ltd.

14.5   TYPES OF HEAT TREATMENT FLUIDS

14.5.1   Quenching Oils

Many heat treatment processes use mineral oil-based quenchants. Before mineral oils became common, vegetable, fish and animal oils, and in particular sperm whale oil, were used for quenching operations. Following the introduction of mineral oils in the 1880s, many advances have been made in the development of quenching oils to provide highly specialised products for specific applications. A wide range of quenching characteristics can be obtained through careful formulation and blending. Selected wetting agents and accelerators added to paraffinic (API Group I) base oils achieve specific quenching characteristics. High-temperature oxidation inhibitors may be included to maintain performance for long periods of continued use. Emulsifiers may be added to enable simple wash-off in water after quenching.

Quenching speed is important because it influences the hardness and depth of hardening that can be obtained. Normal-speed quenching oils have relatively low rates of cooling and are used in applications where the hardenability of the material is high enough to enable the required mechanical properties to be obtained even with slow cooling. Highly alloyed materials and tool steels are typical examples of where a normal-speed quenching would be used. Medium-speed quenching oils provide intermediate quenching characteristics and are used widely for medium- to high-hardenability applications where dependable, consistent metallurgical properties are required. High-speed quenching oils are used for applications such as hardenability alloys, carburised and carbo-nitrided components or large cross sections of medium-hardenability steels where very high rates of cooling are required to ensure maximum mechanical properties.

The temperature of operation of a quenching oil is important because it will have an influence on the oil's lifetime, the quenching speed, the oil's viscosity and hence its rate of "drag-out" with the quenched component and the distortion of components. Cold quenching oils are designed for general purpose use at temperatures of up to 80°C for applications where distortion during quenching is not a problem. Hot quenching oils are designed for use at higher temperatures of up to 200°C for controlling distortion during quenching, a process also known as marquenching.

This process involves quenching the workpiece into a fluid medium maintained at an elevated temperature, generally 100°C to 200°C. The workpiece is held in the fluid until temperature equilibrium is established throughout the section and then air-cooled to ambient temperature. The control of distortion possible with oil quenching techniques is based on the effect of component mass on cooling rate.

Because hot quenching oils are used at relatively high temperatures, their formulation and physical characteristics are different from those of cold quenching oils. They need to be formulated from very carefully selected base oils that have high oxidation resistance and thermal stability. Generally, they have high viscosities and flash points and contain complex antioxidant packages to provide long life under arduous operating conditions. Selection of the marquenching oil grade is based on the required operating temperature and quenching characteristics. A minimum of 40°C to 50°C should be maintained between the operating temperature of the oil and its flash point.

Heat treatment oils can suffer from problems due to the formation of sludge. The presence of sludge may cause non-uniform heat transfer, increased thermal gradients and increased cracking and distortion. Sludge may also plug filters and foul heat-exchanger surfaces. The loss of heat-exchanger efficiency can cause overheating, excessive foaming and fires. Sludge results from thermal and oxidative degradation of the quenchant. Oxidation reactions lead to polymerised and cross-linked molecules, which are insoluble in the oil. Thermal degradation leads to both light and heavy ends, the latter leading to coke and sludge.

In some applications, a requirement exists for quenching oils which can be easily washed off in plain water. Quenching oils incorporating emulsifying agents are available which do not significantly influence the cooling rate of the oil. These enable simple wash-off in cold water eliminating the need for alkaline cleaners or solvent degreasers.

14.5.2 POLYMER QUENCHANTS

Polymer quenchants are concentrated solutions of organic polymers in water, together with corrosion inhibitors and other additives. These can be further diluted in water to give ready-to-use quenching solutions. Several types of organic polymers are used including, polyalkylene glycol (PAG), sodium polyacrylate (ACR), polyvinyl pyrrolidone (PVP) and polyethyl oxazoline (PEO). These have widely differing properties. Unlimited flexibility of quenching characteristics is possible through selection of the type of polymer, polymer concentration, temperature of the quenching bath and degree of agitation.

The successful application of polymer quenchants depends upon many factors, including the hardenability of the metal, section thickness and surface finish of the component, the type of furnace and quenching system and the physical properties required. Polymer quenchants provide a number of advantages compared with mineral oil quenchants.

Polymer quenchants are non-flammable and hence enable the elimination of the need for protection equipment, such as inert gas curtains or fire extinguishing systems. As a result, component entry into the quenchant becomes less critical and fire insurance premiums are much lower. They provide a cleaner and safer working environment because there is no smoke and fumes during quenching or tempering and no oily surrounds to the quenching bath. Oil impregnation is potentially dangerous and requires the use of costly floor absorbents or cleaners. With polymer quenchants, floors can be kept completely clean and safe by occasional washing with water.

Varying the concentration, temperature and agitation of the polymer solution allows a range of cooling rates to be achieved, thus enabling the treatment of a wide variety of materials and components. Producing a uniform polymer film around the component avoids the steam pocketing and soft spot problems often associated with water quenching after induction hardening. The uniform film also reduces thermal gradients and residual stresses associated with water quenching and can, therefore, give substantial reduction in distortion during the solution heat treatment of aluminium alloys. Additionally, large amounts of water contamination can be tolerated before concentrations (and hence quenching speed) are influenced significantly. This eliminates the soft spot, distortion and cracking problems associated with trace water contamination in mineral oils.

Depending upon the type of polymer and the concentration required, the in-tank costs of diluted polymer quenchants can be considerably lower than those of quenching oils. Because polymer solutions have significantly lower viscosities than quenching oils, drag-out and hence replenishment requirements are reduced. Also, components may not require cleaning before tempering. Residual films of polymers will not char, as with oils, but will decompose fully at high temperatures to form water vapour and oxides of carbon. Components may be tempered directly after quenching, thereby eliminating the need for costly alkali cleaning or vapour degreasing operations. For low-temperature tempering or ageing treatments where the polymer may not decompose completely, the residual film can be removed by simple washing in plain water. Because polymer quenchant solutions have almost twice the specific heat capacity of quenching oils, for a given charge weight, the temperature rise during quenching will be approximately halved.

14.5.3 WATER AND SALT SOLUTION QUENCHANTS

Water is the most readily available fluid that can be considered for quenching and has many advantages for such applications. It is inexpensive, easy to handle, does not involve storage problems, is easy to pump and filter, is non-flammable (so there is no fire risk), is comparatively non-toxic (so there are few health, safety or environmental risks) and no cleaning is required prior to tempering or subsequent processing. It is also easy to control bath temperature by continuous supply of fresh water.

However, water is not a panacea and has disadvantages which restrict its use to specific applications. It can be corrosive to the quenching bath, it supports microbiological growth and has poor quenching characteristics. Corrosion inhibitors and biocide packages can be used to overcome the first two disadvantages. However, the quenching characteristics have severe limitations which can cause problems.

As the water temperature increases, the vapour phase becomes prolonged and the maximum rate of cooling decreases sharply. This can lead to the formation of soft spots on the component surface (particularly during induction hardening) or inadequate hardness in the workpiece. The stability of the vapour phase is dependent upon the surface finish of the component. The vapour film is very persistent on flat smooth surfaces, but breaks up readily with the onset of the boiling stage at sharp corners, rough surfaces, defects or other stress raisers. This variation in stability can produce markedly different cooling rates across the component, resulting in distortion and cracking. Water exhibits very high cooling rates in the convection phase compared, for example, with mineral oil. High cooling rates in the martensitic transformation temperature range (Ms–Mf) can result in high residual stresses, excessive distortion and potential cracking. It is best to avoid high rates of cooling.

The detrimental effects of temperature dependence and vapour-phase stability can be minimised by maintaining the water at a low temperature through effective cooling, vigorous agitation to disperse the vapour blanket and/or the addition of an inorganic salt.

The addition of salts to water quenching systems assists in breaking up the vapour phase. Commonly used materials include sodium chloride (NaCl), typically at a concentration of 10% vol., or sodium hydroxide (NaOH), typically at a concentration of 3% vol.

During quenching, very small crystals of salt are deposited on the surface of the component due to evaporation of the water at the surface. The localised high temperatures cause these crystals to fragment violently, and this creates turbulence which destroys the vapour film and gives very high maximum cooling rates. Unfortunately, sodium chloride solutions are corrosive, and sodium hydroxide presents health and safety problems. Some suppliers of aqueous quenchants offer less-corrosive, non-toxic proprietary salt mixtures.

Although the quenching characteristics of water can be modified by close temperature control, effective agitation and the use of salt additives, high cooling rates still persist in the convection phase. For this reason, wherever possible, water-quenched components should be of simple shape with no sharp corners or stress raisers. In practice, water quenching is generally restricted to plain carbon and low alloy steels, low alloy carburising steels, selected surface hardening applications or components of very large section thickness.

14.6 TYPES OF TEMPORARY CORROSION PROTECTIVES

There are five types of temporary corrosion protectives:

- Soft film.
- Hard film.
- Oil film.
- Strippable.
- Volatile.

Soft film temporary corrosion protectives are either deposited onto the component to be protected from a solvent in the cold, applied by dipping the component in hot liquid protective or wiped or brushed onto the component. Solvent deposited types are usually based on petroleum jelly or lanolin mixtures in either white spirit or special boiling point naphtha. These films are usually thinner than films deposited by other methods.

Hot-dipped types are usually petrolatum-based and produce films raging from jelly-like to relatively waxy. Wiped or brushed types can be grease, lanolin, oil, petrolatum or fat-based. They are softer than the hot-dipped types, to permit cold application by swab or brush. Slushing types are a variant of brushed types, possessing some flow properties at room temperature, so that brush marks produced during application are reduced. Some materials contain solvent, so that they are free-flowing as applied, but stiffen as the solvent evaporates.

Hard film temporary corrosion protectives were developed to permit handling of components after treatment and to avoid contamination of adjacent components. They are deposited in the cold. Films should be tough and neither sticky nor brittle. They can be polymer, resin or bitumen based, depending on the degree of transparency or the colour required or acceptable. Solvents used depend on the solubility of the ingredients, the required drying time and the permitted flammability and toxicity.

Oil film temporary corrosion protectives are either medium- or low-viscosity mineral oils with added corrosion inhibitors and antioxidants. They are applied by dipping, spraying, swabbing or brushing. The resulting films are usually very thin, and they are used on the interior surfaces of assembled components or where solids or solvents cannot be tolerated.

Strippable corrosion protectives are applied by hot dipping or deposition from solvents. The hot-dipped types are often based on ethyl cellulose. Dipping temperatures can be up to 190°C. They provide very thick coatings that offer protection against both corrosion and mechanical damage. Reuse of the coating is often possible. They require special dipping tanks and can be expensive. Solvent applied types can be based on latex, vinyl copolymers or polyethylene. Coatings are much thinner than with hot-dipped types and can become brittle with ageing, so can be difficult to remove. They almost always need to contain added corrosion inhibitors.

Vapour-phase corrosion protectives are generally used as an impregnant or coating on paper or synthetic film. They volatilise into the air around components and inside assemblies, so they require some form of enclosure to maintain the vapour around the component. Care should be taken when using them with non-ferrous metals, paints or plastics, as discolouration may occur. They do not need to be removed.

14.7   TESTING METHODS FOR PRODUCTION ENGINEERING FLUIDS AND PASTES

A number of tests are used to monitor and control production engineering fluids and pastes. Some of these, listed below, are the same as those used for other lubricants, as described in Chapters 6, 7 and 8. Some of them are used only for metalworking and metal forming fluids and pastes and heat treatment fluids, so were not included in the preceding chapters and are consequently described below. The numerous sources of contamination of metalworking fluids and their effects on fluid performance are listed in Figure 14.7.

	Effect On	
Contaminant	**Neat Oils**	**Water-Mix Fluids**
Metallic debris	Surface finish	Surface finish
Micro-organisms	No effect	Biodegradation
Tramp oil	Viscosity change	Stability, additives
Water	Additives, clarity	Stability
Dust and dirt	Surface finish	Stability
Other fluids	Stability	Additives
General debris	Clarity	Stability

FIGURE 14.7   Sources of metalworking fluid contamination and their effects.

Source: Pathmaster Marketing Ltd.

The concentration of a water-mix metalworking fluid in water can be measured using several methods. The most commonly used test measures the fluid's refractive index, which can then be used to determine the fluid's concentration. The speed of light is determined by the medium (material) through which it is travelling. Light travels faster in a vacuum than it does in any other medium, and it changes speed as it passes from one medium to another. This is called refraction. The refractive index of a material is a measure of the change in the speed of light as it passes from a vacuum (or air as an approximation) into the material. As light passes through the liquid from the air, it will slow down and create a "bending" illusion. The severity of the "bend" will depend on the amount of the substance dissolved in the liquid.

Refractometers are used for measuring refractive index. The four main types of refractometers are traditional handheld refractometers, digital handheld refractometers, laboratory (or Abbe) refractometers and in-line process refractometers. In order to measure the concentration of the substance in the liquid, a calibration graph must be established. This consists of measuring the amount of "bend", as shown in Figure 14.8 when looking through the eyepiece of a traditional handheld refractometer, for several different test concentrations and plotting the resulting points on a graph, as shown in Figure 14.9. Digital refractometers measure refractive index automatically, but must also be set up with a calibration curve. The measured refractive index is then compared with the corresponding concentration using the calibration curve.

Refractometer measurements suffer from two important problems. They are unable to distinguish between metalworking fluid and contaminants, so the refractive index will depend on anything in the fluid that is not supposed to be there. Additionally, as the fluid ages, in-use and particulate contents (dirt, dust or metallic debris) increase, the band of light becomes blurred, and its precise position on the

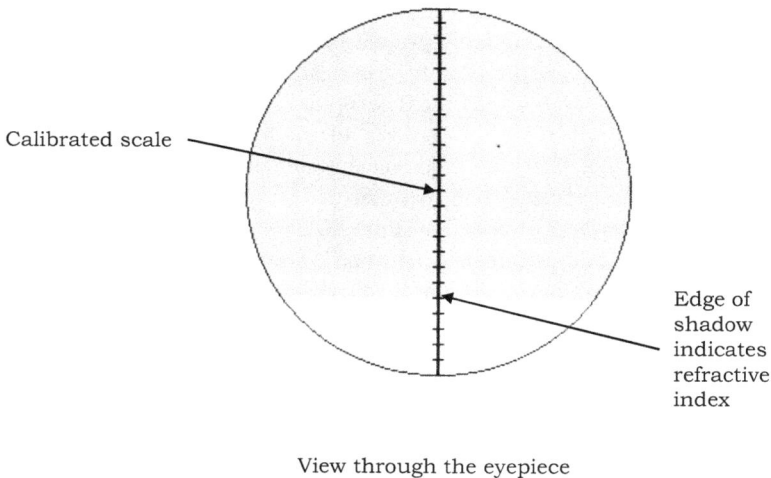

Calibrated scale

Edge of shadow indicates refractive index

View through the eyepiece

FIGURE 14.8   Pocket refractometer.

Source: Pathmaster Marketing Ltd.

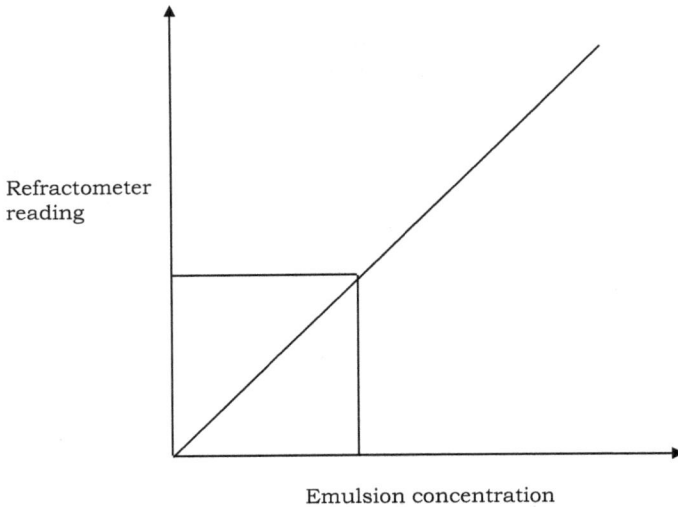

FIGURE 14.9   Pocket refractometer calibration graph.

Source: Pathmaster Marketing Ltd.

refractometer's eyepiece becomes more difficult to determine. Used fluids will tend to give refractometer readings that are stronger than the real value, although this is better than having no determination of concentration at all.

The approximate concentration of a water-mix metalworking emulsion can be obtained by measuring its oil content, using the IP137 test "Soluble Cutting Oil: Oil Content of Dispersions". A known volume of the emulsion is treated with hydrochloric acid in a graduated flask and the volume of liberated oil is measured. However, in addition to the mineral oil originally present in the soluble emulsion, the separated material may also include portions of those emulsifier additives that are decomposed by the acid treatment. A more accurate determination can be achieved by setting up a calibration graph, in the same way as for a refractometer measurement.

If the water-mix metalworking fluid concentrate contains certain anionic surfactants, the concentration of the fluid in water can be assessed using a modification of ASTM D6173 test "Standard Test Method for Determination of Various Anionic Surfactant Actives by Potentiometric Titration". The standard test method is based on a potentiometric titration of common anionic surfactants and blends of anionic surfactant with a hydrotrope. It is solely intended for the analysis of active matter in alcohol ether sulphate, alpha olefin sulphonate, alkylbenzene sulphonic acid, alcohol sulphate, sodium alkylbenzene sulphonate/sodium xylene sulphonate blend (5:1), sodium alkylbenzene sulphonate/sodium xylene sulphonate blend (16:1) and sodium alkylbenzene sulphonate/sodium xylene sulphonate blend (22:1). It has not been tested for surfactant formulations. The anionic surfactant is first dissolved in water, and the pH of the solution is adjusted according to the type of anionic surfactant being measured. In the potentiometric titration, the anionic surfactant is titrated with a standard solution of Hyamine using a surfactant electrode, and the reaction involves the formation of a complex between the anionic surfactant and the cationic titrant (Hyamine), which then precipitates. At the end point the surfactant electrode appears to respond to an excess of titrant with potential change large enough to give a well-defined inflection in the titration curve.

The titration can be adapted to titrate the water-mix metalworking fluid dilution directly and its concentration estimated, again using a predetermined calibration graph.

The pH of a water-mix metalworking fluid can be measured easily and quickly using pH test papers, strips or sticks. The pH of pure water is 7.0. Acidic solutions have lower values of pH, while alkaline solutions have higher values of pH. pH test papers, strips and sticks are marketed in a range of pH values, such as 0 to 14, 0.0 to 6.0, 5.5 to 9.0, 9.5 to 13.0, 7.0 to 10.0 or 5.0 to 8.0. It is better to have a narrower range, which provides greater accuracy.

The acidity or alkalinity of a water-mix metalworking fluid can also be determined using the standard acid number and base number tests (ASTM D664, IP 177, ISO 6619, ASTM D974, IP139, ISO 6618, DIN 51558T1, AFNOR T60-112, IP276, ASTM D2896 or ASTM D4739) which are described in Chapter 6. However, these methods will not give the pH of an aqueous solution or emulsion, although they can be used to monitor and control the fluid in-service.

The pH of an aqueous fluid can be measured using the ASTM E70 "Standard Test Method for pH of Aqueous Solutions with the Glass Electrode". The test method specifies the apparatus and procedures for the electrometric measurement of pH values of aqueous solutions with the glass electrode. It does not deal with the manner in which the solutions are prepared. It indicates that pH measurements of good precision can be made in aqueous solutions containing high concentrations of electrolytes or water-soluble organic compounds or both. It should be understood, however, that pH measurements in such solutions are only a semiquantitative indication of hydrogen ion concentration or activity. The measured pH will yield an accurate result for these quantities only when the composition of the medium matches approximately that of the standard reference solutions.

The method is claimed to provide an accurate measurement of the hydrogen ion concentration and thus is widely used for the characterisation of aqueous solutions. In general, the test will not give an accurate measure of hydrogen ion activity unless the pH lies between 2 and 12 and the concentration of neither electrolytes nor nonelectrolytes exceeds 0.1 mol/litre. The method is also more useful for use with aqueous solutions, rather than soluble-oil emulsions, because the oil and surfactants in emulsions can coat the surface of the electrode and may affect the measurements.

A slightly alkaline pH of a water-mix fluid has been found to slow the growth of bacteria and fungi, as illustrated in Figure 14.10. A more acidic pH is likely to encourage microbial growth. The ideal pH for a water-mix metalworking fluid is between 8.5 and 9.0. Although Figure 14.10 indicates that this level of alkalinity is acceptable to an operator's skin, in a correctly managed production engineering workshop, an operator's skin should not be anywhere near a water-mix metalworking fluid.

Because water-mix fluids are more corrosive to metals than either neat metalworking fluids or lubricating oils, different tests for corrosion need to be used.

```
                              0    Acidic

                              1

                              2

                              3
              _____
                    ⇧      4

                              5    _____
                                      ⇧          Favours microbial
   Acceptable to               6                      growth
   operator's                       ⇩
      skin                     7
                              8      ⇩
                                        _____
              ⇩              9                  Preferred range for
                                    _____          emulsions
      _____     10    ⇧

                              11

                              12

                              13

                              14   Alkaline
```

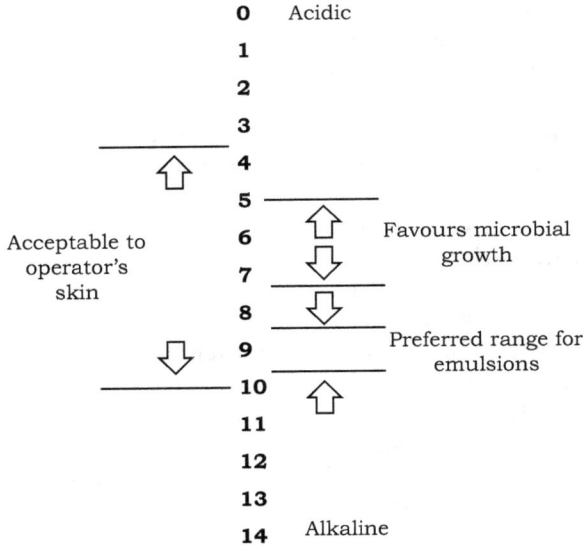

FIGURE 14.10   Water-mix metalworking fluids and pH.

Source: Pathmaster Marketing Ltd.

Corrosion of cast iron can be assessed using either the ASTM D4627, IP125 or IP 287 test methods.

In the IP125 test, "Aqueous Cutting Fluid Corrosion of Cast Iron", steel millings are placed on the cleaned surface of a cast iron plate and the metalworking fluid being tested is poured on to them. The surface of the plate is ground to a uniform smooth finish free from chatter marks and burnishings. The plate and millings are placed in a test chamber, maintained at $18.5°C \pm 5.5°C$ and with a humidity maintained at $52\% \pm 5\%$. After 24 hours, the millings are removed and the surface of the plate is examined for corrosion. The results are recorded in terms of staining and pitting. The proportion of the plate's area that is stained ranges from 0 (no staining), 1 (less than 10%), 2 (10% to 25%), 3 (25% to 50%), 4 (50% to 75%) to 5 (greater than 75%). The intensity of staining ranges from 0 (nil), 1 (hardly perceptible), 2 (slight staining), 3 (heavy staining) to 4 (surface damage, not including pits). The degree of pitting is recorded as the number of pits.

In the ASTM D4627 test, "Standard Test Method for Iron Chip Corrosion for Water-Miscible Metalworking Fluids", cast iron chips are placed in a petri dish containing a filter paper and diluted metalworking fluid. The dish is covered and allowed to stand overnight. The amount of rust stain on the filter paper is an indication of the corrosion control provided by the fluid.

The standard indicates that the results obtained are a useful guide in determining the ability of water-miscible metalworking fluids to prevent or minimise rust under specific conditions. There is usually a relationship between the results of the test and a similar ability of the subject metalworking fluid to prevent rust on nested parts or in drilled holes containing chips. It must be understood, however, that conditions, metal types and other factors found in practice will not correlate quantitatively with the controlled laboratory conditions. The procedure may not be able to differentiate between two products with poor rust control due to the wide spacing between test dilutions.

A similar test is the IP287 method "Rust Prevention Characteristics of Soluble Oil Emulsions: Chip/Filter Paper Method". The method is intended to assess the ability of water-mix metalworking fluids to prevent rusting of machines and components during manufacturing operations. It is suitable for use with conventional soluble oils (transparent and opaque) using anionic emulsifiers as the principal emulsifier system. It is also suitable for use with soluble oils which contain extreme pressure additives and indications are that it is suitable for synthetic metalworking and chemical fluids. The method is intended to arrange oils in order of expected performance but is not meant to specify the safe working dilution.

Cast iron chips are placed on filter paper and wetted with test fluid. The area of stained paper is assessed after 2 hours. This is repeated at various dilutions, and the point at which there is a significant increase in area stained determined. This is known as the break-point. The result is expressed as the dilution (ratio of water to oil) at the break-point.

From the examination of the series of filter papers, the dilution at which the area stained significantly increases, either visually or from the change in slope of a graph obtained by plotting the dilution against the area stained, is evaluated. The dilution (ratio of water to oil) at which there is a significant change is reported as the break-point. The graphical assessment should be used where difficulty is experienced in determining the break-point visually or when the break-point is to be reported between the dilution steps taken in the test method. In these cases the percentage of area staining should be determined by placing a transparent grid over the test area. The area stained after the break-point with one fluid could be less than the area stained before the break-point with another test fluid.

The point at which slight staining first occurs is not necessarily the break-point. Further dilutions can show gradual increases in the stained area, followed by a large increase at some higher dilution. An example of the pattern that could be obtained is 10:1 dilution, no staining, 15:1 dilution, trace of staining, 18:1 dilution, 0.5% area stained, 20:1 dilution, 1.5% area stained and 25:1 dilution, 50% area stained. In this example, the break-point is 20:1 dilution.

Three tests can be used to measure the foaming (or frothing) tendency of a water-mix metalworking fluid. The IP 312 method "Frothing Characteristics of Emulsifiable Cutting Oil" involves shaking the test fluid in a 250 millilitre (ml) measuring cylinder and recording the volume of froth after specified times. One hundred millilitres of test fluid is poured into the measuring cylinder which is stoppered and then shaken vigorously for 15 seconds. The cylinder is stood upright and the volumes of froth are recorded immediately after shaking and after 5, 10 and 15 minutes. The volumes at the four times are reported to the nearest 2 ml.

Two other tests have been used in the past to determine the foaming characteristics of water-mix metalworking fluids. The ASTM D3601 "Standard Test Method for Foam in Aqueous Media (Bottle Test)" covered the measurement of the increase in volume of a low-viscosity aqueous liquid (less than 3 cSt at 40°C) due to its tendency to foam under low-shear conditions. The test fluid is placed in a bottle, which should be no more than half full of fluid, and it is shaken at some specified, reproducible rate. The initial foam height is noted immediately after shaking stops, and the time is recorded for the foam to collapse to some predetermined level. The initial foam height and either the collapse time or the residual foam height after a specified waiting period should be used to compare foaming tendencies of various fluids in both soft and hard water. (Water-mix metalworking fluids tend to foam less in hard waters.)

The ASTM D3519 "Standard Test Method for Foam in Aqueous Media (Blender Test)" covered the measurement of the increase in volume of a low-viscosity aqueous liquid (less than 3 cSt at 40°C) due to its tendency to foam under high-shear conditions. The blender test is a very severe, although not unrealistic method, simulating the agitation a fluid could receive as it is whirled around by a grinding wheel, cutting tool or pump impeller. Two hundred millilitres of metalworking fluid dilution is placed in the jar of a kitchen blender. The mix is agitated at approximately 8,000 rpm for 30 seconds, and the foam height is measured immediately after the blender is switched off. The time is recorded for the foam to collapse to 10 mm in height. If more than 10 mm of foam remains after 5 min, the residual foam height is then recorded.

Both ASTM test methods were withdrawn in May 2013, due to lack of interest in their continued use. The ASTM D892 (IP 146, ISO 6247) foaming test described in Chapter 6, in which air is blown into the fluid to generate foam is designed for lubricating oils, so has little relevance for water-mix metalworking applications.

The stability of a water-mix metalworking fluid concentrate can be assessed using the IP311 "Thermal Stability of Emulsifiable Cutting Oil" test, which is used to determine the thermal stability of emulsifiable cutting oil over the range of temperatures at which oil would normally be stored. The concentrate's high temperature stability is evaluated using approximately 75 ml of the test oil, poured into each of two sample bottles, which are then loosely stoppered and placed in an oven at 50°C ± 1°C for between 15 and 20 hours. At the end of this time, the bottles are removed from the oven, and the oil is examined immediately for any signs of turbidity, separation or gelling. If any doubt exists as to the presence of turbidity, separation or gelling, the concentrate should be re-heated to 50°C and immediately tested for emulsion stability using method IP 263 (described in Chapter 6) with an appropriate water and dilution ratio. An emulsion of the untreated oil prepared in the same manner can be used for reference.

The concentrate's low temperature stability is evaluated in a similar way, except that the bottles are stored in a refrigerator at 0°C ± 1°C for the same time as for the high temperature stability. The concentrate is also examined for any signs of turbidity, separation or gelling. Again, if any doubt exists, the bottles should be cooled to 0°C and immediately tested for emulsion stability using method IP 263. For both temperatures, the appearance of the concentrate is reported, together with the results of the IP263 emulsion stability test, if this was required.

The machining performance of water-mix metalworking fluids can be evaluated using a tapping torque test. The ASTM D5619 "Standard Test Method for Comparing Metal Removal Fluids Using the Tapping Torque Test Machine" described a laboratory technique to evaluate the relative performance of metal removal fluids using a non-matrix test protocol, using the tapping torque test machine. The test method was withdrawn in January 2016, without a replacement, due to lack of interest in supporting the standard with updated/compliant precision information.

The interest in this test was because it is probably the only bench-scale metal cutting test available. Torque values are measured as a tap cuts threads into a predrilled hole in a metal specimen, which can be made of different metals. The average torque value of five tests is calculated. Test results may be expressed either as a simple torque force value or as a percent efficiency, the ratio of the average torque value of a reference fluid to that of the test fluid. The same tap is used on both the reference fluid and the test fluid.

However, the repeatability and reproducability of the test are suspect. The composition and dilution of the water-mix metalworking fluid are only two of many factors that can affect the tapping torque test results. Other factors include the quality of the tap, whether it is a "cut" or "form" tap, the precise size of the predrilled and reamed hole relative to the tap size (sometimes expressed as thread percentage) and the metal and its hardness of the metal (which can vary across the metal specimen). Form taps require higher forces than cut taps and often show greater differentiation between different fluids than do cut taps. This is because, as explained previously, cut taps make threads by cutting into the wall of the hole and removing chips of metal during the process while form taps do not (or should not) generate any chips. Form taps push the metal and force it to flow into the required shape.

Numerous experiments have found that very careful selection of test conditions is required in order to generate reliable conclusions. It is also important to note that other mechanical tests, such as four-ball, Falex, Timken and others described in Chapter 8, used to evaluate the wear or extreme-pressure lubricating performance of oils and greases, are unsuitable for water-mix metalworking fluids because they are too severe. It has been found over many years, by both formulators and users of water-mix metalworking fluids, that tests in metalworking machines in workshops are the only practical methods of evaluating a fluid's machining performance.

The resistance of diluted water-mix metalworking fluids to microbial attack can be assessed using the ASTM E2275 "Standard Practice for Evaluating Water-Miscible Metalworking Fluid Bioresistance and Antimicrobial Pesticide Performance". The practice provides laboratory procedures for rating the relative bioresistance of metalworking fluid formulations, for determining the need for biocide addition prior to or during fluid use in metalworking systems and for evaluating biocide performance. (General considerations for biocide selection are provided in Practice ASTM E2169.)

The practice addresses the evaluation of the relative inherent bioresistance of water-miscible metalworking fluids, the bioresistance attributable to augmentation with antimicrobial pesticides or both. It replaces ASTM D3946 (which was withdrawn in April 2004) and ASTM E686. The relative bioresistance is determined by challenging metalworking fluids with a biological inoculum that may either be characterised (comprised of one or more known biological cultures) or uncharacterised (comprised of biologically contaminated metalworking fluid or one or more unidentified isolates from deteriorated metalworking fluid). Challenged fluid bioresistance is defined in terms of resistance to biomass increase, viable cell recovery increase, chemical property change, physical property change or some combination of these.

The practice is applicable to antimicrobial agents that are incorporated into either the metalworking fluid concentrate or end-use dilution. It is also applicable to metalworking fluids that are formulated using non-microbicidal, inherently bioresistant components.

The factors affecting challenge population numbers, taxonomic diversity, physiological state, inoculation frequency and biodeterioration effects in recirculating metalworking fluid systems are varied and only partially understood. Consequently, the results of tests completed in accordance with ASTM E2275 should be used only to compare the relative performance of products or biocide treatments included in a test series. The practice indicates that test results should not be construed as predicting actual field performance.

As with many other lubricants, portable test kits have been available for many years for the on-site monitoring and control of water-mix metalworking fluids. The kits can include a pocket refractometer, a capillary micro-viscometer, pH test strips, microbial dip slides and a portable particle counter.

Many of the tests used to monitor and control neat metalworking fluids and metal forming fluids and pastes are the same as those used for other lubricants. Viscosity is measured using ASTM D445, acid number using ASTM D664, IP 177 or ISO 6619, steel corrosion using IP 135 and ASTM D665, copper staining using IP 154 and ASTM D130 and water content using Karl Fischer titration, ASTM D6304. Sulphur content is measured by ICP-AES (ASTM D4294, IP 336), chlorine content by XRF (ASTM D6443) and metals contents by ICP-AES (ASTM D4951). Particulate content is evaluated using ISO 4406. Anti-wear and extreme-pressure properties can be measured using the four-ball (ASTM D2783), Timken (ASTM D2782, IP 240 or DIN 51434) or Falex (ASTM D3233) methods.

One test that might be considered applicable to neat metalworking fluids is the ASTM D3705 "Standard Test Method for Misting Properties of Lubricating Fluids". The test provides a guide for evaluating the misting characteristics of oils for use in industrial mist lubrication systems. The standard states that the degree of correlation between the test and service performance has not been fully determined. Unfortunately, the standard is intended to evaluate oils that are intended to be used as mists, to provide both lubrication and cooling of metal parts in machines.

It has been established by many suppliers and users of metalworking fluids that defining a standard method by which to accurately measure and monitor mists generated during machining operations is very difficult. The majority of large metalworking machines are enclosed now and have mist extraction incorporated as well as time delay locks to prevent doors being opened prior to completion of machining. A number of companies also market equipment that can be retrofitted to existing machines. Additionally, it has become apparent that each metalworking fluid behaves differently as its composition changes according to the environment in which it is being used, and this includes the ability to form mist. This applies to both water-mix fluids and neat oils.

The most common methods employed to monitor mists generated by metalworking fluids use extraction systems to collect droplets and aerosols on filter papers and to weigh the collected material. One system uses a novel cascade impactor that separates coarse, fine and ultra-fine particles operated alongside IOM air samplers within the breathing zone of machine workers and was supported by the collection of data using a DataRam particle counter. The difficulty with collecting metalworking fluid mists on filter papers is that many of the collection systems were designed to collect dust particles and oil and/or water of filter papers is likely to result in eventual blockage unless the collection times are controlled effectively. Additionally, the majority of particle counters used for occupational hygiene assessment are portable devices also designed for determining the concentration of dry dust particles. Models can vary greatly in respect to the particle size classes measured and the ease with which they become blocked when used for liquid particle measurement. Studies are ongoing to develop aerosol monitoring particle counters to ascertain whether they could be used to measure metalworking fluid mist against occupational health guidance limits. However, in some studies similar mist concentrations are reported when a particle counter is used compared with gravimetric analysis. It has been noted that, in some respects, particle counters may give a more accurate measurement of metalworking fluid mist as any loss of particles due to poor recovery from the sampler or during analytical stages is negated. It could be concluded that there may be a potential role for particle counters in the measurement of metalworking fluid mist and in particular the rapid assessment of high concentrations possibly due to poor extraction of mists.

The tests used for neat metalworking fluids can also be used for metal forming fluids and pastes. However, as with water-mix metalworking fluids, it has been found over many years that the actual performance of many of these fluids and pastes can only be evaluated reliably in production engineering workshop conditions. Almost all of these tests are customer-specific and require a great deal of specialised mechanical equipment. Most of them are used in the formulation of metal forming products and tend not to be used for fluids and pastes in-service.

Tests used for mineral oil and polymer quenchants include kinematic viscosity, flash point, acid number, Conradson carbon number, RPVOT oxidation, water content and elemental content. The test methods are those described in Chapters 6 and 7. Tests used for water and salt solution quenchants are the same as those used for water-mix metalworking fluids, such as concentration, corrosion, pH, foaming and stability.

One additional test for mineral oil and polymer quenchants is precipitation number, used to determine the relative amount of sludge. The ASTM D91 "Standard Test Method for Precipitation Number of Lubricating Oils" was originally developed to measure the naphtha-insoluble materials present in semi-refined or black oils. Fully refined petroleum oils normally contain no materials that are insoluble in naphtha. The precipitation number is found by adding naphtha to the oil and determining the precipitate volume after centrifuging. The standard indicates that the test can be used for other lubricating oils, so it is used for quenchants as an indication of the amount of sludge present. The remaining life of a quenchant can be estimated by comparing the relative propensity of sludge formation in new and in-service oil.

Quenching speed is probably the most important test for quenchants of all types. It is highly desirable to measure quenching characteristics on site. Specialised, fully portable and self-contained equipment have been developed for this purpose. The ASTM D3520 "Standard Test Method for Quenching Time of Heat-Treating Fluids

(Magnetic Quenchometer Method)" determines the time for cooling a chromised nickel ball from approximately 885°C to approximately 354°C when quenched in 200 ml of test fluid in a metal beaker at 21°C to 27°C. The quenching time is recorded by a digital timer which is energised by a photoelectric cell from light produced by the ball at 885°C and which is stopped when the ball becomes magnetic (Curie Point, approximately 354°C) and is attracted by a magnet to the side of the beaker, tripping a relay to stop the timer. The test apparatus is known as a "GM Quenchometer" ("Nickel Ball") and requires a separate oven or furnace able to heat the ball to at least 885°C.

According to the standard, the results obtained by the test method are useful as guides in selecting fluids with respect to quenching speed characteristics desired for metal quenching applications. However, although the test method has been found useful for some water-based quenchants, the statistical significance of the test has been established only by round-robin testing of petroleum-based fluids.

The GM Quenchometer has been used to classify quenching oils for many years, but has only limited value. The test does not provide any information regarding the cooling pathway, which must be known if the quenchant's ability to harden steel is to be determined. Consequently, GM Quenchometer quenchant characterisation is increasingly being replaced by the ASTM D6200, IP 414 or ISO 9950 cooling curve analysis methods.

The ASTM D6200 "Standard Test Method for Determination of Cooling Characteristics of Quench Oils by Cooling Curve Analysis" provides a cooling time versus temperature pathway which is directly proportional to physical properties such as the hardness obtainable upon quenching of a metal. According to the method, the results obtained may be used as a guide in heat treatment oil selection or comparison of quench severities of different heat treatment oils, new or used. The test is designed to evaluate quenchants in a non-agitated system, and there is no correlation between the test results and the results obtained in agitated systems.

The test uses a 160 mm long Inconel 600 (nickel-chromium-iron alloy) probe at 850°C, with a thermocouple inserted to 30 mm from its end, quenched into 2 litres of test oil at 60°C (or some other temperature), with no agitation. The thermocouple and its associated software are used to measure different variables, including maximum cooling rate, temperature at maximum cooling rate, temperatures ate the start of boiling and convection and times to 600°C, 400°C and 200°C. This enables the plotting of cooling curves. It is worth noting that while cooling curve analysis provides an invaluable tool for monitoring and troubleshooting quenching oil performance, physical property characterisation is still required to identify the causes of the cooling behaviours. However, it has been found by several experimenters that the test method is relatively useful for the evaluation of new oils, it is much less useful when used for testing quenchants in use.

The IP 414 and ISO 9950 test methods (which are identical) are very similar to ASTM D6200 and produce very similar cooling curves. During the last 40 years, a number of companies (including suppliers of mineral oil and polymer quenchants) have developed proprietary test methods for evaluating the quenching characteristics of products, for use either in test laboratories and/or on site.

ASTM D7646 "Cooling Characteristics of Aqueous Polymer Quenchants for Aluminium Alloys by Cooling Curve Analysis" provides a cooling curve versus temperature pathway. The results obtained by the test may be used as a guide in quenchant selection or comparison of quench severities of new or used quenchants. The test method employs aqueous solutions in a non-agitated system. There is no correlation between these test results and the results obtained in agitated systems.

The test uses a silver rod probe assembly, which is heated to 500°C in a furnace and then quenched in an aqueous polymer quenchant solution. The temperature inside the probe assembly and the cooling time are recorded at selected time intervals to establish a cooling temperature versus a time curve. The resulting cooling curve may be used to evaluate quench severity. The presence of contaminants, such as oil, salt, metalworking fluids, forging lubricants and polymer degradation, may affect the cooling curve results obtained by the test.

For temporary corrosion protectives, tests are only used to evaluate their possible corrosion protection in-use. The most commonly used equipment is described in ASTM B117 "Standard Practice for Operating Salt Spray (Fog) Apparatus". The practice covers the apparatus, procedure and conditions required to create and maintain the salt spray (fog) test environment. Suitable apparatus which may be used is described in Appendix X1 of the practice. The practice does not prescribe the type of test specimen or exposure periods to be used for a specific product, nor the interpretation to be given to the results. It provides a controlled corrosive environment which has been used to produce relative corrosion resistance information for specimens of metals and coated metals exposed in a given test chamber.

However, prediction of performance in natural environments has seldom been correlated with salt spray results when used as stand-alone data. The practice indicates that correlation and extrapolation of corrosion performance based on exposure to the test environment provided by the practice are not always predictable. The reproducibility of results in the salt spray exposure is highly dependent on the type of specimens tested and the evaluation criteria selected, as well as the control of the operating variables. In any testing programme, sufficient replicates should be included to establish the variability of the results. Variability has been observed when similar specimens are tested in different fog chambers even though the testing conditions are nominally similar and within the ranges specified in this practice.

The apparatus required for salt spray exposure consists of a fog chamber, a salt solution reservoir, a supply of suitably conditioned compressed air, one or more atomising nozzles, specimen supports, provision for heating the chamber and necessary means of control. The size and detailed construction of the apparatus are optional, provided the conditions obtained meet the requirements of the practice.

14.8   MONITORING AND CONTROL METHODS

14.8.1   METALWORKING FLUIDS

Monitoring metalworking fluids is essential for several reasons. These include determining and, if necessary, correcting fluid condition, characteristics and hence performance, to maintain and extend fluid life and to maintain and extend tool life, to help maintain workpiece quality during production runs and to obtain maximum value for money from the fluid.

Metalworking fluid characteristics and properties change with time for several reasons, including contamination, fluid drag-out, fluid misting, water evaporation, additive drag-out, additive depletion or system imbalance.

Sources of contamination of metalworking fluids and their effects on either the fluid or the metalworking process were summarised earlier in Figure 14.7. Evaporation of water from a water-mix metalworking fluid, which happens continually even when the fluid is not being used, will slowly increase the hardness of the water as a consequence of the slowly increasing concentration of chloride and sulphate ions in the water. This is likely to lead to the formation of soaps, coarsening of emulsions, emulsion instability and/or reduced corrosion protection. In order to adjust the properties of the fluid, it is necessary to add water to replace the evaporated water. Consequently, when making-up the water content of a water-mix metalworking fluid, only distilled or deionised water must be used.

A low concentration of a water-mix metalworking fluid is likely to lead to an intolerance of hard water, emulsion instability, higher microbial activity, a reduced pH, poor tool life and/or poor workpiece finish. Conversely, a high concentration is likely to cause de-fatting of operators' skins, sensitisation of operators' skins, lifting of machine tool paints or coatings, damage to machine tool seals and/or poor rejection of tramp oil. (Tramp oil is oil that has either separated from a water-mix emulsion or has leaked into the water-mix fluid from hydraulic, gearbox or other oil lubricated system.)

Water quality is particularly important when preparing water-mix metalworking fluids for use. Hard water results in fluid instability and the formation of scums and soaps, while soft water causes foaming and poor machining. High chloride ion concentration causes corrosion of tools, workpieces and machines, high sulphate ion concentration causes fluid instability, corrosion and staining of machined components, high nitrate ion concentrations may lead to problems with operator health and safety, and high phosphate ion concentrations are likely to encourage microbial growth.

The presence of bacteria, fungi and/or yeasts in water will encourage microbial growth, leading to all manner of problems, discussed later. There are numerous sources of microbial contamination of water-mix metalworking fluids, including the air (and, therefore, rainwater), dust and dirt, swarf and workshop debris, machined components, operators' foods and drinks and used workshop fluids.

The effects of bacteria on water-mix metalworking fluids are numerous. They include higher fluid spoilage and waste, reduced machining efficiency, reduced component corrosion protection, corrosion of machines, bad smells, unpleasant working conditions and higher total costs.

Monitoring water-mix metalworking fluids tends to be slightly more complex and variable than for other lubricants. Each production engineering workshop is likely to have unique performance requirements as a result of the large number of process and environmental variables discussed previously. Consequently, flagging limits for each type of fluid used in the workshop are likely to have to be established by experience. One of the first signs of deterioration in a water-mix fluid is appearance and smell. Increasing cloudiness or opacity and unpleasant odours are likely signs of either or both contamination and microbial degradation. A reduction in pH may also be a sign of microbial degradation and a warning of corrosion of tools, workpieces and machines. Changes in concentration will affect machining performance, additive drag-out and corrosion. An unstable emulsion will affect metalworking performance and corrosion protection, as will foaming. Tramp oil on the surface of a metalworking fluid in a machine tool reservoir is likely to lead to anaerobic conditions in the bulk of the metalworking fluid, which is likely to encourage bacterial and/or fungal growth. Tramp oil will also affect emulsion stability and additive drag-out, since some of the additives are likely to be more soluble in oil than in water. Changes in water hardness, particularly higher concentrations of chloride, sulphate and phosphate ions will affect emulsion stability, corrosion and microbial growth.

Fortunately, monitoring and controlling metalworking fluids have become easier in recent years, due to significant improvements in production engineering workshop practices. These include:

- Automation: Computer control, less human intervention and enclosed machining environments.
- Consistent working conditions: More even workshop temperatures, less dust and dirt, less fluid contamination and better quality water.
- Electronics: Sensors for machines and sensors for fluids, such as pH and temperature.
- Economics: Waste minimisation, for example, from leaks, tramp oil and reclaimed swarf, fluid reconditioning and fluid re-use.

In a metalworking fluid emulsion, the water phase will contain more of the corrosion inhibitors, coupling agents and biocides while the oil phase will contain more of the extreme-pressure additives and metal passivators. The interfaces between the water and oil will attract the emulsifiers, load-carrying additives and anti-foam additives.

Control of a metalworking fluid emulsion relies on controlling either the whole dual-phase system or one of the two phases. Because of the differing components in the two phases, the effect of adding a single concentrate to only one phase is that it is only possible to adjust one of them and the other phase will be "out of balance". Fortunately, whole fluid concentrates for re-adjustment are more stable, tend to have long storage life and may not need to contain unnecessary emulsifiers or coupling agents. A small number of concentrates are usually required for most applications. Controlling the fluid using phase control provides a number of advantages, including maintaining the fluid in balance, reducing total fluid costs, achieving lower foam, maintaining good tramp oil rejection and swarf separation and extending operating lifetimes.

Controlling microbial contamination involves using biostable fluids, using good quality or deionised water, maintaining fluid pH and concentration, minimising fluid contamination by workshop dirt, debris and wastes, removing tramp oil, filtering fluids and using biostats or biocides only when necessary. An illustrative relationship between the numbers of bacteria present in both a biostable and a conventional water-mix metalworking fluid over a period of 12 months is shown in Figure 14.11. Note that the numbers of bacteria are shown in logarithmic form, as 10^6/ml or 10^9/ml. An illustration of the effect of biostability on pH for both types of fluids after a period of 12 months is shown in Figure 14.12.

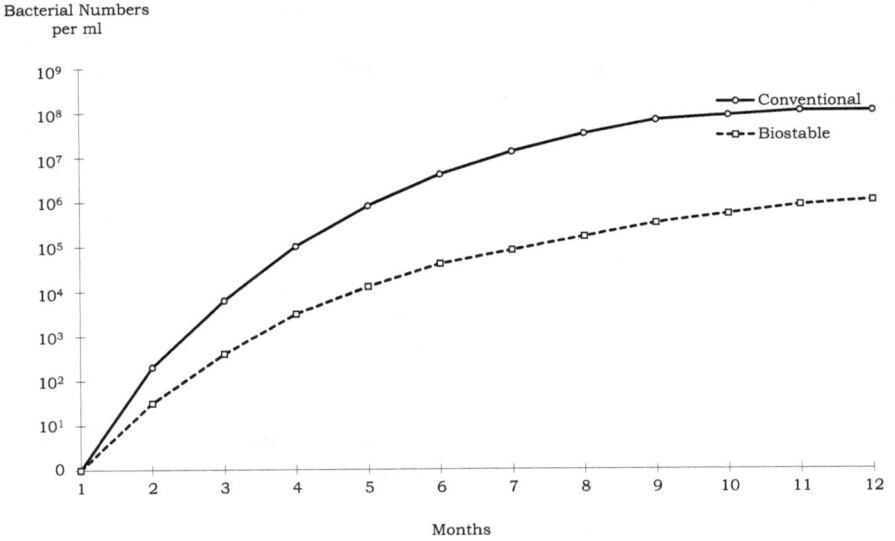

FIGURE 14.11   Biostability and bacteria.

Source: Pathmaster Marketing Ltd.

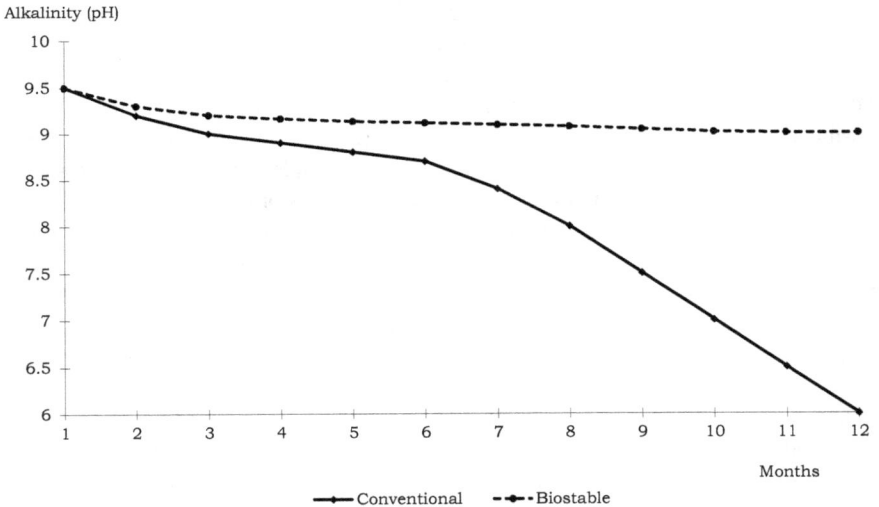

FIGURE 14.12   Biostability and pH.

Source: Pathmaster Marketing Ltd.

Problems arise with emulsion metalworking fluids in use with water quality and its effects on stability, machining performance and biostability, with tramp oil and with health and safety, mainly due to odours and the use of biocides. Monitoring and control of emulsion metalworking fluids involve many parameters and are relatively expensive. These fluids are difficult to recycle or reuse, and their disposal is complex and costly.

Water solution metalworking fluids are also difficult to monitor and control. Again, water quality affects machining performance, foaming and increasing ionic strength. High rates of water evaporation lead to deposits, which may cause machined components to stick together. The compatibility of these fluids with hydraulic fluids, gear oils and slideway lubricants needs to be considered.

Monitoring neat metalworking fluids involves attention to viscosity, additive levels, acid number, water content, particulate content and metals content. Control of neat metalworking fluids requires the effective and efficient removal of swarf, filtration to remove fine particulates, using additive packages to replenish load-carrying and extreme-pressure properties and adjusting the fluid's viscosity as required. Off-site reconditioning can be used to remove acidic oxidation products, to adjust viscosity, colour and clarity and to re-spike additive components. Controlling machining performance requires attention to the concentration of "lubricants", meaning mineral oils, synthetic oils, load-carrying additives and extreme-pressure additives. Changes to these are caused by oil drag-out, oil and additive depletion on metal surfaces and the functioning of additives during machining. Control involves adjust the concentration(s) after identifying which "lubricant(s)" has/have been depleted.

The problem areas for neat metalworking fluids include odour (particularly from sulphurised additives), colour (in terms of workpiece visibility), health and safety (from misting generation and fire risk) and suspended metals, which can risk dermatitis in operators' skin and may be difficult to filter.

14.8.2 METAL FORMING FLUIDS AND PASTES

Monitoring and controlling metal forming lubricants are particularly difficult to do by testing their properties during use. Instead, their properties must be evaluated by looking at their performance during use.

The problems associated with metal forming fluids and pastes are numerous and solving them can be complicated. The main performance problems are:

- Scoring: Either the lubricant has been applied incorrectly, the additive concentration is too low, there is a need to use cushioning additives, there is a need to have improved surface wetting, a more viscous lubricant is needed and/or the lubricant film strength needs to be increased.
- Wrinkling: Either the lubricant has been applied incorrectly, there is too much slip, so the coefficient friction needs to be increased or the wrong lubricant viscosity is being used.
- Breakage: Either the lubricant has been applied incorrectly, the lubricant's viscosity is too low, the workpiece is too heavy for the selected lubricant or the lubricant film strength needs to be improved.

- Poor concentricity: Either the lubricant has been applied incorrectly or the lubricant is too viscous.

If a new metal forming lubricant enables the production of high-quality parts, but the quality then begins to deteriorate either gradually or suddenly, then either the metal has changed over time, the punches and dies have become worn or damaged or the metal forming lubricant has deteriorated and must be replaced.

14.8.3 HEAT TREATMENT FLUIDS

In use, the performance of quenching oils can change for numerous reasons including oxidation, contamination or depletion of additives. Quenching oils should be monitored periodically for their chemical and physical properties, a requirement often dictated by heat treatment specifications or quality control procedures. The examination should determine the condition of the oil and its suitability for continued use. The results of the tests should enable any necessary corrective action to be taken. Several tests can be used.

Changes in the viscosity of a quenching oil may indicate oxidation and thermal degradation or the presence of contaminants. In general, viscosity increases as oil degrades, and this may result in changes in quenching speed. Increasing the quenching bath temperature to compensate for an increased viscosity is likely to accelerate the rate of oxidative or thermal degradation.

The flash point of a quenching oil is important because it is related to the maximum safe working temperature. As a general guideline, the maximum oil temperature in the quenching bath should be 40°C to 50°C below the Cleveland Open Cup flash point of the oil. Changes in flash point in service usually indicate contamination, the presence of dissolved gases or thermal degradation of the oil. When the flash point of an in-service oil becomes too low, it needs to be changed in order to achieve the correct quenching rate for the metals and components being treated.

Contamination of quenching oils with water must be avoided at all costs. As little as 0.05% of water in quenching oil influences quenching characteristics significantly and may cause soft spots, distortion or cracking of treated components. At concentrations of 0.5% or more, foaming during quenching is likely, and this can give rise to fires and explosions.

The acid number of a quenching oil is an indication of the level of oxidation. The formation of oxidised constituents in a basic mineral oil decreases the stability of the vapour phase and increases the maximum cooling rate, thereby increasing the risk of distortion and cracking. The use of specially formulated quenching oils reduces the possibility of these problems occurring.

The oil's saponification number is an indication of the presence of unstable or unsaturated hydrocarbons in the oil which may oxidise and form sludge. It is also an indication of oxidation in the oil. Saponification number can also be used to measure the level of fatty additives that may be present in the oil.

As noted previously, quenching speed or cooling curve analysis is the most important control check for any quenching fluid, since this is the critical performance characteristic.

For polymer quenchants, once the most suitable type of polymer, required concentration, bath temperature and agitation conditions have been selected, it is important to establish a monitoring system to ensure consistent performance in service.

Several methods can be used for measuring polymer concentration. Refractive index is a simple method, suitable for day-to-day control of the quenching bath. Refractive index is measured using a hand-held refractometer, in the same way as for water-mix metalworking fluids, so the concentration is obtained from a calibration curve. The disadvantage of this method is that the refractive index reading can be influenced by contamination in the system which may give rise to an erroneous indication of polymer concentration.

The kinematic viscosity of polymer solutions can also be used to monitor concentration. However, care must be taken with the interpretation of results because polymers may degrade in service leading to viscosity changes.

Quench-rate testing is the most relevant technique, because it measures the actual quenching speed of the polymer solution and takes into account any changes in performance resulting from contamination or degradation. On-site assessment using portable equipment such as the IVF Quenchotest is preferable since it also takes into account bath conditions.

Contaminants can reduce the life of a polymer quenchant or change its performance. Particulate matter such as scale or soot has little effect on quench-rate but can hinder concentration control by making refractive index difficult to measure and may affect the cleanliness of the quenched component. Contamination by metalworking fluids, corrosion inhibitors or hydraulic oils can provide nutrients for biological growth and may also prolong the vapour phase of the quenching process. Cross-contamination with different types of polymer quenchants can adversely affect quenching performance and should therefore be avoided.

As with all water-based systems, polymer quenchants are susceptible to microbiological contamination. Bacteria can give rise to unpleasant odours and cause the depletion of corrosion inhibitors. Accumulated deposits of fungi can block filters and nozzles in spray quenching equipment and may impair the efficiency of cooling systems. Biocides can be incorporated into polymer quenchant formulations to minimise these effects. In addition, continuous circulation of quenchant to maintain aerobic conditions, particularly when the quench tank is not being used, will assist in preventing biological growth. Occasional additions of biocide may be required to control biological growth, as with water-mix metalworking fluids.

Accumulations of inorganic salts from supply water or salt bath heat treatment operations will influence refractive index and hence affect concentration control. These may also change quenching characteristics. Contamination with ammonia (for example, from carbo-nitriding atmospheres) may affect quenching characteristics and corrosion protection.

Polymer quenchants, like quenching oils, may degrade in service at a rate dependent upon the type of polymer, the severity of application and conditions of use. However, all types of polymer quenchants can provide reliable, consistent performance which can be monitored effectively by a combination of refractive index measurement, kinematic viscosity determination and quench-rate testing.

14.8.4 Temporary Corrosion Preventatives

Temporary corrosion protectives cannot be monitored in-service, as there is insufficient material to do so. However, the metal components that are supposed to be being protected must be monitored closely to check for signs of corrosion and deterioration or failure of the protective coating.

The most effective way to do this is to put limits on the times between each stage of the production engineering process for which temporary corrosion protection is required. Date and time of production should be assigned to each batch of components, together with the date and time for the next stage in the process. The time intervals will depend on the metal, method of production, ambient temperature and humidity in the production engineering workshop and precise shape of the component at each stage of the manufacturing. This can only be done by experience. As with all the other monitoring procedures described in this book, written or computerised records of all the variables involved is highly likely to produce quantifiable benefits.

14.9   PRESENTATION OF RESULTS

As with the other types of lubricants described in this book, the effective recording, analysis and presentation of test results and condition monitoring programmes are vitally important. The range and complexity of production engineering processes and lubricants mean that almost all workshop situations are comparatively unique. As a result, flagging limits for most fluids and pastes almost always have to established by experimental experience. This means recording and plotting individual results against a reference standard.

As described in several places in this chapter, it is not possible to test some production engineering fluids and pastes in-service, but their properties and performance need to be monitored by evaluating the end result of the process. On occasions, deterioration is not due to the lubricant but to some other factor. Careful failure mode and effect analysis needs to be used to identify the cause(s) of the deterioration.

Graphs and trend analysis will enable the determination of flagging limits for each production engineering process and lubricant in each workshop. It should be obvious that this may take some time, possibly several years.

One method that can be very useful in this situation is statistical process control (SPC). This is a method of quality control that uses statistics to monitor and control a process, thereby helping to ensure that the process operates efficiently, producing more on-specification products with less waste (rework or scrap). SPC can be applied to any process where the output of on-specification and non-conforming products can be measured. SPC must be practiced in two stages, the first of which is the initial design of the production process and the second of which is the regular use of the process. The period to be examined must be defined in the second stage. The time period will depend on the people, machines, materials, methods, movements and environmental conditions and the wear rates of the equipment used in the manufacturing process.

An advantage of SPC over other methods of quality control is that it emphasises early detection and prevention of problems, rather than the correction of problems after they have occurred. In addition to reducing waste, SPC can lead to a reduction in the time required to produce the product. SPC makes it less likely the finished product will need to be reworked or scrapped.

The origin of statistical process control began with the realisation that data from physical processes seldom produced a normal Gaussian or "bell" distribution curve. Data from measurements of variation in manufacturing did not always behave in the same way as data from measurements of natural phenomena. It was concluded while every process displays variation, some processes display variation that is natural (or "common") to the process, and these processes can be described as being suitable for statistical control. Other processes additionally display variation that is not present in the causal system of the process at all times, and these are described as "special" sources of variation.

In production engineering, quality is associated with conformance to specification. Unfortunately, no two products are ever exactly the same, because any manufacturing process contains many sources of variability, as noted previously in this chapter. Traditionally, the quality of a manufactured product is evaluated by post-manufacturing inspection. SPC uses statistical methods to measure the performance of the production process in order to detect significant variations before they result in the production of sub-standard articles. Any source of variation at any point of time in a process will be either common or special and SPC will be able to identify which.

Obviously, SPC will be much easier with computer records and automated output of results in tables and graphs. Now, in many larger production engineering workshops, automated machines with on-line sensors are connected via the Internet to production engineering lubricant suppliers, so that lubricant and machine test results can be sent automatically to the supplier, as part of a "Total Fluid Management" (TFM) contract. This enables the lubricant supplier to exercise real-time monitoring and control of production engineering lubricants. The scope and operation of TFM contracts are outside the scope of this book, but readers are encouraged to investigate their advantages and disadvantages if they have not already done so.

14.10 SUMMARY

A large number of engineering processes can be used to make metallic components. These processes either remove metal or shape metal, most usually in three dimensions, but sometimes in only one or two dimensions. The large number of engineering processes are likely to require a large number of lubricating fluids. The same shaped component can be made using very different engineering processes. When combined with many types of metals and alloys, the production possibilities are enormous.

Testing, monitoring, control and correction of metalworking and production engineering fluids are completely different from oil and machine condition monitoring for other lubricants. Test methods and monitoring and control techniques for metalworking and production engineering fluids can range from simple and easy to very difficult and complex. Control of metalworking and production engineering fluids can require a detailed knowledge of chemistry, biology, metallurgy, physics and mechanics.

In almost all cases, the monitoring and control methods for production engineering fluids and pastes are customer- and application-specific. As a consequence, the flagging limits for these lubricants are likely to be slightly different for each workshop.

15 Condition Monitoring of Automotive and Industrial Greases

15.1 INTRODUCTION

The ASTM defines a grease as "A solid to semifluid product of dispersion of a thickening agent in a liquid lubricant. Other ingredients imparting special properties may be included".

Greases are most usually used where a lubricant is required to maintain its position in a mechanism, particularly where opportunities for frequent re-lubrication may be limited or commercially unjustifiable. This may be due to the type of mechanism, the mode of action, the type of sealing or a need to minimise ingress of contaminants into the mechanism.

The main disadvantages of greases are their inability to remove heat rapidly and the lack of flushing action to remove wear particles or contamination from the mechanism. Conventional soap-based greases have a relatively limited operating temperature range, although more specialised greases can operate in more extreme conditions, but at a higher cost.

Approximately 90% of all rolling bearings are lubricated by grease. Monitoring greases is therefore quite important for many types of industrial machines. It is therefore important to explore and explain the specific requirements for monitoring the condition of greases and the machines in which they are used, in comparison with lubricating oils.

15.2 TYPES OF GREASES AND SOLID LUBRICANTS

Four main types of greases are used:

- Soap thickener.
- Clay thickener.
- Inorganic thickener.
- Polymer thickener.

It is very important to note here that soap-based greases MUST NOT be mixed with either clay-thickened or inorganic-thickened greases. This is because these types of thickeners are chemically incompatible and their interaction is highly likely to destroy the thickeners' abilities to thicken.

Soap-based greases are made by reacting an acid with a base, in oil, in such a way as to create a solid or semi-solid matrix in the oil carrier. For example, a lithium soap

DOI: 10.1201/9781003245254-15

grease is made by reacting 12-hydroxy stearic acid with lithium hydroxide in a base oil, to form lithium 12-hydroxy stearate soap. The manufacturing process involves heating with stirring and time, shock chilling with more base oil to form a fibrous structure, milling and homogenising, adding additives, de-aerating and finally filtering. Types of soaps are lithium and lithium complex, calcium and calcium complex, aluminium and aluminium complex, sodium and mixed soaps. Barium soaps used to be made, but are not now due to their potential health, safety and environmental problems.

The base oils used for soap-based greases include paraffinic and naphthenic mineral oils, diesters, polyol esters, polyalphaolefins, phosphate esters, silicones and fluorinated polyethers and rapeseed, linseed and cotton seed vegetable oils. "Complex" soap greases are made by co-crystallising longer-chain fatty acids (stearic) with short-chain acids (azelaic) or inorganic salts (carbonates, sulphonates). These greases have a much "tighter" grease structure, so more are physically stable, have significantly higher drop points and improved shear stability.

Making other types of greases is much less complicated. The four main types of thickener, bentone (clay), silica, polytetrafluoroethylene (PTFE) and polyurea (polymer), are simply mixed into a base oil in the proportions required to achieve the required softness or hardness. The thickener is dispersed into the base oil, a gelling agent is added, the mixture is sheared until homogenous, additives are added, the grease is de-aerated and finally filtered.

The base oil provides the hydrodynamic lubrication properties of a grease, although the thickener system is more important for rheological properties than was thought previously. The thickener does not simply act as a "sponge" for the base oil. The main properties required of greases are:

- Consistency.
- Drop point.
- Load carrying.
- Water resistance.
- Corrosion protection.
- Oxidation resistance.
- Leakage tendency.
- Roll stability.
- Compatibility.
- Temperature performance.
- Bearing tests.

Solid lubricants are not like greases. They are intended to form a physical barrier between moving metallic or non-metallic surfaces. However, they are used in some of the same applications in which greases are used, although in generally more extreme operating conditions. Many types of solid lubricants, shown in Figure 15.1, have been used over the years, although most of these are not used now due to health, safety and/or environmental issues. For example, lead compounds, barium compounds and cadmium compounds are toxic and damaging to the environment. The most commonly used solid lubricants are molybdenum disulphide,

Layer-lattice compounds		
Molybdenum disulphide	Graphite	Calcium fluoride
Tungsten disulphide	Graphite fluoride	Barium fluoride
Tantalum disulphide	Tungsten diselenide	
Polymers		
Polytetrafluoroethylene	Nylons	
Polytrifluorochloroethylene	Acetals	
Polyvinyl fluoride	Polyimides	
Ultra-high MW polyethylene	Phenolics	
Powdered metals		
Lead	Tin	Zinc
Silver	Aluminium	Gold
Inorganic powders		
Lead oxide	Boron nitride	Lead sulphide
Zinc oxide	Boron trioxide	Barium oxide
Molybdic oxide	Cadmium oxide	

FIGURE 15.1   Types of solid lubricants.

Source: Pathmaster Marketing Ltd.

graphite, PTFE, nylons, acetals, ultra-high-molecular-weight polyethylene, pow-dered tin and zinc oxide. As noted earlier, some of these are used as thickeners in greases.

Solid lubricants are required to provide low friction and wear, to have good film-forming ability, ductility, good adhesion to the substrate, film continuity and low shear strength. They should also have high thermal, oxidation and hydrolytic stability, high softening and melting points, high thermal conductivity and diffusiv-ity and low thermal expansion. They should be chemically inert, have appropriate electrical conductivity if required, have no abrasive impurities and have low toxicity and good environmental compatibility.

There are several advantages to using solid lubricants in specific applications. They have almost no tendency to flow, creep or migrate, so should remain in place for long periods. There is little tendency to contaminate adjacent systems or products. Some are usable in high vacuums, such as on spacecraft. Some are usable at very high or very low temperatures. They are often inert to other chemicals and are gener-ally more stable to radioactivity.

However, they have just as many disadvantages. They are difficult or impos-sible to feed or replenish, have a limited life due to finite wear rates and have lim-ited sliding speeds due to poor thermal conductivity. They have different thermal expansion from metals, so clearances between moving surfaces can be lost with temperature changes. Friction is generally higher than with liquids. They are more complicated to apply, have no cooling properties and only limited corrosion protec-tion abilities.

Unlike greases, solid lubricants are usually used only in very specific applica-tions. In general, monitoring solid lubricants is either impossible or impractical. As with many other machines, items of equipment or systems, the performance of a solid lubricant can only be monitored by monitoring the machine or system, often using vibration or noise analysis and/or thermography.

15.3 GENERAL CONDITION MONITORING
FOR GREASES

Although greases are required to stay in a mechanism, they do need to be able to move within the mechanism and to be pumped into the mechanism in many applications. For a grease to lubricate moving parts, it must flow, but under the proper circumstances.

Greases are often described by their consistency, or penetration, as outlined in an earlier chapter. Penetration, as indicated by the NLGI (National Lubricating Grease Institute) or ELGI (European Lubricating Grease Institute) number, is a measure of a grease's hardness or softness, rather than its viscosity. As such, consistency is a physical parameter used by grease manufacturers, equipment manufacturers and users to specify, recommend and purchase a grease for their applications. Therefore, the importance of consistency to the end user is very high.

However, it is obvious that two greases having the same NLGI number are not necessarily the same. It has been shown that some greases pump harder, are stickier or look harder or runnier than other greases with the same penetration number.

While penetration has been the standard for measuring consistency, newer methods have been developed that could provide a more complete picture of grease properties. Other physical properties of greases and equipment should also be considered, including pumpability, bearing speed, operating temperature, grease bleed and environmental conditions.

Two types of flow must be considered when selecting a grease: flow from the grease pump into the machine and flow within the lubrication contact zone. In turn, flow is controlled by two other properties: cohesion and adhesion. NLGI defines cohesion as the molecular attraction of the molecules within a grease and adhesion as the forces that cause two substances, such as grease and metal, to stick together. Both can affect how difficult or easy it is to make a grease begin to flow.

The first challenge is to get the grease moving by overcoming the yield stress, which is the minimum amount of stress needed to make a plastic-like material flow. In general, flow is affected by the dynamic viscosity of the lubricant, which is the viscosity under conditions of shear. However, grease viscosity is difficult to measure because most greases do not flow easily, due to their cohesive and adhesive properties.

Another important property is pumpability. A grease will not function if it cannot be transported to the lubricating point. Centralised lubricant systems are often used to improve efficiency, convenience and safety. Various factors affect pumpability, including temperature, distance to be pumped, pump input pressure and the inner diameter of the pipe.

Operating temperature also has a significant effect on grease flow. The standard penetration test is performed at 25°C, but this is not always a typical operating temperature. Also, studies have shown that grease flow changes with temperature in a nonlinear way, but the NLGI rating system is linear.

Oil bleed from the grease must also be considered. For most users, bleed is when a grease becomes thin enough (or the dynamic viscosity is low enough) for it to flow out of a bearing. The technical definition of grease bleed is when the base oil and additives release from the grease. This condition can create storage problems and can result in the grease appearing to be thicker than it should if the oil is not reincorporated.

Environmental conditions affect grease consistency, both in storage and in service. Contamination is one important environmental factor. Contamination from another type of grease can be very serious, as some greases are not compatible with other greases. Two greases get mixed when an operator pumps the wrong grease into the application. Other environmental contaminants include dirt, water, chemicals and wear metals. Depending upon concentration, any of these can affect grease consistency, some dramatically. Therefore, consideration should be given to selecting grease ingredients such that consistency can be optimised to manage contamination.

As part of an overall condition monitoring programme for greases, ASTM has developed ASTM D7918 "Flow Properties and Evaluation of Wear, Contaminants, and Oxidative Properties of Lubricating Grease by Die Extrusion Method and Preparation". Changes in the wear, contamination, consistency and oxidative properties of a lubricating grease or deviations from the new grease can be indicative of problems with the lubricated component. Problems can include the mixing of incompatible thickener types, excessive wear or contaminant levels or significant reductions in levels of oxidation inhibitors. The test methods in ASTM D7918 also make it possible to develop trends that can be used to predict failures before they occur and allow for corrective action to be taken. The standard provides guidance on evaluating in-service grease samples, NLGI grades 00 to 3.

A grease sample of known volume is measured to determine the density of ferrous material in the sample. The grease sample held with a defined geometry sample holder is placed into a temperature-controlled instrument and extruded onto a substrate as a thin ribbon or strip of grease. The grease is tested at three different rates to reflect the non-Newtonian nature of greases. Testing at several different rates creates a series of step changes that are then compared with an unused baseline grease that also has been tested under the same conditions. While the flow properties are being measured, the grease is simultaneously deposited onto a thin film substrate containing substrate segments. Each substrate segment contains approximately 0.25 g of grease. The individual substrate segments are used for further testing of wear, contamination and oxidative properties. The substrate is removed from the instrument and processed further to obtain information related to the chemistry and content of the grease sample, including linear sweep voltammetry and grease colorimetry.

15.4   GREASE LUBRICATION FOR BEARINGS

Four primary components determine grease lubrication for rolling bearings:

- Grease selection.
- Application method.
- The volume of grease to be delivered.
- The frequency with which the grease is applied.

Many different methods can be used for specifying these values. Opinions can vary significantly as to which approach is best. For this reason, the best approach may be to use more than one method and develop a strategy that defines default values which are then fine-tuned based on feedback from visual inspections, operating temperature, acoustic monitoring and others. Due to the variability of operating conditions and machine design, it can be very difficult to be truly precise without introducing a "condition-based" component to the method. Condition-based lubrication should be part of any effective monitoring programme.

When developing a grease lubrication strategy, a good starting point is to select the right grease. Simply selecting the best quality grease is not as important as choosing the correct grease for a given application. Too often, grease selection is over-simplified and the key properties are overlooked. Grease selection is actually more complicated than lubricating oil selection.

Proper grease specification requires all of the components of lubricating oil specification, including base oil viscosity, additive requirements and base oil type, together with other special considerations for grease selection, including thickener type and concentration, consistency, dropping point and operating temperature range. OEM recommendations coupled with operating experience can be very useful.

Several different methods are used applying grease. It can be applied through centralised application systems, single-point automatic application systems, hand packing or a manual grease gun. A compelling argument can be made for the superiority of continuous application systems, but for many applications, these systems are impractical.

Manual grease application, when done correctly, is a very effective method and does provide advantages over automatic systems. One requirement of manual application is that the technician is normally in close proximity to the lubricated component. This allows for inspections to be made in conjunction with the re-lubrication activity. In addition to sensory observations (sight, smell and sound), instrument inspections, such as acoustic monitoring and temperature readings, may be used to provide component condition information and to fine-tune the lubrication activity. The addition of acoustic monitoring to a well-developed greasing strategy will help take the lubrication programme to a world-class level.

One of the more important and often badly carried out parts of condition monitoring is re-lubrication volume. Many industrial users of grease do not know how much grease to apply when re-greasing a bearing. Several acceptable methods can be used, although one of the most common is that recommended by SKF, in which grease replenishment volume is defined by:

$$Gp = 0.005 \times D \times B$$

where Gp is grease replenishment amount in grams, D is bearing outside diameter in millimetres and B is total bearing width, also in millimetres. This method generally provides good results but does not necessarily take all factors into account. For example, it does not account for differences in bearing housings and application points. Instead, it assumes that the grease is added at the optimum location. Also, it is not always possible to know the amount of and the condition of old grease in the housing at the time of reapplication.

For these reasons, it may be advantageous to modify the calculated values with a condition-based approach. The most advanced condition-based technique is the use of acoustic instrumentation to optimise the re-lubrication volume. By establishing a baseline value and determining a statistically appropriate limit, the volume of grease added can be optimised.

The component of the greasing strategy that has the most variability is the frequency of re-lubrication. Many factors must be considered in order to be even reasonably precise in determining the best application frequency. These factors include operating temperature, seal type and condition, particulate contamination, moisture, vibration and grease quality. Although there are several methods for calculating frequency, some of which take many of these factors into consideration, they can still generate significantly different values. A grease re-lubrication chart with associated correction factors is published by SKF.

Optimising a grease lubrication strategy can be difficult. The best approach is probably to combine the latest technology with proven traditional methods, combined with experience and knowledge. To keep it simple, it involves using the right lubricant in the right place, with the right application method in the right amount and at the right time.

15.5   MONITORING AUTOMOTIVE GREASES

Automotive greases, like automotive engine, gear and transmission oils, are almost always used in small amounts, so sampling and monitoring them is completely impractical.

The three main applications for automotive greases are in rolling bearings, constant velocity joints and truck and bus chassis lubrication systems. In the first two applications, the equipment is "sealed for life" and does not need re-lubrication. If the equipment fails, it is simply replaced. There is very little monitoring of either rolling bearings or constant velocity joints, except on large off-highway machines, where vibration analysis and operating temperature measurements can be used to detect possible lubrication problems.

With chassis lubrication systems, new grease is added at periodic intervals, usually using an automatic pressurised grease delivery cartridge. Surplus grease coming out of the lubrication points simply drips into the environment.

15.6   MONITORING INDUSTRIAL GREASES

In the past, bearings were re-greased as a function of time. The grease quantities and lubrication intervals were calculated numerically. Now, more attention is being paid to condition monitoring of greases and bearings in order to determine re-greasing or grease renewal.

Greases, like oils, contain a variety of additives. Antioxidant levels are of particular interest in identifying the useful life of a grease. Differential scanning calorimetry (DSC) is a modern analytical method for measuring the onset of oxidation in used grease (ASTM D5483). When compared with the new reference grease, the test can be used to determine the remaining useful life (RUL) of a grease. This test is analogous in the information it seeks, if not in methodology, to the RPVOT test (described in Chapter 6) commonly used to determine the RUL of turbine oils and other lubricating oils.

The viscosity of grease is often misunderstood. The viscosity typically listed on a new grease data sheet is usually the kinematic viscosity of the oil used in making the grease, measured using ASTM D445, IP 71, ISO 3104, DIN 51562 or AFNOR T60-100. The kinematic viscosity of the base oil is important in ensuring the correct grease containing the correct grade of oil is used for lubrication purposes. However, the viscosity of the grease itself can be measured, using ASTM D1092. Since a grease is non-Newtonian, only the apparent viscosity is measured, as described in an earlier chapter.

Rheology measurements of grease may begin to replace both the penetration and the apparent viscosity measurements. Rheology is the study of the deformation and/or flow of matter when it is subjected to strain, temperature and time. A rheometer only requires a few grams of sample to perform the analysis, yielding much more information than the penetration or the apparent viscosity measurements. This makes the rheology measurement an ideal test for small amounts of used grease.

The dropping point of a grease establishes its maximum usable temperature, which is usually set at 50 to 100°C below the experimentally determined dropping point. Dropping point can help to establish if the correct grease was supplied or is in use and to determine if a used grease is good for continued service.

Many bearings fail prematurely due to contamination. Grease contamination can come from common environmental contaminants such as dirt and water and from cross-contamination from other grease types, as explained earlier. Contamination from water or other grease types can be identified by Fourier transform infra-red (FTIR) spectroscopy. FTIR can also measure gelling agent type and concentration, as well as with oxidation by-products.

FTIR can identify the presence of water in greases. However, it is not sensitive to low levels. Water in greases in the ppm range can be measured using a variation of ASTM D6304. This method allows for the distillation of water using a distillation tower at 120°C into a titration vessel where it is soublised in toluene and sparged with nitrogen. The toluene/water mixture is then titrated using Karl Fischer Reagent as per ASTM D6304. The levels of detection using this method are in the low ppm range.

Occasionally, if cross-contamination with a different type of grease is suspected, elemental analysis (after acid digestion) can be used to check. For example, if a grease that is supposed to be an aluminium complex grease has become contaminated with a calcium sulphonate complex grease, both aluminium and calcium will show up in ICP-AES spectroscopy, described in Chapter 6.

The methods for establishing criteria, and methods for alert and alarm values, vary according to the experience of the user. Alert values are those considered to be above or below the norm, while alarm values are those beyond a safe operating level. Absolute values, also referred to as fixed or hard number, may be assigned to any characteristic. These values are based on the equipment type and grease type and grade. In some cases, fixed values may be obtained from the OEM. In cases where there are no recommended values, the fixed limit may be set using the experience of the laboratory with the specific lubricant and machine combination. It is important to remember that hard number alert values are a place to start a programme, which contain many unknown factors.

During the initial phase of the programme, it is not uncommon for the alarm values to remain unchanged and invalidated for a substantial period of time. If the alarm value set is not appropriate for the machine in its unique configuration, the risk of machine failure remains higher than acceptable.

For some tests, such as oil versus gelling agent, it is more appropriate to set values on a percentage change rather than standard deviation. An advantage of this type of alert is that it does not require valid statistical populations if the baseline is considered. Many percentage alarms can be converted to absolutes when the baseline value is a known quantity or the test has published typical values.

Statistical analysis of wear metals is effective on mature databases. This requires a statistically valid population, typically 30 data points or more. It is therefore normally based on similar equipment in a group rather than a single piece of equipment. Once sufficient historical data for the single machine is available, statistical analysis may be applied to the machine alone.

An example of setting alert and alarm limits for wear metals in greases is shown in Table 15.1.

It is easy to see in Table 15.1 which of the samples have high iron, aluminium, copper and silicon contents. The alarms set are based on 18 samples from various parts inside the bearing cavity from different wheel bearings. Usually, as in this case, a data evaluator must make an initial judgement about what is considered normal. After sorting the data set by iron, it is clear there is a break at 144 ppm. Considering all of the samples lower than 144 ppm of iron as normal, the basis for the analysis is established:

- Okay samples: The average of all normal data is added to the standard deviation (STDEV) of all normal data. These samples are considered OK.
- Abnormal samples: Twice the STDEV of all normal samples is added to the Average. These samples are considered ALERT.
- Critical samples: Three times the STDEV of all normal samples is added to the Average. These samples are considered ALARM.

Conventional methods for analysing wear debris are ferrographic analysis and elemental analysis. While the quantitative estimation of wear debris is difficult in a used grease sample using elemental analysis, because of the difficulties of obtaining a representative sample, ferrographic analysis, which by its very nature is a qualitative technique, is ideal in determining the active wear mechanism and severity of the problem in grease-lubricated bearings. Ferrographic analysis on used greases is carried out by extracting the wear debris from the sample and analysing it visually using an optical microscope, in a way similar to how ferrography is used for used oil samples.

Setting alert values-based trend analysis, or on the slope of the curve, for a specific wear metal above a predetermined minimum threshold value can be accomplished after the initial three sets of data are entered into the database. The logic behind

TABLE 15.1

Establishing Alarm Limits for Metals Contents of Industrial Greases

Sample	Fe	Al	Cu	Si
	Element Content, ppm			
1	8	7	3	17
2	18	6	80	17
3	11	0	0	18
4	31	5	34	18
5	40	0	120	28
6	12	4	28	29
7	89	0	32	34
8	0	0	0	39
9	163	0	24	40
10	84	0	10	43
11	32	0	0	48
12	234	98	540	323
13	345	87	340	830
14	300	43	234	167
15	378	16	253	230
16	628	28	8	1,110
17	144	29	458	230
18	68	0	0	10
Average	36	2	28	27
Standard Deviation	31	3	39	12
STDEV + Average	67	5	67	40
2 × STDEV + Average	98	8	100	51
3 × STDEV + Average	129	11	145	63

Source: Noria Corporation, with permission.

the three histories is simply that it takes a minimum of three points to calculate a curve. While this can provide additional information to the analyst, it relies heavily on obtaining correct and consistent operating time values, normally hours. It is also more likely to be invalidated by other variables such as inconsistent sampling techniques.

Acoustic monitoring has for many years been used successfully to monitor the condition of electrical systems and identify leaks in vacuum, compressed air, steam and other fluid transfer operations. In recent years, more and more maintenance professionals have come to rely on this technology to also monitor the condition of mechanical components and even monitor bearing lubrication condition. The amount of noise produced by a lubricated bearing can be a useful indicator of the effectiveness of the lubricating film.

Rolling element or anti-friction bearings typically employ an elasto-hydrodynamic lubricating film. In this type of lubrication, the loaded surfaces elastically deform, and the load is carried by a film of oil sufficiently thick to prevent the interacting surfaces from contacting one another. At a microscopic level, the finished surfaces in the bearings present irregularities or bumps often referred to as asperities. When they collide, it generates noise which can be measured by acoustic monitoring devices.

In a properly lubricated bearing, these collisions should be few and, thus, generate a relatively low noise level, but as the grease in a bearing is "used up", the oil film begins to dissipate and the collisions become more frequent and create more noise. While it is certainly possible to hear this phenomenon with a stethoscope, the acoustic instruments allow it to be quantified and provide an objective interpretation of the sound levels. Based on the normal or baseline noise levels for a particular bearing, limits can be established that alert the technician to the precise time the bearing requires re-lubrication and even indicate when to stop applying grease to prevent over-lubrication.

A more modern alternative to penetration for determining changes in the consistency of used greases is thermal gravimetric analysis (TGA). TGA measures the mass of a substance in relationship to temperature and is used to determine the loss of material with increasing temperature. The analysis can be carried out in an inert atmosphere such as nitrogen or a reactive atmosphere such as oxygen. Typically, a few milligrams of the sample is weighed and heated under controlled conditions. The weight loss at specific temperatures allows the technician to evaluate the oil/gelling agent ratio as compared with new (unused) grease, as well as the presence of volatile compounds such as water, allowing any significant change in gelling agent chemistry to be determined.

An innovative sensor has been developed to enable online condition monitoring of greases in rolling bearings. The sensor is ideal for monitoring critical plant and machinery located in difficult-to-access areas, such as wind turbines and automated assembly lines. The sensor has been developed jointly by The Schaeffler Group, Freudenberg Dichtungs- und Schwingungstechnik and Klüber Lubrication. It incorporates what is claimed to be a unique electronic evaluation system, which enables the condition of the grease to be monitored while the bearings are operating. The sensor is positioned directly in the grease in the bearing.

Schaeffler believes this is a significant breakthrough, as the schedule for replacing rolling bearing grease can be planned into maintenance schedules precisely. Any changes in the condition of the grease can be detected early, before any damage might be caused to the bearings. With preventive maintenance regimes, the operating life of the grease is critical, particularly if it is less than the expected life of the bearing. Schaeffler claims that the new grease sensor enables grease to be replaced according to the actual operating requirements of the bearing and not according to any pre-defined time period.

The sensor, which has a diameter of just 5 mm and a length of 40 mm, is able to detect four parameters of the grease: water content, cloudiness (opacity), wear (thermal or mechanical) and temperature. From these, the sensor's electronic evaluation system uses complex software algorithms to generate an analogue signal (4 to 20 mA), which then displays the condition of the grease. By setting alarm thresholds (limit values), digital signal outputs can also be generated, indicating whether the grease quality is "poor" or "good", ranging from 100% for as-new, to a theoretical

0% for an unusable grease. A user can decide at which point grease re-lubrication or replacement should be carried out.

The sensor operates by using the optical, near-infrared reflection principle. This method, developed in conjunction with the Fraunhofer Institut for Electronic Nano Systems (ENAS) in Germany, is based on an infrared process used by laboratories to measure the quality of grease, but has been adapted for online measurements in rolling bearings. The know-how involves both the set-up of the sensor and how the measurements are evaluated.

Evaluating the measurements involves rotationally symmetrical irradiation of the grease by the sensor, at an angle of 45° using specific wavelengths within the infrared spectrum. The sensor head is embedded directly in the grease during this procedure. The reflected light is measured perpendicular to the grease, which enables any shadow effects or surface anomalies to be excluded completely. The reflected light is then evaluated in terms of the quality of the grease. In terms of sensor set-up, the optimum measurement point will vary depending on the application. The measurement depth of the sensor extends only a few millimetres into the grease. There must be grease directly in front of the sensor for measurement. Air inclusions can lead to incorrect measurements. Tests carried out on the rolling bearing lubricant test rig FE8 in accordance with DIN 51819-1 have shown that the sensor must not record grease in direct rolling contact. The areas adjacent to the raceway also contain highly homogeneous grease conditions. As a result, comparable measurement results can be obtained.

Schaeffler, Freudenberg and Klüber have validated the measurement method for around 95% of greases currently available. A further solution is currently being developed for integration of the sensor in rolling bearing seals. The sensor is now being used in bearings made by FAG.

15.7 SUMMARY

Modern methods of analysis for used grease samples from industrial and large off-highway equipment or systems are rapid, sophisticated and require only a fraction of the sample volume necessary in the past. Sound, cost-saving, maintenance decisions can be made using grease analysis as the basis for preventive and predictive programmes.

Monitoring greases used in automotive applications is generally impractical, generally due to the small volumes used. Monitoring solid lubricants is either impossible or impractical, mainly due to the very specific or extreme applications in which they are used.

16 Lubricant Condition Monitoring Programmes, Their Implementation, Benefits and How to Avoid Problems

16.1 INTRODUCTION

Lubricant condition monitoring and machine condition monitoring can be expensive, complex and time-consuming. They are not suitable for all mechanical equipment. Fortunately, for large and expensive machines, items of equipment and systems, they can be extremely cost-effective.

Unfortunately, there are numerous significant issues that must be addressed in order to run an effective condition monitoring programme. Avoiding problems is very worthwhile and provides significant benefits when planning, implementing and running a condition monitoring programme.

This final chapter sets out why and how to define, plan and implement an achievable lubricant condition monitoring programme. It also describes and discusses the benefits of using the results of a lubricant condition monitoring programme to instigate effective predictive maintenance on all machines, items of equipment and lubricated systems.

16.2 DEVELOPING A CONDITION MONITORING PROGRAMME

The goals of a lubricant monitoring programme will determine the amount of useful data which needs to be acquired in order to perform the proper analysis, so as to be able to make meaningful judgements about the maintenance of lubricants and equipment.

At the beginning of a condition monitoring programme, a decision must be made as to whether the programme will aid decision-making regarding the condition of the equipment by looking at just the lubricant or by looking at the conditions of the lubricant and the equipment. These different goals require different approaches. Many condition monitoring programmes begin by just looking at lubricants. However, it eventually becomes apparent that a number of aspects of the behaviour of the equipment need to be included.

DOI: 10.1201/9781003245254-16

Therefore, to avoid (or at least, minimise) future problems, it is wise to start by monitoring both lubricants and the specific machines they lubricate. A common misconception in the evolution of a condition monitoring programme is that the transition between lubricant monitoring and machine condition-based maintenance is the difference between the types of testing that are done on the samples. This is not necessarily true. Of course, the type of test that is run on a sample will provide specific information either on the condition of the lubricant or on the condition of the equipment it came from. The difference is the location the lubricant was extracted from the equipment, the method that was used to obtain the sample and the tools that were used.

Integrating lubricant analysis into an established maintenance programme can yield significant returns in the form of more reliable and longer-lasting equipment. However, in a recent survey in the United States, 43% of oil analysis programmes left half of the equipment unsampled and 36% did not adjust the preventive maintenance activities based on the results. Changing a reactive or preventive maintenance programme to one that predicts and avoids wear or failures requires the right people, processes and technology, but the rewards are more than justified.

Many maintenance programmes have an uncertain start when lubricant analysis is first included. Either the value is tested using a pilot programme or a starting location is selected that will discover the best way to fit lubricant analysis into a maintenance programme, before rolling it out to the entire plant or company. Sometimes, it is believed to be easier and more feasible to select one or two types of equipment to test at the beginning. Adding the task of collecting oil samples regularly from all equipment is not realistic for most maintenance programmes, particularly if staff and budgets are already stretched.

The first step is identifying where lubricant analysis is likely to achieve the greatest financial returns. Focusing efforts on increasing the reliability of particularly "critical" and "problem" equipment helps time-starved maintenance engineers and production supervisors to get on top of the work load. It also provides a bonus of demonstrating to management the largest benefits.

During the last 30 to 40 years, a number of machinery, equipment and system maintenance methods have been designed and introduced. These have been slowly amalgamated into a conventional set of terminologies. Many machine, equipment and system maintenance decisions are based on an assessment of what has come to be called "Overall Machine Criticality" (OMC). This assessment includes lubricant condition monitoring and machine condition monitoring. Critical equipment should be checked more frequently than less critical equipment. Based on the definition of "critical", this refers to the machines with the highest importance to the plant or company. Obviously, knowing how to define an asset as "critical" is essential and there are many approaches to this. Some plants and companies use a simple 1 to 10 scale and subjectively assign numbers to each item of equipment.

The OMC assesses criticality in the context of lubrication. It is calculated as the multiplied product of the "Machine Criticality Factor" (MCF), which relates to the consequences of machine failure, and the "Failure Occurrence Factor" (FOF), which corresponds to the probability of failure. A more detailed discussion of these factors, and the OMC, can be found at www.machinerylubrication.com and an article "Mac hinery-criticality-analysis".

Using the OMC, a machine's candidacy for lubricant analysis is influenced by factors such as:

- Whether the machine is exposed to failure-inducing conditions, such as high loads, speeds, shock, contamination and so on.
- Whether the machine is suffers from chronic problems.
- Whether the consequences of failure are high, in terms of safety, downtime, repair costs, environmental effects and others.
- Whether failures can be lubricant-induced, for example, from degraded or contaminated oil.
- Whether failures can be revealed by the lubricant, for example, wear debris from shaft misalignment.
- Whether early detection is important.

In addition to a machine being critical to a plant's operation, there are some cases when the lubricant is also critical. This is assessed as the "Overall Lubricant Criticality" (OLC). The OLC defines the significance of lubricant health and lon-gevity as influenced by the probability of premature lubricant failure and the likely consequences for both the lubricant and the machine. The method of determining the OLC tends to be rather subjective, although it is grounded in solid principles in applied tribology and machine reliability.

The "Lubricant Criticality Factor" (LCF) defines the specific economic conse-quences of lubricant failure separate from machine failure consequences. The LCF is influenced by the cost of the lubricant, the cost of downtime to change the lubricant, flushing costs and system disturbance costs. For example, machines that use large volumes of expensive, premium lubricants will understandably have high LCF val-ues. Studies have shown the true cost of an oil change can far exceed ten times the apparent cost, in terms of labour and oil costs.

The "Degradation Occurrence Factor" (DOF) defines the probability of prema-ture lubricant failure. The conditions that influence this probability are:

- Lubricant robustness: Synthetic lubricants and other chemically and ther-mally robust lubricants lower the DOF.
- Operating temperature: Lubricants exposed to high operating temperatures, including hot spots, can experience accelerated oxidation and thermal deg-radation, and these conditions will raise the DOF.
- Contaminants: Water, dirt, metal particles, glycol, fuel, refrigerants, process gases and others can greatly shorten lubricant service life, thereby raising the DOF.
- Lubricant volume and top-up rate: Lubricant volume relates to the amount of additives available to reduce lubricant degradation, the estimated runtime to complete additive depletion and the density of contaminants. In normal ser-vice, it can take some time for the additives in systems containing thousands of litres of lubricant to be used up. The top-up rate refers to the introduction

of new additives and base oil(s). New additives replenish depleted additives, and new base oil dilutes pre-existing contaminants. High oil volume and a high makeup rate will reduce the DOF.

Machines that are good candidates for lubricant condition monitoring have high OMC or OLC values. Even marginal OMC or OLC machines may be suitable for a reduced lubricant analysis programme, with fewer samples and fewer tests. Using this methodology, much of the guesswork is taken out of the first major decision related to any oil analysis programme. Determining the OMC and/or OLC values enables candidate machines to be selected. Then the lubricant sampling locations, sampling frequencies, list of tests, flagging limits and alarms, data interpretation strategies, presentation of results and maintenance procedures can be defined.

Machines, items of equipment or systems that have a history of high rates of failure are good initial candidates for lubricant and machine condition monitoring. Units can be grouped by type, manufacturer, model, application, replacement cost, hours/miles operated or how vital they are to production. Calculating the maintenance costs (especially rebuilds and replacements) and the number of failures will typically identify the machines, equipment or systems that would benefit the most from lubricant analysis.

Presenting the business case for condition monitoring in this way is likely to convince management. Maintenance and equipment costs are unavoidable, but they can be reduced. Experienced managers are willing to listen to solutions to problems and strategies for improvement, especially when they are supported by real-world data. However, the business case must be completely realistic.

Management approval to begin a condition monitoring programme is just the first step. Decisions about sampling locations, sampling frequencies and sampling methods come next. A detailed action plan, selection of appropriate staff, preparation of written procedures and a training programme are required. Exactly who will be responsible for what duties and how their work will be evaluated must be identified. Dates or triggers for staggered rollouts should be set, but leaving spaces to deal with delays and unexpected issues. Although there is nothing wrong with exacting schedules, falling behind can demoralise staff and raise management concerns. Everyone involved with the programme needs to know how it will benefit the maintenance plan before they can start believing it will help them and the company. The training also provides an opportunity for feedback and can identify overlooked obstacles or weaknesses in the plan.

The machines and equipment may need preparation. Retrofitting lubricant sampling ports can expedite the sampling process and free up many maintenance hours. The equipment list needs to be updated in the computerised maintenance management system and laboratory analytical results files. Trend analysis graphs need to be set up.

The best way to accelerate the adoption of a new condition monitoring programme is to measure the progress and share the results. One way to achieve this is to "grade" the progress of each section/location/division and discuss them with all staff involved. Peer pressure and friendly competition can go a long way towards motivating slow adopters. Reporting major cost-savings to the whole company can be very motivating to the staff involved. The reviews also help to establish best practice and identify any issues.

It is important to capture information about maintenance and replacements that were avoided due to lubricant and machine condition monitoring. Much time can be saved when information is documented as quickly as possible. Information management software, or even a simple spreadsheet, can track the hours spent and the cost of parts for each project. The hours spent conducting repairs multiplied by the engineer's labour rate, added to the cost of parts, provide the total cost of an individual repair. The cost of lubricant testing and the labour spent collecting samples provides the monitoring costs. By adding these and comparing the total to the costs of complete machine or system replacement, more involved repairs and/or loss of production, the immediate return on investment (ROI) can be determined.

The final step to integrating lubricant analysis into a maintenance programme is to take advantage of the sample result data as a whole. Basic programmes just review individual results for unexpected wear. Although this is likely to save equipment and reduce downtime, all the past test results contain a lot of information that can be mined for even more savings. Lubricant analysis providers can sort and filter specific results from the data and compile them into a management report. A wide variety of reports are possible and the helpfulness of each depends on each person's position in the maintenance programme.

16.3   PREDICTIVE MAINTENANCE OF MACHINES, EQUIPMENT AND SYSTEMS

Because condition monitoring measures lubricant and equipment parameters to identify trends or changes, the opportunity exists to use the data to predict impending equipment failure. Predictive maintenance analysis uses the measurements to predict when equipment maintenance should be undertaken.

Effective condition monitoring programmes allow preventive action to be taken without unplanned downtime, by identifying and detecting equipment failure modes (FMEA, see Chapter 2) and predicting the rate of progression of deterioration or impending failure. This is called predictive maintenance. The trend analysis, root cause analysis and failure mode and effect analysis methods described in previous chapters enable users of lubricants and equipment to predict the future with some degree of accuracy. The approach can also be described as preventative maintenance.

Predictive maintenance strategies can extend a machine's operating life by addressing issues before they develop into expensive failures, while reducing unnecessary maintenance. Lubricant and equipment condition monitoring programmes are becoming more common as organisations recognise how they can increase reliability and reduce costs.

However, not all failure modes can be found through lubricant analysis, so it is important to start with the initial steps of identifying all failure modes, applying a criticality number and then deciding if lubricant monitoring and testing can help with those modes. Using failure mode and effect analysis (FMEA) to identify all failure modes enables a clear direction on appropriate actions that need to be taken. When used correctly, lubricant analysis can be the earliest indicator of impending machine failure. An effective condition monitoring programme uses root cause analysis and FMEA to select the correct equipment to test and the right tests to use to evaluate specific failure modes.

Trend analysis and flagging limits are intended to alert maintenance engineers that action needs to be taken. The aim of a trend or a limit is to make it easy to spot when something unusual is occurring. OEMs sometimes provide information on how to set alarm or condemning limits for lubricants or equipment, although OEM recommendations may not include all of the parameters needed and may not be useful for a specific application. If alarm limits are not set by OEMs, equipment maintenance engineers or plant supervisors can establish them either by experience or statistically, generally using trend analysis. ASTM Guides D7669 and D7720 are particularly useful in this regard.

Alarm limits tend to be static, while trend analysis is able to identify abnormalities. Combining both approaches is synergistic. Statistical process control (explained and discussed in Chapter 14) can be used to establish alarm limits. These should initially be developed by reviewing a statistically acceptable population of relevant data together with data associated with failures, if this is available. However, it is very important to recognise that the operating conditions for machines, equipment or systems may not be identical, so that statistics may not always prove useful. The author has experiences of identical machines installed at the same time in fairly close proximity in manufacturing plants, suffering different rates of deterioration. Extensive investigations were unable to determine why this was so. It is therefore important to establish trend lines for each machine, item of equipment or system. The flagging limit for one machine may be slightly different from an apparently identical machine. It is possible that there is no such thing as "normal" operating conditions. Over-reliance on alarms shifts focus from identifying underlying trends that might truly predict a failure before it occurs.

Predictive maintenance methods are intended to help determine the condition of machines, equipment or systems so as to establish when maintenance should be performed. The approach has the potential for cost savings over routine or time-based preventive maintenance, because tasks are performed only when warranted. It is viewed as condition-based maintenance carried out as suggested by estimations of the degradation state of a machine, item of equipment or system. Its main aim is to allow convenient scheduling of corrective maintenance and to prevent unexpected equipment failures. This leads to optimum equipment lifetime, increased plant safety, fewer accidents with negative impacts on the environment, optimum spare parts inventory and handling and, thereby, lower costs. Predictive maintenance differs from preventive maintenance because it is based on the actual condition of equipment, rather than average or expected life statistics.

16.4 LUBRICANT SAMPLING ISSUES

Lubricant sampling and extraction are perhaps the most important, highly variable and easily overlooked step taken prior to the analysis of the sample. It is also the easiest to make consistent. The methods and procedures used for sampling will determine the amount of useful data that can be acquired from the sample, as illustrated in Chapter 3.

Before deciding the appropriate location(s) and way(s) to extract samples from the equipment, the desired end result must be defined. It is then possible to work backwards to ensure the outcome can be achieved in practice. The end result can be determined by first examining the reason(s) for the lubricant and condition monitoring programme. It is then possible to make decisions on the location(s) for taking samples and what equipment and methods to use.

For example, a sample taken from a hydraulic system reservoir will provide only a limited amount of information on the condition of the system. Samples from a number of locations may be needed. Reservoir samples (primary samples) can provide excellent data on the homogeneous properties of the lubricant, including acid number, remaining useful life, viscosity and additive properties. A primary sample is all that will be needed if the only aim of the lubricant monitoring programme is to monitor the health of the lubricant. It is easy and inexpensive to take only a primary sample and, in some lubricated items of equipment, a primary sample is the only type that can be obtained. For example, a bearing housing or small gearbox will only have one primary sample port.

Secondary samples taken at specific strategic locations throughout a system will provide much more information about its condition. On a hydraulic system, a primary sample port is likely to be on a main return line and secondary sample ports will be downstream of major components such as pumps, motors, coolers, valves and filters. Secondary samples support what may be found in a primary sample and help to determine the cause of a problem.

A number of rules apply to taking samples, so as to avoid (or at least, minimise) problems.

Samples should only be taken from operating machines. Lubricants in "cold" systems MUST NOT be sampled. This rule goes beyond simply starting the machine to take the sample. When at rest, anything heavier than the oil will begin to settle. It takes only 2 minutes for a 20 microns (μm) particle of Babbitt bearing metal to settle 15 cm in an ISO 22 bearing oil. The principle behind oil analysis is to capture a "snapshot" of the system at the time of sampling. The timing of the sampling should be when the system is under the greatest amount of stress. Typically, the best time to sample a system is when the system is under normal working load and normal operating conditions. This can be a tricky task when sampling from a system that continuously cycles during normal production, such as the hydraulic system on an injection moulding machine. It is under these conditions that the samples will best represent the machine conditions most likely to cause accelerated wear.

Oil samples should be taken upstream of filters and downstream of machine components. These may be the same locations. Filters are designed to remove wear debris and contaminants, so sampling downstream of them provides less value. However, taking a sample before and after a filter for a simple particle count will reveal how well the filter is operating currently. Obviously, the particle count before the filter should be higher than after the filter. If it isn't, it is time to change the filter. Condition-based filter changes can be very important for sensitive systems. Filters are inexpensive compared with the cost of machine failure.

Written procedures should be established for each machine, item of equipment or system sampled. Sampling methods or locations MUST NOT be changed. Everything in oil analysis and machinery lubrication should have a detailed procedure to specify what and how things need to be done. Each maintenance point in the plant should have specific and unique procedures detailing who, what, where, when and how. Lubricant sampling procedures are no different. The sample location, the amount of flush volume, the frequency of sampling, the timing within a cycle to sample and what tools and accessories to use on that specific sample point based on lubricant type, pressure and amount of fluid required need to be specified.

It is important to ensure that sampling valves and sampling devices are thoroughly flushed prior to taking the sample. Dirty sampling equipment or reused sample tubing should NOT be used. Cross-contamination has always been a problem in oil sampling. Flushing is an important task that is often overlooked. Failure to flush the sample location properly will produce a sample with a high degree of variability. Flushing prior to sampling needs to account for the amount of dead space between the sample valve and the active system multiplied by a factor of 10. If there is a pipe 30 cm long between the sample valve and the system and it holds 30 millilitre (ml) of oil, a minimum of 300 ml of flushing will be needed before taking the sample. Flushing the dead space also will flush the other accessories such as the sample valve adapter and new tubing.

Samples must be taken at appropriate and consistent intervals. Sampling when "time permits" is likely to lead to problems, particularly if a supervisor, engineer or operator "forgets" to take samples at the agreed time. Some of the people responsible for taking oil samples rarely see the results of the analysis. One of the most powerful aspects of lubricant analysis is identifying a change in the baseline of a sample and understanding the rate at which the change has occurred. For example, a sample of new oil should have zero parts per million (ppm) of iron when tested as the baseline. As regular sampling and analysis continue, the iron level may increase. An increase of 10 or 12 ppm per sample might be considered critical. However, if the frequency is not consistent, what is considered normal becomes very subjective. If the frequency of sampling is 12 months, a rise in iron of 12 ppm isn't a major cause of concern. If the frequency is weekly, a rise in iron of 12 ppm is very concerning. Setting up the appropriate sampling frequency and adhering to it will allow for precise analysis and sound maintenance decisions.

Samples MUST be sent to the lubricant analysis laboratory immediately after sampling. Waiting more than 24 hours to send samples for analysis may also cause problems with the accuracy of the results. It may also result in delays in taking emergency corrective action. Lubricant sampling is very like taking a snapshot of the system at a point in time. The health of a lubricated system can change dramatically in a very short period of time. If a problem is developing in a system, the earlier it is detected, the less catastrophic potential it may have. Responding quickly to a problem will not only allow time to plan for a repair, but the repair is likely to be less significant or costly.

When installing oil sampling ports on equipment, it is wise to incorporate permanent sample port identification tags. These tags help to minimise confusion regarding the actual location of the sample port on the equipment and ensure that samples are drawn from the correct location. Sample port identification tags also help to ensure that the correct label is fixed to the sample bottle before it is sent to the laboratory.

Taking samples consistently refers only to the practice of meticulously following a prescribed procedure. It does not take into account the accuracy or precision of the procedure. It is believed (mistakenly) by some analysts that because oil analysis data is commonly trended, once a baseline is established and samples are taken consistently, problems will be revealed in trend-line movements. The fallacy with this belief is that it ensures only repeatability, not data quality or accuracy.

There are many ways to get useful and trendable information from oil samples taken consistently from many different locations. This is particularly true for properties of the oil that remain homogeneous throughout the body of the oil, such as viscosity, additive concentration and oxidation stability. Although a lot can be learned from such data, it is unlikely to provide the greatest amount of useful information. The major concerns with a trending regime are false positives (nuisance alarms) and false negatives (missed alarms) that undermine the condition monitoring programme and erode confidence because sampling practices have been limited to consistency but with no regard for accuracy. There are a number of ways in which this can occur.

Samples taken consistently from the bottom of reservoirs and sumps will show higher (and unrepresentative) concentrations of bottom sediment and water as compared with system live zones. When oil physical properties, contaminants and wear metals are alarmed, it is assumed that blended overall concentrations are being measured, not concentrates in collection bowls, filters and reservoir bottoms.

Samples collected consistently from the turbulent zones of reservoirs and sumps provide trendable information only on homogeneous oil properties. However, wear metals and many contaminants become hidden from view by extraction or dilution. This is because these insolubles, which are commonly ingressed or are generated at the working end of the equipment (hydraulic components, bearings, gears and others), are then deposited in the large tank of oil that is cleaner, or worse, are removed by return-line filters in the case of many high-pressure hydraulic systems. Even if there is no return-line filter, once wear particles, water and solid contaminants enter the reservoir, their concentration will immediately and progressively change due to dilution, settling and off-line (kidney loop) filtration.

Samples taken consistently on the feed-line of large circulating oil systems are typically the same oil with the same precision problems as the reservoir samples described above. This is also true for samples taken from off-line circulating systems, such as filters, heat exchangers or coolers. As such, the actual concentrations of wear metals and contaminants are often hidden from view. This is an easy way to obtain a false negative.

Occasionally, lubricant users prefer to take samples consistently downstream of pressure-line, off-line or return-line filters. Apparently these users are not interested in analysing the presence of particulate matter in the oil, such as the size of particles that filters typically remove, preferring sampling convenience over sampling accuracy.

Taking an oil sample from a dead zone is the same as sampling the wrong machine. Dead-zone oil, gauge-line extensions, regenerative loops and standpipes are stagnant and typically possess properties different from working oils.

There are numerous sampling procedures commonly used that are not really best practice. These include drop-tube sampling (using a vacuum pump), inadequate flushing and using dirty sampling hardware and sample bottles. Although these procedures may be used consistently, they will also consistently fail to optimise the quality and precision of the sample taken. Often these methods are used simply for convenience, in a misguided attempt to save valuable time, at the expense of valuable data and ultimately valuable equipment.

Consistency alone does not ensure quality sampling. It is not possible for even the very best oil analysis laboratory to extract quality data from an unrepresentative oil sample. The adage, garbage in, garbage out is just as true in oil analysis as it is in any other subject of endeavour. With oil sampling, the goals are always to choose a location that maximises the density of the data in the sample bottle and to choose a procedure that minimises the disturbance of the data. Understanding this is not a matter of intuition or intelligence. It can be done only through proper instruction of the person collecting the samples and by following documented best-practice sampling procedures.

16.5   CONTAMINATION CONTROL

In many contamination control programmes, a common approach is to assign a "blanket" target cleanliness level for all machines and equipment. Assigning target cleanliness levels in this manner is certainly not consistent with the goal of component-specific objectives. For example, having a target ISO cleanliness code of –/16/13 for industrial gearboxes suggests that the company has a desire to have a "very clean" status. While such cleanliness is likely to be needed for high criticality machines or components, it may not be necessary to spend the same amount of time and effort on components that are lower on the criticality ranking.

It is feasible for some equipment to be considered critical simply due to component replacement and maintenance costs, yet not be so critical in terms of overall process costs. Equipment that falls in this category may very easily function at what would be a "clean" level with an ISO cleanliness target of –/18/15. It is much easier to achieve and maintain a cleanliness level of –/18/15 than it is to achieve and maintain a –/16/13 cleanliness level.

Solid particle contaminants vary in hardness, friability and ductility, depending on the composition of the particle. Common contaminants such as rust and black iron oxides have a Mohs hardness rating of 5 to 6 (on a scale of 1 to 10, with 10 being the hardest). Environmental dust, mostly silica, has a hardness rating of 2 to 8. Quarry dust has a rating of 5 to 9. From production engineering processes, tool steel has a hardness rating of 6 to 7. Silicon carbide and aluminium oxide have a hardness rating of 9. Diamond is at the top of the scale with a hardness rating of 10. The size, hardness and friability or ductility of the particle influence the amount of damage that the particle can cause.

Cleanliness levels for gearboxes and hydraulic systems were discussed in earlier chapters. To avoid problems with contamination control in operating gearboxes, for example, it is recommended that a 3 μm filter is used during factory testing to remove any contamination left after assembly or added during testing. After the factory test, the gearbox should be drained and flushed, and a new filter should be installed (if there is an online system). If oil does not meet cleanliness requirements after the factory test has been performed, then the gearbox assembly cleanliness should be improved.

If oil samples do not meet cleanliness requirements during service, there may be one or more failure modes in progress, or seals, breathers or maintenance procedures need to be improved. Contaminants may also be generated internally. These particles are usually wear debris from gears, bearings, splines or other components resulting from micropitting, macropitting, adhesion, abrasion or fretting corrosion wear modes.

Lubricant-borne solid particles much larger than elastohydrodynamic (EHL) film thickness can be entrained between gear teeth and between bearing rollers and raceways due to rolling action. Debris is subjected to enormous pressure under contact. Brittle particles fracture into smaller pieces, with some particles embedding in gear teeth and bearing surfaces, and other smaller fragments passing through the contacts. Ductile particles larger than the film thickness are able to pass through the contacts by the combined effects of flattening of particles and denting of surfaces.

Debris dents cause loss of EHL film thickness and lead to stress concentrations at shoulders around dents. Cyclic contacts at these sites generate pressure spikes, plastic deformation and tensile residual stresses that eventually initiate micropits, which may grow into macropits. Hard friable particles such as titanium carbide pulverise into small fragments and promote abrasion, whereas hard ductile particles such as tool steel create deep dents with high shoulders that promote fatigue.

16.6 ASSESSING ANALYTICAL RESULTS AND TREND ANALYSIS

It is important to ensure that all the key tests are used for each type of lubricant being monitored. If some of the key tests are not performed, interpretation of the results could be made on the basis of tests that are unable to define the actual condition of the lubricant. Lubricant analysis reports must also include important information relating to the machine application and the type of lubricant. This is essential for proper interpretation. Other common mistakes include:

- The same set of tests is used for different types of machines, equipment or systems.
- The person assessing the lubricant analysis report lacks the appropriate knowledge or understanding of lubricants and lubrication.
- The lubricant analysis report is received several months after sampling.
- The laboratory does not follow the strict test procedures defined by ASTM, ISO, IP, DIN, AFNOR or other standards.
- There is no quality assurance for meeting the required standards.
- The report contains insufficient information to make any decisions.
- No interpretation or recommendation by the lubricant analysis laboratory is included in the results.
- Cross-contamination between various lubricant samples leads to inaccurate results.

It is important that the person interpreting the test results is qualified and/or suffi-ciently experienced to do so. Most analytical laboratories are able to conduct tests for various forms and states of solids, liquids and gases, including lubricants. However, they may not have someone who is expert in lubricant analysis and interpretation. Ideally, such a person will be a certified Laboratory Lubricant Analyst (LLA) by the International Council for Machinery Lubrication (ICML).

To achieve the maximum benefits from in-service lubricant analysis, a holistic approach is essential. Lubricant analysis is considered the second most important predictive maintenance technique (after vibration analysis), but it is also one of the most neglected condition-based maintenance (CBM) technologies.

At the beginning of a new lubricant condition monitoring programme, it is likely that a number of abnormal or critical sample reports will be observed. Particular attention should be paid to the main problem areas of water contamination, high levels of particulates, high levels of wear metals, improper oil top-ups and poor oil condition.

The sample report will show water results under either "water" or "H$_2$O". The flagging limits should have been set at the start of the programme, in discussion with the OEM, lubricant supplier and/or testing laboratory. With most machines, equip-ment or systems the contamination limit for water will be 0.1 %wt. Limits will range as low as 0.03 %wt for turbines and as high as 0.2 %wt for gearboxes. Some compres-sors using synthetic oils may tolerate water contents as high as 1.0 %wt.

Dirt will appear in a sample report under silicon (Si), together with the other elemental data for wear and additives. Secondary samples taken upstream of filters should be tested for particle counts. Attention should be paid to the ISO cleanliness code, as well as the particle counts by micron size. Abnormal or severe silicon levels and/or particle count results indicate a problem with contamination. Again, the labo-ratory will be using typical industry limits for silicon and oil cleanliness, which is acceptable when starting the oil analysis programme. Standard silicon alarm levels for most equipment are approximately 25 parts per million (ppm). ISO cleanliness codes for filtered systems are generally around 19/17/14.

Improper oil top-ups are slightly more difficult to detect. Changes in the elemental additive levels (phosphorus, zinc, magnesium, boron, calcium, sulphur and others) are likely to provide clues. Changes in oil viscosity that are plus or minus 10% from the oil specification are another indicator. Elemental additive levels can fluctuate as much as 25%, so a laboratory will look for other elements that should not be present or the lack of an element that should be present in the oil. Some laboratories have very sophisticated algorithms that not only compare the used oil to the new baseline but can also determine the fluid type and compare it to the generic fluid type for the oil that has been specified. FTIR spectra can be used to identify chemicals or fluids that should not be present in the oil. The laboratory can then provide an alert that a different type of fluid has been added.

The most blatant types of improper oil top-ups or incorrect oil usage are when the viscosity varies drastically from the specification. For example, an ISO 320 viscosity gear oil should be being used in a gearbox, but the analytical test result indicates that the sample's viscosity is 104 cSt at 40°C, indicating a possible top-up with hydraulic, compressor or circulating oil.

For most lubricated plant machinery, a common monitoring test is the oil's acid number (AN). An increasing AN signifies oil degradation, and once the AN is over the limit for the oil, an oil change should be scheduled as soon as practicable. Large systems, such as steam turbines, require more advanced testing, such as rotating pressure vessel oxidation testing (RPVOT), demulsibility, rust properties, foaming characteristics and air release, to determine if the oil is suitable for continued use.

As the condition monitoring programme progresses, more data will accumulate from the regular analytical test reports. It is very important to plot the results over time for each group of tests for each machine. These graphs, as discussed in Chapter 5, will begin to identify trends. Some machines will continue to operate satisfactorily with few changes to lubricant or machine considerations. The test results for other machines will begin to show increasing changes to one or more parameters. Action needs to be taken when these trends begin to approach the "abnormal" flagging limits. The action may be increasing the frequency of sampling and testing, additional tests or even scheduling maintenance as indicated by the earlier discussion about predictive maintenance. It should be possible to identify problems well before an "alarm" flagging limit is reached and to plan for maintenance, overhaul or replacement before a machine fails. As noted earlier, predictive maintenance is often more cost-effective than preventive maintenance.

Unfortunately, it is not uncommon for staff in production facilities who are responsible for making decisions about the operation of critical machines to fail to take the appropriate action required by a lubricant analysis report. On too many occasions, oil analysis reports are filed in the records without any proactive or predictive action being taken.

16.7 SOLVING AND CORRECTING PROBLEMS

Lubricant-related problems are opportunities for improvement in the lubrication and maintenance programme. Most solutions are of low cost and provide a high rate of return on investment. Many of the recommendations can be implemented within a short amount of time and don't require a large capital investment. All should be implemented as quickly as possible, to reduce maintenance costs.

Air breathers are easy to implement and low cost for preventing water and particulates from entering lubricated machinery. Air breathers can reduce moisture levels in lubricants even when oil analysis results show 0.2% or less water contamination. Desiccant air breathers dry the air that enters the machinery during operation and also dry the headspace in reservoirs, moving moisture out of the oil. The result is drier oil. In addition, air breathers have a rated micron filter that cleans the air of dust and dirt, leading to cleaner oil.

For very large systems, dry gas blanketing may be an effective option, especially when there is a readily available source of inert gas present, such as in chemical plants, petrochemical plants, refineries and pharmaceutical plants. For example, feeding dry nitrogen into a turbine reservoir can create a positive pressure that prevents the introduction of contaminants. The dry gas causes moisture to move out of the oil and into the headspace, where it is exhausted externally.

For systems with major water contamination issues (0.5% or more water in the oil), a more involved solution will be required. All machine hatches and inspection ports must be properly sealed. Upgrading the seals may be necessary. To remove water contamination between 0.3% and 2.0% on smaller systems (less than 50 litres of oil), an off-line filtration cart fitted with water adsorption filter media should be considered. If there is too much water, there is a risk of spending a lot of money on filter elements. Consultation with the filter cart provider can help assessment of the situation. If the water contamination issues are chronic (from a leaking cooler, for example) and the system is large (more than 500 litres of oil), a vacuum dehydrator or bypass centrifugal filtration system could be needed.

Particulate contamination can be managed easily with proper lubrication, drain ports and off-line filtration. An appropriately sized off-line filtration cart may be more cost-effective than retrofitting either in-line filters or a bypass filtration loop. Hydraulic filter carts are fairly straightforward and inexpensive. Gearbox applications require heavy-duty equipment and time spent to ensure that the filter cart has the proper specifications for the application.

The addition of lubrication and drain ports to machinery that will be part of an off-line filtration programme is advised, as these ports feature connections to allow maintenance technicians to easily connect a filter cart and perform oil top-ups and changes without having to remove fill and drain ports.

Controlling oil top-ups can be managed easily. Installing lubrication porting on equipment makes top-ups easier and provides the right kind of dispensing equipment to empower the lubrication technicians to do the task correctly. Oil identification tags should be attached to the lubrication ports. Colours and/or symbols help identify the lubricant to be used. Dispensing equipment is available in a variety of colours to match.

If a sample report indicates a poor oil condition (not contamination), an oil change should be scheduled as soon as is practicable. For more expensive oil changes, it may be prudent to invest in advanced oil testing to determine if an oil change is required immediately or whether the task can be put off for three months or more. This is an example of predictive maintenance. Unlike contamination, in 99% of cases when the oil condition is a problem, the oil will need to be changed. For a better indication of what is happening with the oil, membrane patch colorimetry (MPC) testing for varnish potential and RULER tests to determine the exact amounts of anti-oxidants remaining in the oil are recommended, but beware of the issues explained in Chapter 6.

Most plants and facilities with a dedicated reliability programme use maintenance software for data storage, trending and reporting, but it can also be useful to integrate with lubricant analysis results for a comparison trend. The most important benefit of lubricant analysis is detecting wear metals at an early stage. Wear metals may appear first and then vibration, followed by eventual equipment failure, or vibration may appear first followed by wear metals. By integrating lubricant analysis results and recommendations with vibration analysis, greater insights into machine condition can be deduced. Additionally, the likelihood that a maintenance specialist will react quickly to investigate any problems will be enhanced.

The following actions for improving a lubricant condition monitoring programme should help lead to fewer unexpected equipment failures:

- The routine oil analysis timeline should be updated regularly. This can be communicated to the laboratory analysts in order to obtain their comments.
- The laboratory should be requested to present the results in a format that can be imported into the maintenance software alongside the traditional reports.
- The alarms triggered on laboratory reports for machine and oil condition should be counted for each component, machine, item of equipment or system. A list of these should be recorded so they can be tracked over time.
- A scoring system should be implemented, so samples that prompt more than one alarm and locations that trigger repeat alarms earn more points.
- Repeat problem machines should rise to the top of the maintenance priority list, especially if they are the result of wear metal particles or severe water contamination.
- When a lubricant analysis report recommends vibration or ultrasound analysis or thermography for a machine or item of equipment, goals of reducing the alarms over time should be set, taking into account the maintenance team's workload.
- The "abnormal" and "alarm" thresholds that the laboratory, OEM and/or lubricant supplier has set for the machines should be checked regularly, to ensure they are still correct. This applies to all lubricant parameters.

Obviously, this list will require some site-specific fine-tuning and customisation, but hopefully these steps will add structure to the routine condition monitoring programme and help to minimise unexpected downtime or equipment or system failures.

16.8 SUMMARY

Lubricants can be regarded as information messengers of numerous failure modes and root causes of machine, equipment or system failure. It is difficult for a machine to be in trouble without the lubricant showing it first. For most laboratories, the number of non-conforming samples from lubricant analysis will sometimes exceed 20%. That is, more than one in every five samples has a reportable condition that requires a corrective response. For this reason, it is very important to be prudent about which machines are selected for lubricant analysis as well as the sampling methods and frequencies.

Establishing an effective lubricant and equipment condition monitoring programme is not easy and its initial stages can be time-consuming. It is often best to start small, establish the methods, reporting and corrective actions and then gradually expand the programme as the cost-benefits become clear and management support is established.

Predictive maintenance can be a more cost-effective approach than preventive maintenance. Some of the main components that are necessary for implementing predictive maintenance are data collection, trend analysis, early fault detection, fault detection, time to failure prediction, failure mode and effect analysis, maintenance scheduling and resource optimisation.

Correct lubricant sampling methods and comprehensive testing are vitally important for an effective condition monitoring programme. Identifying, solving and correcting problems require detailed knowledge and considerable experience of both lubricants and machines.

Glossary

AA	Atomic absorption (spectroscopy)
ACC	American Chemistry Council
ACEA	Association des Constructeurs Européens d'Automobiles
ACR	Sodium polyacrylate (quenchant)
ACTIA	Association of Technical Cooperation for the Food Industry
ADEME	French Agency for Environment and Energy Management
AES	Atomic emission spectroscopy
AFNOR	Association Française de Normalisation
AGMA	American Gear Manufacturers Association
AN	Acid number
ANSI	American National Standards Institute
API	American Petroleum Institute
ASIC	Application-specific integrated circuit
ASLE	American Society of Lubrication Engineers
ASME	American Society of Mechanical Engineers
ASTM	American Society for Testing and Materials
AT	Automatic transmission
ATF	Automatic transmission fluid
ATIEL	Association Technique de l'Industrie Européene des Lubrifiants
AW	Anti-wear (additive)
BMEP	Brake mean effective pressure
BN	Base number
BPT	Borderline pumping temperature
BRT	Ball rust test
CAD	Computer-aided design
CBM	Condition-based maintenance
CBT	Cummins high-temperature corrosion bench test
CCS	Cold cranking simulator
CEC	Co-ordinating European Council
CEN	Comité Européen de Normalisation
CFC	Chlorofluorocarbon (refrigerant)
CIGRE	Conference Internationale des Grandes Reseaux Electriques a Haute Tension (France)
COC	Cleveland open cup (flash point test)
COFRAC	French Accreditation Committee
CPA	Colorimetric patch analyzer
CRM	Customer relationship management
CSTB	Scientific and Technical Center for Construction (France)
CTI	Center Network Industrial Technology (France)
CV	Constant velocity (joint)
CVT	Continuously variable transmission

DI	Detergent inhibitor (pack)
DIN	Deutsche Institut für Normung
DOF	Degradation occurrence factor
DSC	Differential scanning calorimetry
EC	European Commission
ECHA	European Chemicals Agency
EDM	Electrical discharge machining
EELQMS	European Engine Lubricants Quality Management System
EFTA	European Free Trade Association
EHC	Electro-hydraulic control
EHM	Engine Health Management
ELGI	European Lubricating Grease Institute
ELID	European lubricants industry directory
EMD	Electro-Motive Division (Progress Rail division of Caterpillar)
EP	Extreme-pressure (additive)
EPA	Environmental Protection Agency (US)
EPR	Engine pressure ratio
EU	European Union
FMEA	Failure mode and effect analysis
FMECA	Failure modes, effects and criticality analysis
FOF	Failure occurrence factor
FTIR	Fourier transform infrared (spectroscopy)
FVA	Forschungsvereinigung Antriebstechnik
FZG	Forschungsstelle für Zahnräder und Getriebebau; Technische Hochschule Munchen
GC	Gas chromatography
HCFC	Hydrochlorofluorocarbon (refrigerant)
HDDEO	Heavy-duty diesel engine oil
HDPE	High-density polyethylene
HFC	Hydrofluorocarbon (refrigerant)
HSE	Health, safety and the environment
HWBF	High water-based hydraulic fluid
IBC	Intermediate bulk container
ICM	Intelligent combustion monitoring
ICML	International Council for Machinery Lubrication
ICP	Inductively coupled plasma emission (spectroscopy)
IEC	International Electrotechnical Commission
ILMA	Independent Lubricant Manufacturers Association (US)
ILSAC	International Lubricant Standardization and Approval Committee
IMO	International Maritime Organisation
INERIS	National Institute for Industrial Environment and Risks (France)
IP	Institute of Petroleum (The Energy Institute)
ISA	International Society of Automation
ISO	International Organization for Standardization
ITS	Intertek Testing Services
ITU	International Telecommunications Union

JASO	Japanese Automobile Standards Organisation
JSA	Japanese Standards Association
JSAE	Society of Automotive Engineers of Japan
KV	Kinematic viscosity
LCF	Lubricant criticality factor
LCIE	Laboratoire Central des Industries Électriques (France)
LLA	Laboratory Lubricant Analyst
LMOA	Locomotive Maintenance Officers Association (US)
LNE	Laboratoire National Metrology and Testing (France)
LNG	Liquefied natural gas
LVF	Linear variable filter
MCF	Machine criticality factor
MPC	Membrane patch colorimetry
MRV	Mini rotary viscometer
NAS	National Aeronautical Standards (US)
NLGI	National Lubricating Grease Institute (US)
NOx	Nitrogen oxides
OD	Optical density
ODI	Oil drain interval
OEM	Original equipment manufacturer
OIT	Oxidative induction time
OLC	Overall lubricant criticality
OMC	Overall machine criticality
PAG	Poly alkylene glycol
PAO	Poly alpha olefin
PCMO	Passenger car motor oil
PCV	Positive crankcase ventilation
PDSC	Pressure differential scanning calorimetry
PEO	Polyethyl oxazoline
PET	Polyethylene terphthalate
PIB	Polyisobutylene
PLOQ	Pooled limit of quantitation
PMC	Pensky Martins Closed Cup (flash point test)
PP	Polypropylene
PRO	Pumpability reference oils (ASTM)
PTFE	Polytetrafluoroethylene
PVC	Polyvinylchloride
PVP	Polyvinyl pyrrolidone
R&D	Research and Development
RBOT	Rotating bomb oxidation test
RCA	Root cause analysis
RIC	Radial internal clearance (of a bearing)
RMS	Root mean square
ROI	Return on investment
RPVOT	Rotating pressure vessel oxidation test
RULER	Remaining useful life evaluation routine

SAE	Society of Automotive Engineers (US)
SGEO	Stationary gas engine oil
SOx	Sulphur oxides
SPC	Statistical process control
STLE	Society of Tribologists and Lubrication Engineers (US)
TBS	Tapered bearing simulator
TEOST	Thermal-oxidation engine oil simulation test
TFOUT	Thin-film oxygen uptake test
TLC	Thin layer chromatography
TLT	Tribology and Lubrication Technology (Magazine)
TOST	Turbine oil stability test
UCVD	Ultra clean vacuum device
UEIL	European Union of Independent Lubricant Manufacturers
UKLA	United Kingdom Lubricants Association
UTAC	Union Technique de l'Automobile, Cycle and Motorcycle (France)
UTE	Union Technique de l'Électricité (France)
VI	Viscosity index
VII	Viscosity index improver
XRF	X-ray fluorescence (spectroscopy)
ZDDP	Zinc dialkyl dithiophosphate or zinc diaryl dithiophosphate

Index

For Product Safety Concerns and Information please contact our EU
representative GPSR@taylorandfrancis.com
Taylor & Francis Verlag GmbH, Kaufingerstraße 24, 80331 München, Germany